The Global Geometry of Turbulence

of Turbulence

Impact of Nonlinear Dynamics

NATO ASI Series

Advanced Science Institutes Series

A series presenting the results of activities sponsored by the NATO Science Committee, which aims at the dissemination of advanced scientific and technological knowledge, with a view to strengthening links between scientific communities.

The series is published by an international board of publishers in conjunction with the NATO Scientific Affairs Division

A	**Life Sciences**	Plenum Publishing Corporation
B	**Physics**	New York and London
C	**Mathematical and Physical Sciences**	Kluwer Academic Publishers
D	**Behavioral and Social Sciences**	Dordrecht, Boston, and London
E	**Applied Sciences**	
F	**Computer and Systems Sciences**	Springer-Verlag
G	**Ecological Sciences**	Berlin, Heidelberg, New York, London,
H	**Cell Biology**	Paris, Tokyo, Hong Kong, and Barcelona
I	**Global Environmental Change**	

Recent Volumes in this Series

Series B: Physics

The Global Geometry of Turbulence

Impact of Nonlinear Dynamics

Edited by

Javier Jiménez

School of Aeronautics
Universidad Politécnica
Madrid, Spain

Springer Science+Business Media, LLC

Proceedings of a NATO Advanced Research Workshop on
The Global Geometry of Turbulence,
held July 8–14, 1990,
in Rota, Spain

Library of Congress Cataloging-in-Publication Data

NATO Advanced Research Workshop on the Global Geometry of Turbulence
(1990 : Rota, Spain)
    The global geometry of turbulence : impact of nonlinear dynamics /
edited by Javier Jimenez.
        p.   cm. -- (NATO ASI series. Series B, Physics ; v. 268)
    "Proceedings of a NATO Advanced Research Workshop on the Global
Geometry of Turbulence, held July 8-14, 1990, in Rota, Spain"--T.p.
verso.
    "Published in cooperation with NATO Scientific Affairs Division."
    Includes bibliographical references and index.
    ISBN 978-1-4613-6670-6   ISBN 978-1-4615-3750-2 (eBook)
    DOI 10.1007/978-1-4615-3750-2
    1. Shear flow--Congresses. 2. Turbulence--Congresses.
3. Nonlinear theories--Congresses. 4. Global differential geometry-
-Congresses.   I. Jimenez, J. (Javier) II. Title. III. Series.
QC150.N38 1990
532'.0527--dc20   91-20142
                                                            CIP

ISBN 978-1-4613-6670-6

© 1991 Springer Science+Business Media New York
Originally published by Plenum Press, New York in 1991
Softcover reprint of the hardcover 1st edition 1991

SPECIAL PROGRAM ON CHAOS, ORDER, AND PATTERNS

This book contains the proceedings of a NATO Advanced Research Workshop held within the program of activities of the NATO Special Program on Chaos, Order, and Patterns.

PREFACE

The aim of this Advanced Research Workshop was to bring together Physicists, Applied Mathematicians and Fluid Dynamicists, including very specially experimentalists, to review the available knowledge on the global structural aspects of turbulent flows, with an especial emphasis on open systems, and to try to reach a consensus on their possible relationship to recent advances in the understanding of the behaviour of low dimensional dynamical systems and amplitude equations.

A lot has been learned during recent years on the non-equilibrium behaviour of low dimensional dynamical systems, including some fluid flows (Rayleigh-Benard, Taylor-Couette, etc.). These are mostly closed flows and many of the global structural features of the low dimensional systems have been observed in them, including chaotic behaviour, period doubling, intermittency, etc.. It has also been shown that some of these flows are intrinsically low dimensional, which accounts for much of the observed similarities. Open flows seem to be different, and experimental observations point to an intrinsic high dimensionality. However, some of the transitional features of the low dimensional systems have been observed in them, specially in the intermittent behaviour of subcritical flows (pipes, channels, boundary layers with suction, etc.), and in the large scale geometry of coherent structures of free shear flows (mixing layers, jets and wakes). Moreover, the recent availability of detailed direct numerical turbulence simulations allows the investigation of the structure of some of these flows in unprecedented detail, as well as the study of simpler ones (e.g. two dimensional turbulence) that are usually beyond experimental methods. Also, a lot of recent work is related to the representation of turbulent systems in terms of deterministic models, especially in the framework of vorticity dynamics. These models also entail a drastic reduction of dimensionality, and many of them result in a Hamiltonian representation, which connects again directly with the theory of dynamical systems. In a related development, simplified descriptions of the flows in terms of amplitude equations, such as the Ginsburg-Landau or Kuramoto-Sivashinski equations, are beginning to show the underlying reason for many of the observed similarities.

It appeared, therefore, a good moment to bring together specialists in dynamical systems, open and closed flows and structural Physicists, to inquire how much of this structure is dependent on low dimensionality and how much is more general, as well as which should be the consequences for future studies both in turbulence and in dynamical systems theory.

The Workshop was initially organised on rather classical lines, with some sessions devoted to experimental and theoretical "engineering" turbulence and transition, dealing essentially with three dimensional shear or homogeneous flows, while others were reserved for more traditionally "physical" or "mathematical" subjects, such as closed flows, pattern selection mechanisms and dynamical systems. The hope was to prepare the ground for future meetings in which these two schools of thoughts could be more closely linked. Happily, this organisation proved to be inadequate, and is not reflected in these Proceedings. It turned out that the two methodologies were already closer to each other than we suspected. Many workers in classical turbulence, especially those in transition, had already been applying for some time tools coming from nonlinear dynamics, and some of the researchers in this latter field had began to tackle the problems of the systems with high dimensionality which are implicit in free turbulent flows.

Some differences persist, however, the main ones being those between the group which would not consider as turbulence anything that cannot be derived explicitly from the three dimensional Navier Stokes equations, and the one which prefers to begin with the study of model equations, even if their relation to more realistic flows is not immediately apparent. This division has been taken as the main organisational backbone of these Proceedings. Beyond that, papers have been just classified into experimental and theoretical. The further subdivision of the former into three subclasses merits some comments. **Experiments of observation** are the classical ones and were, of course, included in the Workshop as such from the start. The **Numerical experiments** are a later development, and have only recently begun to appear as a separate group in scientific conferences. A Panel discussion was organised to try to elucidate whether they should be considered as experiments or as something else, and their inclusion here reflects the general, although by not means unanimous, conclusions of that panel. Finally the third subclass, **Control experiments**, was not anticipated at all, but emerged during the Workshop as a surprisingly coherent group of experiments whose immediate aim was to control turbulence, but which were being used much more generally as a means of understanding the turbulent structure by disturbing it. They have been put here together to highlight the similarity of their methodologies. It is also this group that showed the greatest influence of nonlinear dynamical concepts in experimental turbulence.

A few authors were invited to prepare review lectures on several subjects of interest, and their contributions can be found at the beginning of their respective groups. We feel that they add significantly to the usefulness of the book. Another important part of the Proceedings are the three Panel discussions included at the end. They were structured as open colloquia for the whole conference, conducted in each case by a three person panel, and they constitute, in some way, the conclusions of the Workshop. They were recorded, and are included here with minimal editing.

The interaction of nonlinear science and "engineering" turbulence is an obvious, and highly desirable, consequence of the recent advances in both areas, and the papers in these Proceedings show that it is already well under way. We feel that they represent a comprehensive snapshot of how the process stood on the summer on 1990. Let us hope that future Workshops can report further advances in the many problems that remain unsolved.

Javier Jiménez
School of Aeronautics
Universidad Politécnica
Madrid, Spain

January 1991

ACKNOWLEDGMENTS

It is a pleasure to acknowledge the help of the multiple institutions and persons which made this Workshop possible. Most of the funding was provided by the NATO Scientific Affairs Division, through its Programme on Chaos, Order and Patterns. I want to thank especially its director, Dr. G.A. Venturi for his constant help and understanding during the organisation of the Workshop. Additional funding was also provided by the European Offices of the U.S. Air Force Office of Aerospace Research and Development, of the U.S. Army Research Office, and of the U.S. Office for Naval Research.

The help of many other individuals was also indispensable, beginning, of course, with the members of the Scientific Committee, with the Chairmen of the different sessions, and with the members of the Discussion Panels. A. Jiménez and N. Rosell acted as secretaries before and during the meeting and, as usual, carried most of the weight of the organisation by themselves. M.L. de la Riva was responsible for the social events, and I am sure that the participants will remember her efforts with special pleasure. I am grateful for the help of J.C. Agüí, J.A. Zufiria, R. Corral, J.A. Hernández and C. Martel for their help in the day to day running of the Workshop. The interaction with B. Kester, of the NATO ASI Publication Office, and with J. Curtis, of Plenum, was consistently helpful. Finally, the staff and management of the Hotel Playa de la Luz, in Rota, made our stay during the Worshop a week to remember. I think that I interpret the feeling of most of the participants if I transmit to them our appreciation for their efforts and for their beach.

CONTENTS

SHEAR FLOWS: EXPERIMENTAL OBSERVATIONS

SHEAR FLOWS: CONTROL EXPERIMENTS

SHEAR FLOWS: NUMERICAL EXPERIMENTS

CLOSED FLOWS: EXPERIMENTS

THEORETICAL MODELS: THE N-S AND RELATED EQUATIONS

THEORETICAL MODELS: AMPLITUDE AND MODEL EQUATIONS

PANEL DISCUSSIONS

SHEAR FLOWS:

EXPERIMENTAL OBSERVATIONS

THE MIXING TRANSITION IN FREE SHEAR FLOWS

Anatol Roshko

California Institute of Technology
Pasadena, California 91125

INTRODUCTION

The term "mixing transition" denotes an increase in molecular mixedness observed in a shear flow which has earlier experienced the conventional (momentum) transition from laminar flow. First defined by Konrad (1976), from measurements of concentration in a free shear layer, the transition has been described and measured by a number of other methods, in particular by flow visualiztion, by measurement of chemical reaction product and by hot-wire anemometry, in aqueous as well as gaseous flows. In this presentation we review some of the measurements and try to assess what insight they may give on several questions that occur. 1. What is the relation of the mixing transition to Reynolds number and to other events: the momentum transition; vortex pairing; development of streamwise vortex structure? 2. How much does interfacial area increase during the transition? 3. How fast does this occur? 4. Does chaotic advection play a role? The answers are tentative and incomplete.

BACKGROUND

The term "transition" has traditionally been used to denote the departure from steady "laminar" to unsteady "turbulent" conditions in shear flows, more precisely to the corresponding increase of momentum transport which results in an increase of skin friction (in boundary layers) or of spreading rate (in free shear flows). Turbulent transport of advected heat or other scalars is also increased after such a transition; the term "increased mixing" is used. However, this is not generally accompanied by increased "mixedness" or homogenization, which can be accomplished only by molecular diffusion and is completed later in the flow development. It was found that part of the increased mixedness occurs fairly abruptly, at some time after the initial, momentum transition, and that it is accompanied by an increase in small-scale structure. The term "mixing transition" was coined to denote it. ("Mixedness transition" might be more appropriate.)

The phenomenon occurs in many shear flows starting from initially laminar conditions but was first identified and seriously studied in mixing layers. A mixing layer is created when two streams of fluid meet downstream of a dividing partition, or "splitter plate"; the plane interface between the two fluids becomes perturbed by the rapidly growing Kelvin-Helmholz instability and rolled up into the interior of the resulting vortices as illustrated in the shadowgraph in Figure 1 of a mixing layer between two gases. Another example, for water, may be seen in Figure 7 of Koochesfahani and Dimotakis (1986), in which a laser sheet has illuminated a *cross section* of such vortex structures. In water, molecular diffusivity is very small and the two fluids remain practically "unmixed", except for a very thin diffusion layer at the highly convoluted

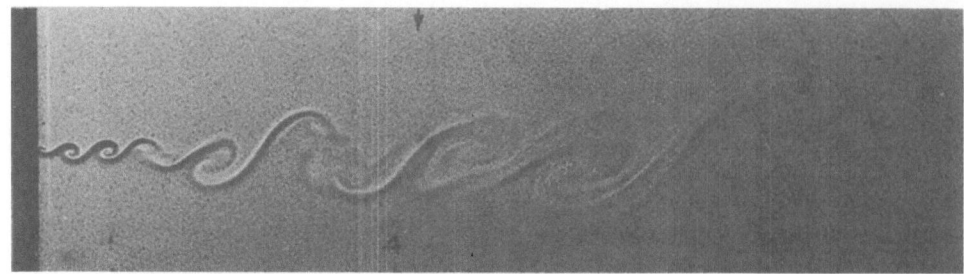

Fig. 1. Mixing layer between N_2 at 10m/s and equal-density $He-A$ mixture at 3.8m/s. $p = 1$atm. $\Delta U \theta_0/v \doteq 80$. The arrow indicates the beginning of mixing transition as defined by Fig. 2; the numbers indicate values of Rx/λ_0. Photo by Konrad.

interface. A concentration probe placed inside this region will register a signal in which the values of concentration alternate mainly between one fluid or the other, with only small indications of mixture at the interfaces. If the two fluids are chemically reactive then a correspondingly small amount of chemical product is formed at the interfaces. At higher Reynolds number, or further downstream in the development, the mixedness dramatically increases, as indicated by the large increase of regions with values of concentration between the unmixed-fluid values (color coded in Figure 9 of Koochesfahani and Dimotakis). Because the diffusivity is so small, the implication is that the *interface area* per unit volume has also increased greatly. The increase in interface area has been attributed mainly to development of three dimensional, small-scale "eddy" structure through a hierarchy of instabilities and not to effects of chaotic advection as have been observed in strictly two dimensional flows, e.g., in the numerical experiment of Aref and Jones (1989) or the analytical/numerical study of Rom-Kedar et al. (1990). At this time it is not clear how one would distinguish between the two mechanisms for the flows discussed here.

DEFINITION

The first measurements of this change, or "transition", in mixedness were made by Konrad (1977), using the Brown-Rebollo (1972) concentration probe in the $He-N_2$ mixing layers of Brown and Roshko (1974). It is the prominent feature in Figure 2 where a measure of the

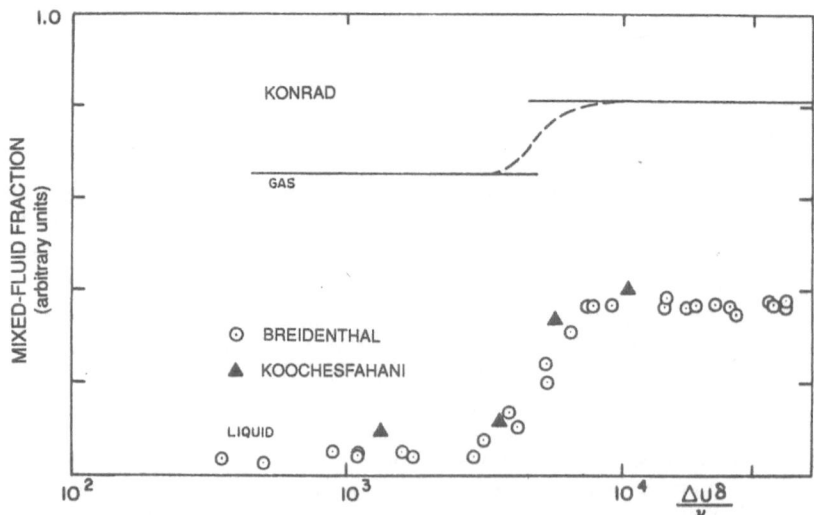

Fig. 2. The mixing transition.

mixedness, or mixed-fluid fraction is plotted as a function of large-structure Reynolds number $\Delta U \delta/\nu$; here $\Delta U = U_1 - U_2$ is the difference of velocities in the two streams, δ is the vorticity thickness and ν is the kinematic viscosity. The measured mixedness is a time averaged equivalent thickness of mixed fluid or reaction product normalized by the "vorticity thickness", $\delta_\omega = (U_1 - U_2)/(\partial U/\partial y)_{max}$ at the same location. δ_ω is approximately ½ the "visible thickness". Also shown in Figure 2 are results for a mixing layer in water for the same velocity ratio $U_2/U_1 = 0.38$. These were obtained by Breidenthal using a chemical reaction technique and by Koochesfahani and Dimotakis (1986) from measurements of concentration in the experiments noted above.

Konrad had set out to determine whether mixedness increases or decreases with Reynolds number. The direction is not obvious *a priori*. One expects the interface area to increase with increasing Reynolds number. On the other hand the thickness of the diffusion layer at the interface will decrease, and it is not clear how the product of thickness and area will change. The results of those experiments may be summarized as follows.

(1) The mixedness does not change continuously with increasing distance (time) in the development, but rather increases from one level to a higher one over a relatively short interval (on a logarithmic scale) called the "mixing transition". The levels are a measure of mixed-volume fraction whose quantitative value depends on normalizing factors. The *relative* values in gas and water are consistent.

(2) In water, the increase in the fraction of mixed fluid after the transition is by a factor of almost 10, corresponding to the "dramatic" increase which is illustrated in the concentration images of Koochesfahani and Dimotakis discussed above. There is some uncertainty because of difficulty in measuring accurately the small lower value. The implication is that interface area has increased by about 1000%. In gas, on the other hand, the increase in mixed-fluid fraction is only about 25%.

(3) However, the transition in gases and in water occurs at the same Reynolds number. This is not unexpected since the vorticity structure at a given Reynolds number is the same in gas and liquid (if the experimental geometries are similar). Thus, the topology and area of the interface (however convoluted), on which the diffusion and reaction are occurring, must be the same in both cases. One must conclude that in the gas the fraction of mixed fluid does not increase much because the diffusion mixed layer on the convoluted interface is already almost space filling.

(4) This is consistent with the large difference in values of mixedness before the mixing transition, where the mixed-fluid fraction in water is about 1/20 that in gas (again, the small values in water make for some uncertainty). One would expect it to be proportional to the square root of the ratios of diffusivities, that is about 1/25, but not strictly since faster merging of the thicker diffusion layers in gas will preclude a strict proportionality of the interface areas.

RELATION TO REYNOLDS NUMBER

When the measurements of mixedness were undertaken by Konrad it was thought that, in the spirit of Kolmogorov scaling, the large-scale Reynolds number, $\Delta U \delta/\nu$, might be the correlating parameter since the development of small scales seemed central. (It should be noted that the local Reynolds number Re increases with distance downstream from separation, ie. with flow evolution time.) On the other hand, in correlating momentum transition in free shear layers, it is customary at typical laboratory conditions to ignore effects of Reynolds number, i.e., to assume that scaling depends only on the initial conditions. For example, Sato (1956) placed the momentum transition, defined as an increase in growth rate $d\theta/dx$, at $x_{tr} = 40$ to $50\,\theta_0$, where θ_0 is the initial momentum thickness. This location is in good agreement with measurements of Freymuth (1966) for the exponential growth of the Kelvin-Helmholtz instability in a free shear layer and his measurements agree well with the predictions from *inviscid* instability theory (Michalke, 1965). Of course, x_{tr} for given θ_0 depends also on the level of the external forcing, either natural as in Sato's experiments or acoustically imposed in

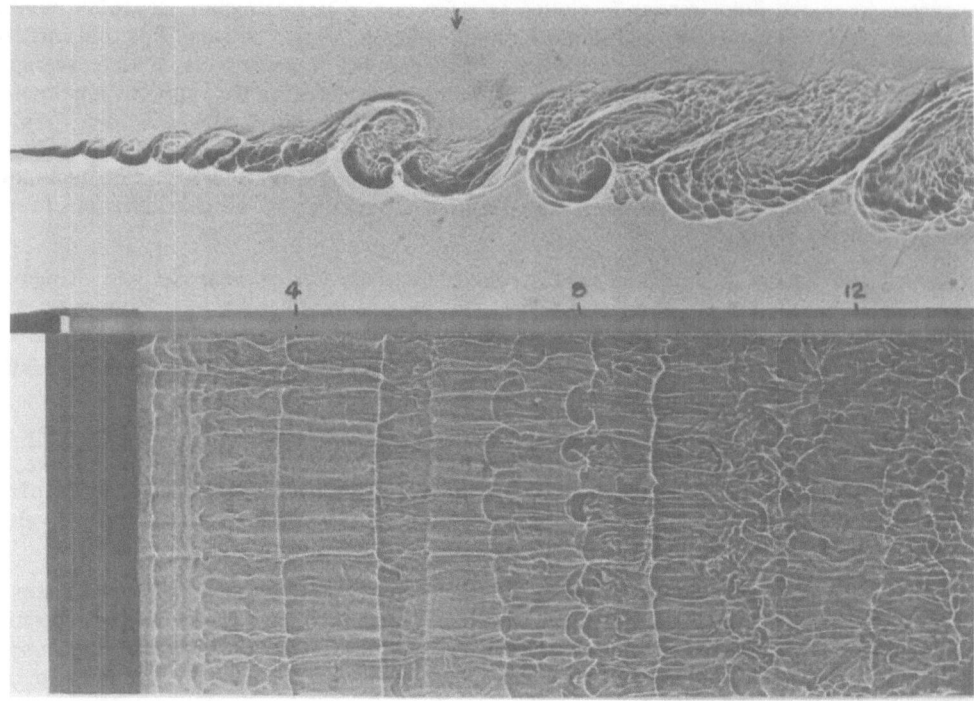

Fig. 3. Mixing layer between *He* at 10m/s and N_2 at 3.8m/s. $p = 4$atm. $\Delta U \theta_0/\nu \doteq 54$. Marking as in Fig. 1. Simultaneous plan and edge views. Photo by Konrad.

Freymuth's. The values given are nominal for laboratory conditions. Bradshaw (1966) also used x/θ_0 as the parameter against which to plot the development of momentum mixing, e.g., the Reynolds stress. Downstream of Sato's transition point this stress increases exponentially, as in Freymuth's measurements, saturates and oscillates around values considerably higher (by 2 or 3) than the steady-state, "fully-developed-turbulence" value reached at $x = 500$–$1000\ \theta_0$, to which it then relaxes at $x = 500$–$1000\ \theta_0$. One might say that the overall momentum transition extends from about $x = 50\ \theta_0$ to $1000\ \theta_0$. The mixing transition, as defined in Figure 2, is embedded in the overall momentum transition.

Starting with Breidenthal (1978), later investigators of the mixing transition began to consider x/θ_0 as the correlating parameter. Breidenthal noted however that it may still be necessary to take also into account the Reynolds number, especially at its lower values.

RELATION TO DEVELOPMENT OF STREAMWISE VORTEX STRUCTURE

Figure 1, an edge-view shadowgraph in which the light has traversed a span equal to 40 times the diameter of the first prominent vortex, suggests that the initial structure is two-dimensional, highly correlated spanwise. A view in the direction normal to the plane of the shear layer (Figure 3 for a different flow) indicates that actually there is a spanwise perturbation, initially too weak to affect the edge view. The streamwise streaks mark the edges of developing streamwise vortex structures, as was illustrated by Bernal (1981), who obtained sectional views (normal to the flow direction) of strongly developed vortex pairs. One of his pictures (Fig. 13 in Bernal and Roshko, 1984) is reproduced here in Figure 4. This section is at a location ($\Delta U \delta_\omega/\nu \doteq 2.4 \times 10^3$) which is near the beginning or ahead of Breidenthal's mixing transition, measured earlier in the same apparatus (cf. Fig. 2). This upstream view shows a section through a spanwise vortex structure which must have resulted from two pairings, as can

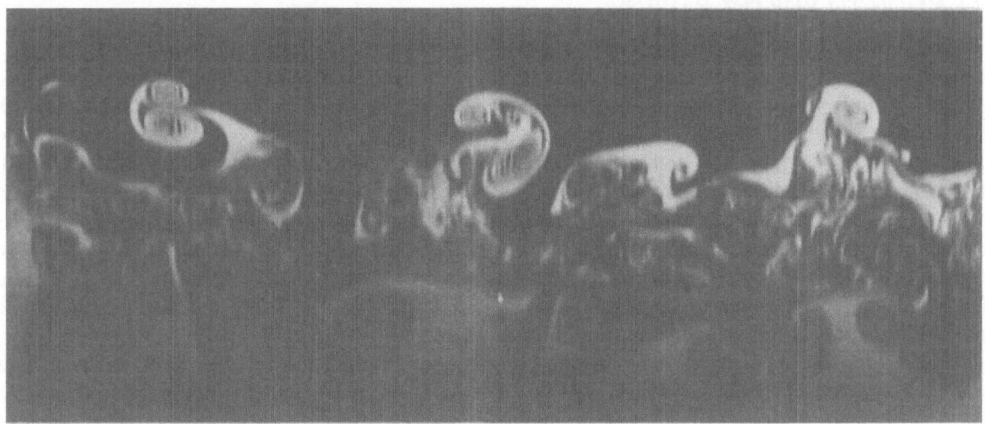

Fig. 4. Section normal to the flow direction. $U_1 = 20$cm/s, $U_2 = 8$cm/s in water; $\Delta U \theta_0/\nu \doteq 60$; $Rx/\lambda_0 \doteq 5$. Photo by Bernal.

be estimated from vortex spacings in the edge or plan views and from the location at $x/\theta_0 \doteq 300$, i.e. $x/\lambda_0 \doteq 10$, where $\lambda_0 \doteq 30\theta_0$ is the most amplified initial wave length. The Reynolds number $U_1\theta_0/\nu \doteq 60$. The streamwise vorticity has rolled up into vortex pairs. At least five layers of interface can be seen in each of the prominent rolls (perhaps not clearly in the reproduced picture). It may also be noted that some of the vortex pairs are tilting.

Using quite different flow visualization, in which mixedness is marked by chemical reaction product, Breidenthal (1981) obtained plan views in water in which the streamwise vortices are seen to evolve by convective straining amplification of an initial, spanwise-periodic distribution of streamwise vorticity. The stretching results in intense streamwise vortex pairs (Figs. 12-15, Bernal and Roshko, 1986) which bridge the space between primary vortices in a "rib" like structure (Hussain, 1986). That is, the view in Figure 3 denotes an organized pattern of secondary, streamwise vortices perpendicular to and carried on the primary, spanwise Kelvin-Helmholtz vortices. From Bernal's pictures it is inferred that the streamwise vortices are already quite strongly developed upstream of the mixing transition, at the relatively low initial Reynolds number $U_1\theta_0/\nu = 60$.

RELATION TO THREE DIMENSIONALITY AND SMALL SCALES

It had early been conjectured that increased mixedness was associated with development of three dimensionality but, as noted in the previous section, the development of organized streamwise vortex structure increases mixedness very little, at first. The large increase associated with the mixing transition must correspond to further increase of material surface area from an increase of small-scale structure. Just how this happens is not clear from the flow pictures. A shadowgraph in which the light has traversed several structures may exhibit a deceptive amount of "small-scale structure". Sectional views are more reliable. Even so, it is not possible to tell from an image of the scalar field, like that of Koochesfahani and Dimotakis, cited above, whether the complex small-scale structure is due to new, smaller vortex structure or to chaotic advection in the field of the pre-mixing-transition structure (Fig. 4).

The first quantitative definition of the development of small-scale three-dimensionality was made by Jimenez et al. (1979) from measurements of the power spectra of velocity fluctuations. The development of three dimensional small scales was associated with the emergence of an inertial subrange, as measured by the logarithmic slope of the spectrum at high wave number. The slope evolved from values of about -4 early in the shear layer to an asymptotic value of -5/3 by $x/\theta_0 \doteq 500$ or $x/\lambda_0 \doteq 16$. This corresponds to the end of the momentum transition suggested by Bradshaw's measurements.

RELATION TO VORTEX PAIRING

Jimenez (1983) began to associate the development of transition events with the location of vortex pairings. He found that the first pairing occurred at about $x = 90\ \theta_0$, i.e., $x \doteq 3\lambda_0$, (where λ_0 is the initial wave length), the second at $x \doteq 5\lambda_0$, the third at $x \doteq 8\lambda_0$ and the fourth at $x \doteq 16\lambda_0$. The end of his small-scale transition at $x/\theta_0 \doteq 500$ $(x/\lambda_0 \doteq 16)$ thus corresponded to the fourth vortex pairing. Jimenez' approach was developed further by Huang and Ho (1990) but with a useful addition, namely, controlled excitation of the shear layer. Using acoustic forcing at a combined frequency which contained the most amplified frequency and its subharmonic they stabilized the locations of the first and second pairings at $Rx/\lambda_0 = 4$ and 8, respectively, where $R \equiv (U_1 - U_2)/(U_1 + U_2)$ is used to scale the development time (Ho and Huerre, 1984). (In Jimenez' experiments, $U_2 = 0$ and $R = 1$.) Their measurements of the power spectrum of velocity fluctuations also indicated an absence of inertial sub range at the first vortex pairing or "merging", where the high-frequency logarithmic slope was -4. The latter increased asymptotically to the value -5/3 at $Rx/\lambda_0 = 8$, implying that the small-scale transition was complete by the second vortex merging, compared to the later time in the experiments of Jimenez. Huang and Ho attribute the slower completion of transition to effects of lower Reynolds number. In their experiment $U_1\theta_0/\nu \doteq 500$ while Jimenez (1983) lists values of 220 and 430.

A more direct examination of small-scale velocity fluctuation was made by Zohar (1990), using a "peak-valley-counting" technique to extract the short-wave-length low-amplitude velocity variations from the overall, large-scale parts of the signal. The associated gradients give estimates of the dissipation, which is found to reach an equilibrium state near the third vortex merging for their particular conditions, without forcing. When the shear layer is acoustically forced, equilibrium is reached earlier, near the second pairing. The conclusion was that forcing the main vortices ("rollers") enhances the development of the fine-scale structure, which develops qualitatively the same as in the free mixing layer, and is completed between the first and second pairings.

DISCUSSION

It is difficult to correlate the observations from the different experiments since not all the relevant parameters were measured in each one. Only in the numerical experiment of Moser and Rogers (1990) were small-scale transition and mixing transition both measured; it was found that the "small-scale transition is accompanied by an increased mixing of a passive scalar which is similar to the mixing transition in experiments". In the earliest laboratory experiments, in which the mixing transition was defined, θ_0 or λ_0 were not measured, but in some cases it is possible to estimate them from the flow conditions or from the pictures. In the measurements of small-scale velocity structure, the scalar mixedness was not (could not) be measured. For the following discussion we have tried to make estimates of parameters, especially θ_0 or λ_0, which were not measured or recorded and we use them for estimating the location of the mixing transition, as defined in Figure 2, in relation to pairing locations and to initial Reynolds number.

In none of the data is mixing transition or small-scale transition found to begin before the first pairing. It seems clear that unlimited increase of the initial Reynolds number would eventually move the small-scale transition to the separation point and into the boundary layer, but the character of the associated mixing transition would perhaps be different from the one discussed here.

Conversely, one can suppose that the mixing transition can be initiated after several pairings if the initial Reynolds number $U_1\theta_0/\nu$ is small enough. The local Reynolds number would eventually be large and, provided the necessary disturbances are present, the streamwise vortices would eventually emerge. This trend is illustrated in the set of flow pictures in Figure 20 of Brown and Roshko (1974) for He/N_2 mixing layers. These depict the flow structure over a range of Reynolds number varying by a factor of four. The (unit) Reynolds number was varied by changing pressure or velocity or both; for the flows depicted the Reynolds numbers have relative values 8, 4 and 2; the geometry and the velocity ratio $U_2/U_1 = 0.38$ are unchanged.

Thus we can estimate that the reference scales θ_0, λ_0 vary by a factor of 2. The positions of the beginning of transition, defined by Konrad's measurements in those flows, are at $x \doteq 0.15L$, $0.6L$ and $1.1L$, respectively, where L is the overall length of the view in the pictures. They correspond approximately to $Rx/\lambda_0 = 6$, 9 and 12. These may be compared to values of 4, 8 and 16 for nominal positions of the first, second and third pairings. (λ_0 has been estimated from the pictures at the smallest Reynolds number and then scaled up to the higher values.) The small-scale structure on the shadowgraph correspondingly moves from $x \doteq 0.1L$ to $x > L$. (It should be noted however that increasing pressure tends to increase the appearance of "small-scale structure" on shadowgraphs.) The estimates for initial values of Reynolds number are $U_1\theta_0/\nu = 76$, 54 and 38 respectively. In Figure 1, for a mixing layer at uniform density and for which $U_1\theta_0/\nu \doteq 80$, the initiation of mixing transition, inferred from Figure 2, is at $Rx/\lambda_0 \doteq 4$.

These speculations and the preceding observations suggest a scenario as follows. Reynolds number affects the mixing transition, not as originally anticipated in terms only of the local, large-scale value, but also through the initial conditions. It follows, as pointed out by Jimenez, that the pairing or merging process plays a role. As in other convective instabilities, the strength of the initial disturbances, natural or forced, also plays a role. Lin and Corcos (1984) developed a model describing the development of streamwise vorticity from an initial distribution amplitude (strength) into coherent rib-like vortex structures like those seen in cross section in Figure 4. This development is governed by a balance between stretching in the strain field of the primary vortices and diffusion by viscosity. Moser and Rogers (1990) propose that this system of primary and secondary vortices, if the latter are sufficiently strong, can be triggered by the first pairing into "diverging solutions" which initiate the small-scale transition; if not sufficiently strong then the transition may not be initiated until the second pairing. Thus the transition initiation depends on the initial Reynolds number. At sufficiently low values of the latter, initiation can occur downstream of the second pairing, as suggested by the example described in the preceding paragraph. With increasing Reynolds number, from low values, the mixing-transition initiation moves upstream through the pairing locations until it becomes stuck at the first pairing and is complete by the second one, as proposed by Huang and Ho.

The nature of processes at play in the mixing transition is still not clear. Of particular interest here is the question of the origin of the "small-scale structure". The traditional view implies that there is a "hierarchy" of (vorticity) instabilities creating a cascade of "eddies", from large to small ones. Much depends on the implication of the term "eddies". High frequency content in the Fourier spectrum may be associated with velocity interfaces or internal shear layers in the large vortical structures and not with small vortical entities. In the development of the mixing layer the evolution of large eddies, i.e. the primary rollers and the secondary, streamwise vortices, is quite evident. These result from growth and saturation of particular instabilities e.g., the "roll-up" in the Kelvin-Helmholtz instability and what Lin and Corcos termed the "collapse" (of distributed vorticity) to form the streamwise vortex structures. These "primary" and "secondary" structures may be viewed as the manifestation of the first two steps in a hierarchy of instabilities. The first one can be initiated at very low values of $\Delta U \theta_0/\nu$, of order 10, while the second one requires a higher value, of order 10^2, as suggested by the model of Lin and Corcos and the observations of Moser and Rogers. Will the next ones require another large increase in Reynolds number? No further analogous, smaller vortex structures have been seen in flow visualizations. Vorticity can be visualized in displays of computer simulated flows, but even here it may be very difficult to recognize small vortical structures, if they exist. The "small structure" that appears in shadowgraphs may result from disorder in the arrangement of the scalar large structure. May the same be true for the velocity or does small structure there indicate the presence of "eddies", in the sense described above? Probably not. For example, the layered structure in Bernal's sectional view of streamwise vortices may be associated with a layered velocity structure that one would see as a small-amplitude, small-scale fluctuation in the output of a velocity probe across which the vortex had moved, as suggested by Lundgren (1982). He proposed a model for fine structure consisting of slender axially strained spiral vortex solutions of the Navier-Stokes equation. The model produces "a cascade of velocity fluctuations to smaller scale" and results in the Kolmogorov spectrum.

The question, then, is whether with two or three, relatively large structures, which are perturbed from a regular pattern, one might realize a mixing transition by the "chaotic advection" route. It had earlier been conjectured (Corcos, 1979; Roshko, 1981) that such a few dynamical structures may be sufficient to describe the momentum development, where a long-range mechanism is available for momentum mixing. They may also suffice, through the mechanism of chaotic advection, to account for the scalar mixing transition, where intimate mixing at the diffusion scale is needed.

As indicated above, the mixing transition measurements of Figure 2 were made at values of initial Reynolds number $\Delta U \theta_0/\nu$ smaller than 10^2. It seems unlikely that the flows contained organized vortical structures or "eddies" smaller than those described here. The suggestion is strong that the large increase in mixedness was accomplished, or helped, by chaotic advection.

Much of the experimental work used for reference here was supported by the U. S. Office of Naval Research and the Air Force Office of Scientific Research. Discussions with P. Bernal and R. Breidenthal provided helpful clarifications.

REFERENCES

Aref, H. and Jones, S. W., 1989, Enhanced separation of diffusing particles by chaotic advection. Phys. Fluid A 1:470.

Bernal, L. P., 1981, Streamwise vortex structure in plane mixing layers, Ph. D. thesis, Calif. Inst. of Technology; also with A. Roshko, J. Fluid Mech. 1986, 170:499.

Bradshaw, P., 1966, The effect of initial conditions in the development of a free shear layer, J. Fluid Mech. 26:225.

Breidenthal, R. E., 1978, A chemically reacting, turbulent shear layer, Ph. D. thesis, Calif. Inst. of Technology; also J. Fluid Mech. 1981, 109:1.

Brown, G. and Roshko, A., 1974, On density effects and large structure in turbulent mixing layers, J. Fluid Mech. 64:775.

Corcos, G. M., 1979, The mixing layer: deterministic models of a turbulent flow, Univ. California Berkeley Rept. No. FM-79-2; also Corcos, G. M. and Sherman, F. S., 1984, J. Fluid Mech:29.

Freymuth, P., 1966, On transition in a separated laminar boundary layer, J. Fluid Mech. 25:683.

Ho, C-M. and Huerre, P., 1984, Perturbed free shear layers, Ann. Rev. Fluid Mech. 16:365.

Huang, L.-S. and Ho, C.-M., 1990, Small-scale transition in a plane mixing layer, J. Fluid Mech. 210:475.

Hussain, A. K. M. Fazle, 1986, Coherent structures and turbulence, J. Fluid Mech. 173:303.

Jimenez, J., 1983, A spanwise structure in the plane shear layer, J. Fluid Mech. 132:319.

Jimenez, J., Martinez-Val, R. and Rebollo, M., 1979, On the origin and evolution of three dimensional effects in the mixing layer, USA-ERO Rep. 79-6-079, London.

Konrad, J. H., 1977, An experimental investigation of mixing in two dimensional turbulent shear flows with application to diffusion-limited chemical reactions, Ph. D. thesis, California Institute of Technology.

Koochesfahani, M. M., 1984, Experiments on turbulent mixing and chemical reaction in a liquid mixing layer, Ph. D. thesis, Calif. Inst. of Technology; also, with P. E. Dimotakis, J. Fluid Mech. 1986, 170:83.

Lin, S. J. and Corcos, G. M., 1984, The mixing layer: deterministic models of a turbulent flow. Part 3. The effect of plane strain on the dynamics of streamwise vortices, J. Fluid Mech. 141:139.

Lundgren, T. S., 1982, Strained spiral vortex model for turbulent fine structure, Phys. Fluids, 25:2193.

Michalke, A., 1965, Spatially growing disturbances in an inviscid shear layer, J. Fluid Mech. 23:521.

Rom-Kedar, A., Leonard, A. and Wiggins, S., 1990, An analytical study of transport, mixing and chaos in an unsteady vortical flow, J. Fluid Mech. 214:347.

Roshko, A., 1981, The plane mixing layer: flow visualization results and three dimensional effects, in Lecture Notes in Physics No. 136 (The Role of Coherent Structures in Modelling Turbulence and Mixing, J. Jimenez, ed.), Springer.

Sato, H., 1956, Experimental investigation on the transition of laminar separated layer, J. Phys. Soc. Japan 11:702.

Zohar, Y., 1990, Fine-scale mixing in a free shear layer. Ph. D. dissertation, Univ. Southern California.

VORTEX SHEDDING FROM SPHERES AT SUBCRITICAL REYNOLDS NUMBER IN HOMOGENEOUS AND STRATIFIED FLUID

P. BONNETON*, J.M. CHOMAZ* and E.J. HOPFINGER**

*Centre National de Recherches Météorologiques
31057 TOULOUSE Cedex, France
**Institut de Mécanique, Université de Grenoble
38000 GRENOBLE, France

ABSTRACT

The structure of the wake of a sphere at subcritical Reynolds numbers and the frequencies associated with the vortex shedding and oscillatory modes were studied in homogeneous and stratified fluid. Flow visualization techniques and local probe measurement were used to obtain the Strouhal number dependence on the Reynolds and Froude numbers. The interaction between the two instability modes has been characterized. It is found that stratification suppresses the oscillatory mode when Froude number $F < 3$ or 4 depending on the Reynolds number and the vortex shedding mode when $F < 2$. The Froude number is defined with the sphere radius, R, its velocity, U, and the Brünt Väisälä angular frequency, $N = \sqrt{|g/\rho \ d\rho/dz|}$, $(F = U/RN)$.

INTRODUCTION

Wakes of spheres in homogeneous fluids at low Reynolds number ($Re < 400$) are well documented in Batchelor (1967). However, results concerning vortex shedding frequencies and instability modes at larger Reynolds numbers are controversial and relatively few studies have been reported in the literature.

Flow visualization of wakes of spheres in homogeneous fluids by Taneda (1956), Achenbach (1974) and Pao and Kao (1977) have shown a strong dependence of the wake structure on Reynolds number even in the subcritical range ($Re < 3.10^5$). In particular, there exists a transition in the dominant frequency near $Re \approx 6.10^3$ (Cometta, 1957 ; Achenbach, 1974). It was also noted by Achenbach that several instabilities coexist in the regime $Re > 6.10^3$. Studies of the effects of stratification on the wake structure and on vortex shedding frequencies are practically inexistent. Only overall effects on the wake have been considered (Lin and Pao, 1979). It is of importance to know how the different modes of wakes oscillations are affected by stratification because internal wave generation by the turbulent wake is thought to be directly connected with these modes of oscillation.

In this paper, we present some novel results on the wake structure of a sphere and the dependence of the modes of oscillations on Reynolds number. These results were obtained mainly by different types of visualization techniques. Some conductivity probe measurements were also made. First, results will be presented for homogeneous fluid and then the effects of stratification will be discussed in terms of the internal Froude number. The Reynolds number range investigated is $200 < Re < 2.10^4$ and the Froude number range $0 < F \overset{>}{\sim} 8$, where

$F = U/NR$, with U the velocity of the sphere, R its radius and $N = \sqrt{|g/\rho\, d\rho/dz|}$.

EXPERIMENTAL INSTALLATION

We used visualization techniques to have access to the spatial development of the wake, and local probe measurements to analyse more accurately wake instabilities. The visualization experiments were carried out in a tank with glass walls of dimensions 50 cm deep by 50 cm wide by 400 cm long, filled with water or with a linearly stratified salt solution. Three different spheres of diameter 2.2 cm, 5 cm and 7.2 cm were towed through the fluid at constant speed. These spheres were supported by three steel wires 0.1 mm in diameter. The towing speed ranged between 1 cms^{-1} and 50 cms^{-1}, giving Reynolds numbers $200 < Re < 2.10^4$. The Brünt Väisälä frequency N was always 2.02 $rd\ s^{-1}$, and the Froude number range [0.3, 8].

To visualize the flow we covered the sphere surface with fluresceine dye. Either a central plane of the wake was visualized by means of a thin vertical laser light sheet or the whole wake was made visual with a mercury vapour spotlight. We also used in some cases the particle streaks technique with laser plane lighting.

Probe measurements were carried out in a big tank 1 m deep, 3 m wide and 22 m long. We mounted a conductivity probe flush on the surface of the sphere at 90° from the stagnation point. The wave length and vortex shedding frequencies were determined from video films. Probe measurements were made in stratified fluid. Some hot film measurements were also conducted.

WAKE STRUCTURE IN HOMOGENEOUS FLUID

For $Re \in [200, 20000]$, we can distinguish three instability modes of the recirculation zone attached to the sphere : (i) a pulsating mode, associated with a fluctuation of the length of the recirculating zone; (ii) a nonaxisymmetric oscillating or sinuous mode, coming from a rotation of the detachment line; (iii) a Kelvin-Helmoltz mode of the annular vorticity sheet between the mean flow and the recirculation zone. These instabilities are locked or unlocked depending on Reynolds number.

Five different wake regimes can be distinguished:

Regime 1, $Re < 200$: the recirculation zone is axisymmetric and steady (Batchelor, 1967). There is no vortex shedding and only a thin filament of dye drags behind the recirculation bubble.

Regime 2, $200 < Re < 400$: the recirculation zone is still axisymmetric but now it pulsates along the axis of sphere displacement **x**. We note in Figure 1 that there is still no vortex shedding. The apparent disymmetry on the photograph is due to the light absorption by the fluresceine dye.

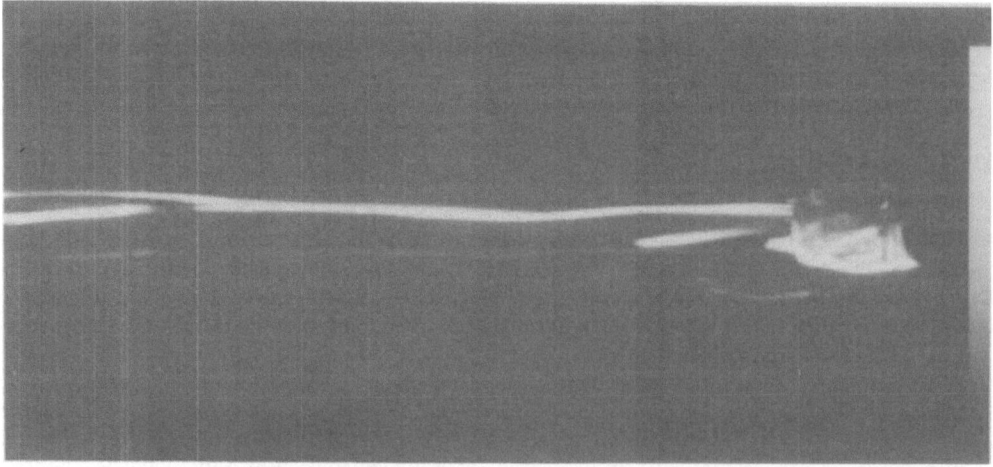

Figure 1. The wake at $Re = 300$, $R = 1.1$ cm visualized by a central vertical laser light sheet. The sphere is moving from left to right.

Regime 3, 400 < Re <800 : Figure 2 shows that the recirculation zone has lost its axial symmetry. This zone oscillates around the **x** axis and creates a helical wake. At the downstream edge of the annular vorticity sheet, periodic vortices are emitted. We observe on Figures 2 and 3 a thin spiral of dye which links periodic vortices. In this case, the helical mode and the Kelvin-Helmholtz mode are locked.

Figure 2. Visualization of the wake in a vertical central plane ; $Re = 500$; $R = 1.1\ cm$. The sphere is moving from left to right.

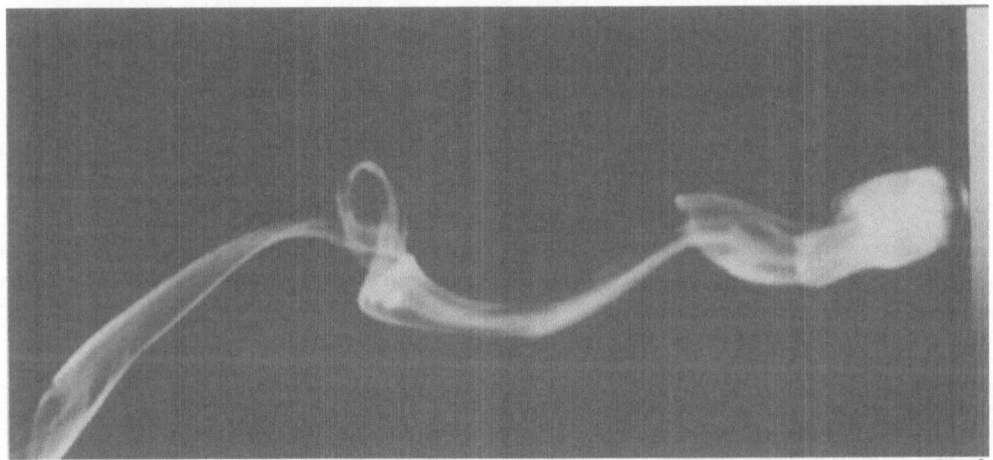

Figure 3. Global visualization with a spotlight ; $Re = 400$; $R = 1.1\ cm$. The sphere is moving from left to right.

Regime 4, 800 < Re <6000 : The annular vorticity boundary layer of the recirculation zone is Kelvin-Helmholtz unstable . Toric, axisymmetric vortices grow on the annular vorticity sheet and then split off from the recirculating zone (Figure 4). The global visualization shown in Figure 5 indicates that the shed vortex rings organize themselves in an irregular manner on a mean spiral. The weak spiral mode is not locked with the vortex ring shedding and corresponds to a lower broad band frequency.

Regime 5, Re >6000 : In this range the nonaxisymmetric oscillation of the recirculating zone becomes dominent as was shown by Achenbach (1974) the vortex ring shedding is still present but of small amplitude compared with the oscillation. The wake has the structure of a turbulent spiral composed of irregular vortex rings. Figure 6 presents a picture of the wake development where the rapid growth of the spiral is visible.

Figure 4. Visualization of the wake in a vertical central plane ; $Re = 4000$; $R = 1.1\ cm$. The sphere is moving from left to right.

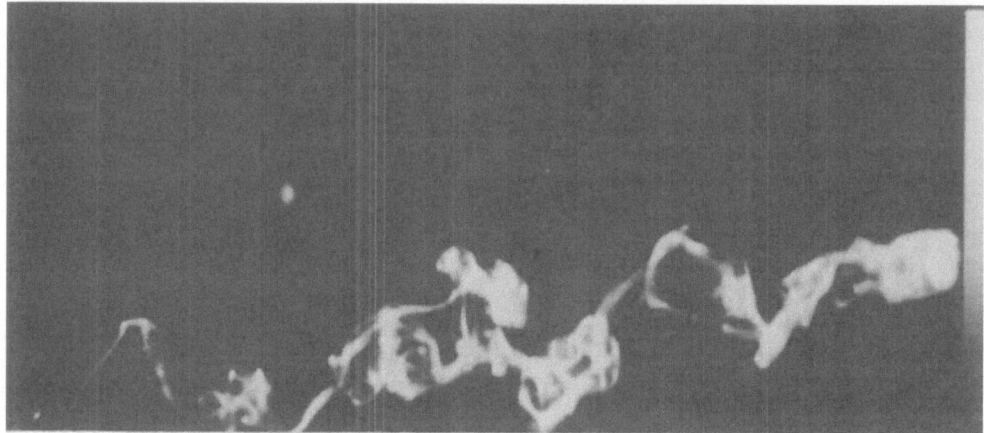

Figure 5. Global visualization with a spotlight ; $Re = 1000$; $R = 1,1\ cm$. The sphere is moving from left to right.

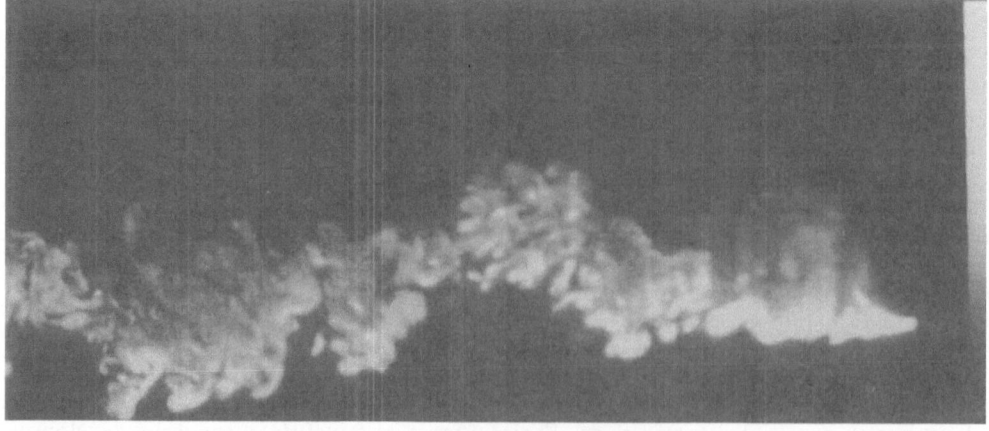

Figure 6. Visualization of the wake in a central plane ; $Re = 6000$; $R = 3.6\ cm$. The sphere is moving from left to right.

Most of the authors reject the idea of a helical wake configuration because a single helix does not respect Thomson's circulation theorem. This theorem states that the circulation around the wake must be zero. This led Pao et Kao (1977) to propose a complex double helix which unwinds in the opposite sense. However a single helix associated with strong vortices does not violate Thomson's theorem and is in agreement with our experiments.

THE FREQUENCIES ASSOCIATED WITH THE INSTABILITY MODES

Kelvin-Helmholtz instability

This instability is present over the full unstable range. The measured wave length of this instability mode non-dimensionalized by $2R$ ($S' = 2R/\lambda$) is shown in Figure 7 as a function of Reynolds number. It is seen that S' has the expected $Re^{1/2}$ dependence. This is because the annular shear layer thickness is determined by the boundary layer thickness on the sphere at separation which varies as $(\nu R/U)^{1/2}$.

We have also determined the vortex shedding frequency f by measuring the time required for a sequence of about 15 vortices to be released. Figure 8 shows the Reynolds number dependence of the Strouhal number $S = 2Rf/U$. The Strouhal number is related with S' through the relation $f = \overline{U}/\lambda$ where $\overline{U} = (U_1 + U_2)/2$ with U_1 and U_2 being respectively the velocities outside and inside the annular shear layer. In the case of a sphere $U_1 \approx U$ and $U_2 \approx 0$, giving $\overline{U} \approx U/2$ ($S = S'\overline{U}/U$). It is seen in Figure 8 that for $Re > 800$ the functional dependence of S on Re is similar to that of S'. At lower values of Re ($Re < 700$) the Kelvin-Helmholtz mode is locked with the sinuous mode and $S \approx 0.17$.

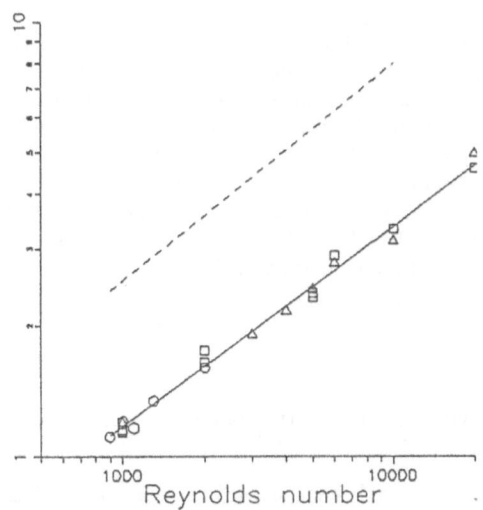

Figure 7. Inverse of the non-dimensional wave-lenght of the Kelvin-Helmholtz instability as a function of Reynolds number. $\bigcirc$, for $R = 1.12\ cm$; $\square$, for $R = 2.5\ cm$; $\triangle$, for $R = 3.6\ cm$; ____ linear interpolation: ($S' = 0.0483\ Re^{0.462}$), − − − slope $Re^{1/2}$.

For $Re > 800$, it seems that there are two distinct levels of Strouhal number. This could be an indication of a multistable behaviour of the wake caused by the interaction of the different instability modes or it could indicate an intermittent behaviour. More experiments will be required to conclude on this point.

Sinuous instability

This instability is mainly visible for regimes 3 and 5. Measured values of $S'' = 2R/\lambda_r$, where λ_r is the wave length of the helical mode, are given in Table 1. The relation with the Strouhal number $S_r = 2Rf_r/U$, measured by others, is via $f_r = (U - U_e)/\lambda_r$ where U_e is the

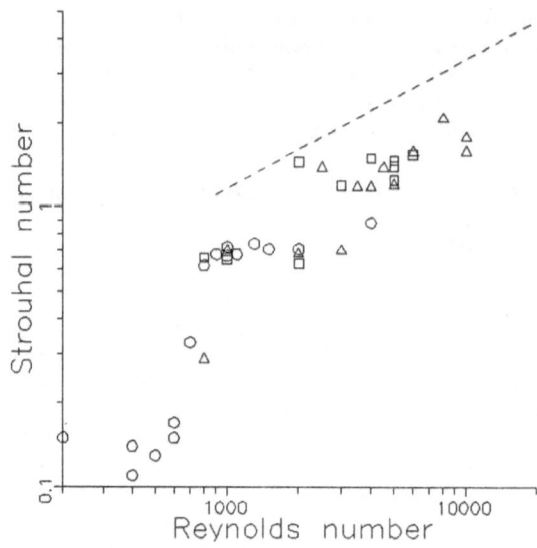

Figure 8. Strouhal number of vortex shedding as a function of Reynolds number. o, for $R = 1.12$ cm ; □, for $R = 2.5\ cm$; △, for $R = 3.6\ cm$; $- - - S' = 2R/\lambda$.

displacement speed of the helical or sinuous mode with respect to the fluid. Hot film measurements have shown that $U_e \approx 0.1U$. The difference between S'' and S_r lies within experimental error and S'' is, therefore, a good approximation of the Strouhal number of the sinuous mode.

Table 1 compares present results with previous ones. It is noted that for $R_e > 400$ there exists a sinuous instability mode with a Strouhal number of about 0.17. This agrees with the results, of Pao and Kao (1977) and Cometta (1957), that is, the low frequency instability is present even when $R_e < 6000$, and is contrary to the conclusion of Achenbach (1974). In fact, hot film measurements indicate that $R_e = 6000$ is associated with a strengthening of the sinuous instability and the emergence of a spectral peak.

Table 1. Values of the Strouhal number of the sinuous or spiral mode for three different Reynolds number regimes.

Regime	Re	S''	$S_{\text{Achenbach}}$	$S_{\text{Pao \& Kao}}$	S_{Cometta}	$S_{\text{Mujundar }et\ al}$
$n°3$	400	$0,13 \pm 0,02$				
	500	$0,13 \pm 0,01$				
	600	$0,17 \pm 0,01$				
$n°4$	1000	≈ 0.2				
	4300			0.14	0.19	
	5000	≈ 0.2				
$n°5$	6000	$0,16 \pm 0,02$	0.120			
	7820		0.135	0.18	0.19	
	10000	$0,17 \pm 0,02$	0.150			0.2
	17400		0.165	0.22	0.19	
	20000	$0,17 \pm 0,02$	0.170			

WAKE STRUCTURE IN STRATIFIED FLUID

In stratified fluid the geometry of the wake is controled by two parameters: the Reynolds number Re and the Froude number $F = U/NR$ which compares inertia forces to buoyancy forces ($N = \sqrt{|g/\rho \; d\rho/dz|}$). The limit $F = +\infty$ corresponds to the neutral case described previously.

Adding the buoyancy force breaks the axial symmetry of the wake. We have covered a wide range of Froude and Reynolds numbers. It appears that according to Froude number, three stratified wake regimes can be identified (fig.9 & 10). When $F > 4.5$ the structure of the flow in the near wake is unaffected by the stratification. The regime $4.5 > F > 2$ corresponds to a transition zone where the sinuous mode (spiral) will disappear first between $F = 2$ and 4, depending on Reynolds number. The shear or Kelvin-Helmholtz mode will disappear around $F = 2$ and the recirculating zone is suppressed between 1.5 and 2. For $0.9 < F < 1.5$ the flow is totally controled by the internal wave generated by the body which is of maximum amplitude in this Froude number range. When $F < 0.9$ the flow in the central horizontal plane is similar to the cylinder wake because of the strong plane symmetry imposed by the buoyancy. The flow structure for $F < 1$ has been described in Bonneton, Chomaz and Perrier (1990) and will not be discussed here. Figure 10 shows side views of particle streaks in the central plane for $F = 1$, 1.2, 1.5, 2, 2.5 and 3.18. As F increases, recirculation zone downstream of the sphere is building up and is clearly visible when $F = 2$. Kelvin-Helmholtz vortices start at 2.5.

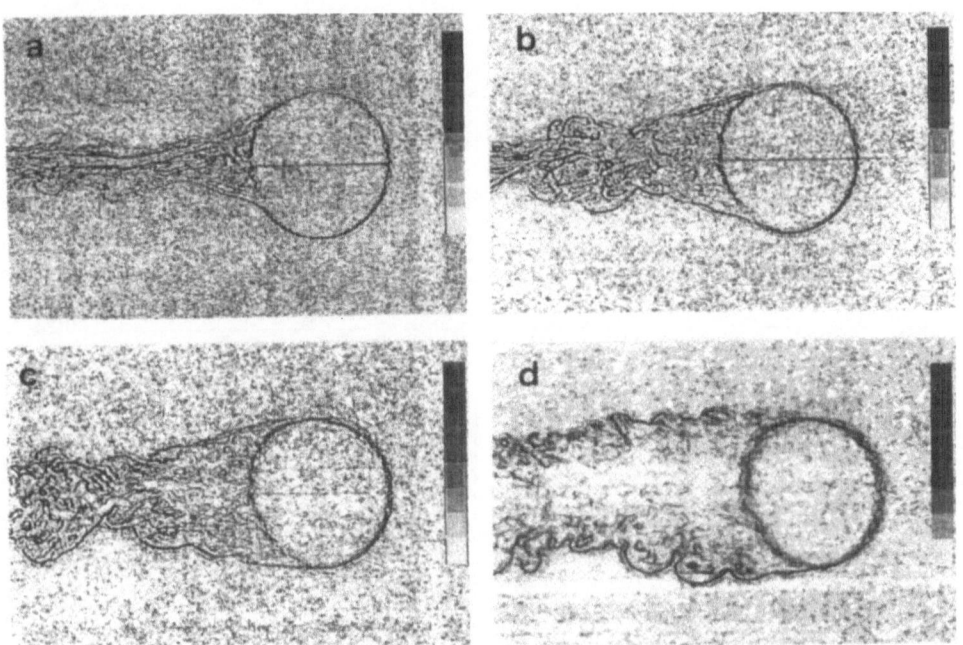

Figure 9. Side view shadowgraph pictures. a), $F=1$; b), $F=1.5$; c), $F=2.0$; d), $F=3.18$; $R = 2.5$ cm ; $N = 1.89 \; rd \; s^{-1}$; $Re = 2360\text{-}7500$. The sphere is moving from left to right.

This is demonstrated by side view shadowgraph pictures shown in Figure 9 which were treated by image processing ; the gradient modulus of the light intensity has been computed to enhance the small structures. Kelvin-Helmholtz waves are visible when $F > 2$ and a spiral oscillation is present when $F=3.18$. To study the disappearance of the spiral mode we performed experiments with a conductivity probe mounted on the sphere surface at 90° angle measured from the stagnation point.

The probe measures the time variation of the detachment point in a vertical central plane. The results are shown in Figure 11 where the Strouhal number is plotted vesus F. Two zones are clearly visible in this figure, one where $F > 4.5$ with a very small scatter of repeated measurement of the Strouhal number and the other between $F=3$ and 4.5 where the Strouhal number is double valued. This indicates the possible existence of a multistable system between

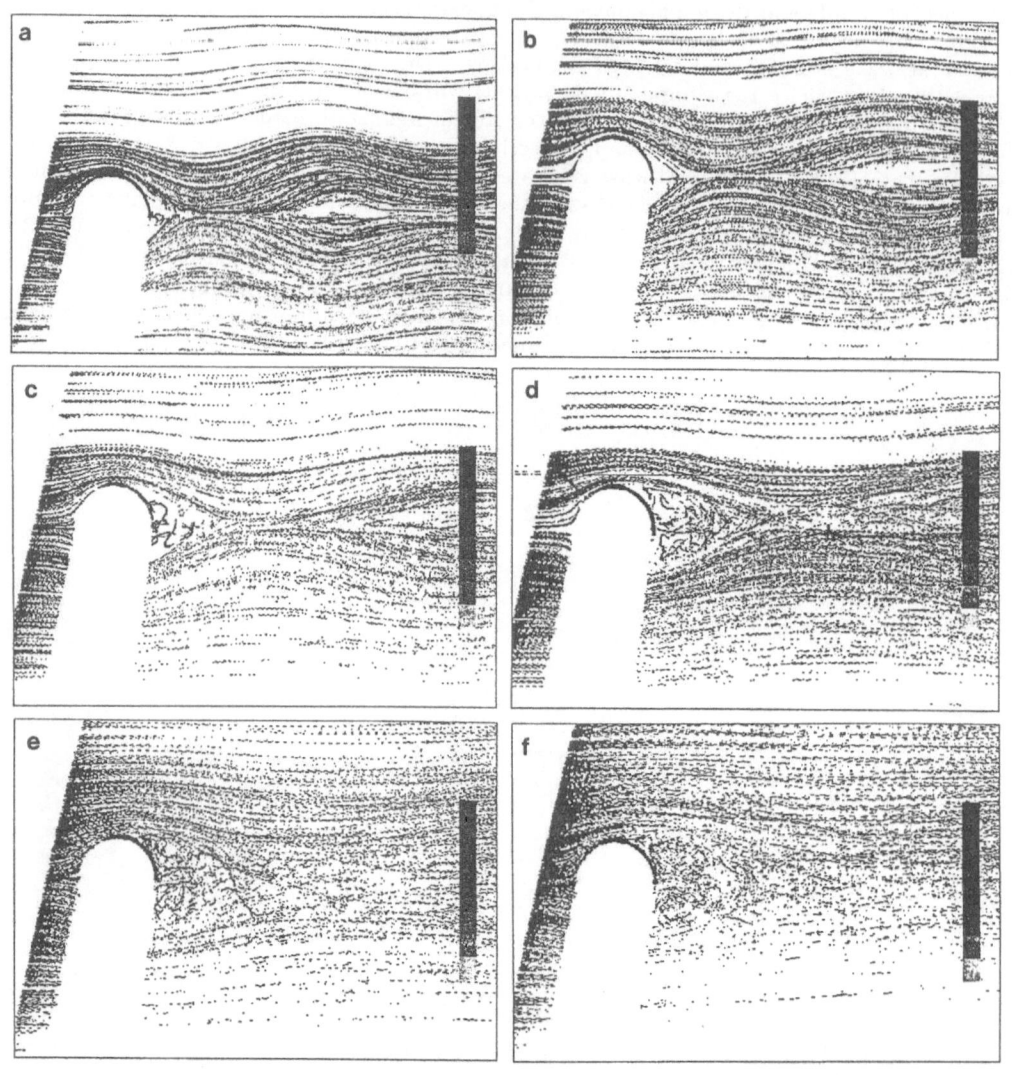

Figure 10. Particle streak photographs for a), $F=1$; b), $F=1.2$; c), $F=1.5$; d), $F=2$; e), $F=2.5$; f), $F=3.18$; $R=2.5\ cm$; $N=2.02\ rd\ s^{-1}$. The sphere is moving from right to left. $Re = F * 2R^2 N/\nu$.

3 and 4.5. It could be also due to the presence of long transients. One has to keep in mind that towing tank experiments give finit time sequences with starting and ending effect. Nevertheless, this scatter occurs in a Reynolds number range [8000, 12000] where no such a phenomenon was observed in homogeneous fluid. A detail study of this transition is in progress using several hot films to determine the coherent motion of the recirculating zone.

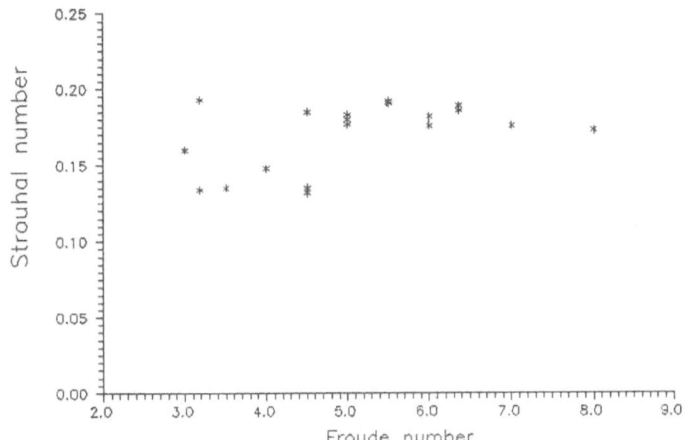

Figure 11. Strouhal number of spiral mode as function of the Froude number for the stratified water $R = 3.6$; $N = 1.17 \ rd \ s^{-1}$.

The experiment was made possible by the team SPEA and we wish to thank for their kind help, enthusiasm and efficiency: B. BEAUDOIN, J. C. BOULAY, A. BUTET, S. LASSUS-PIGAT, C. NICLOT, M. NICLOT, M. PERRIER and H. SCHAFFNER.

This work was supported by contract DRET number 88-126.

REFERENCES

Achenbach, E., 1972 : Experiments on the flow past spheres at very high Reynolds numbers. *J. Fluid Mech.* 54, 565-575.

Achenbach, E., 1974 : Vortex shedding from spheres. *J. Fluid Mech.* 62, 209-221.

Batchelov, G.K., 1967 : An introduction of fluid dynamics, *Cambridge University Press.*

Bonneton, P., J.M. Chomaz and M. Perrier, 1990 : Interaction between the internal wave field and the wake emitted behind a moving sphere in a stratified fluid. *Proc of Engineering Turbulence Modelling and Experiments*, Dubrovnik.

Cometta, C., 1957 : An investigation of the unsteady flow pattern in the wake of cylinders and spheres using a hot wire probe. *Div. Engng. Browns University, Tech. Rep.* WT-21.

Hanazaki, H., 1988 : A numerical study of three-dimensional stratified flow past a sphere. *J. Fluid. Mech.* 192, 393-419.

Lin, J.T., and Y.H. Pao, 1979 : Wakes in stratified fluids. *Ann. Rev. Fluid. Mech.* 11, 317-338.

Pao, H.P. and T.W. Kao, 1977 : Vortex structure in the wake of a sphere. *Physics of Fluids* 20, $n°2$.

Möller, W., 1938 : Experimentelle Untersuchung zur Hydromechanik der Kugel. *Phys. Z.* 39, 57-80.

Taneda, S., 1956 : Studies on the wake vortices (III). Experimental investigation of the wake behind a sphere at low Reynolds numbers. *Res. Inst. Appl. Mech., Kyushu University, Fukuoka, Japan, Rep.* 4, 99-105.

NATURE OF THE GÖRTLER INSTABILITY : A FORCED EXPERIMENT

J.M. CHOMAZ, M.PERRIER

Centre National de Recherches Météorologiques
31057 TOULOUSE Cedex, France

ABSTRACT

We study the instability of a concave boundary layer (Görtler instability) on a concave-convex model, in a water channel. Two different geometries were used either with (no streamwise pressure gradient) or without a counter-profile (favorable streamwise pressure gradient). In the first case Görtler vortices were present and randomly spaced. In the second case no vortex was detectable. The two flows exhibit different responses to localized forcings. We argue that these results may be interpreted by the non-linear convective instability concept, with a subcritical bifurcation when in presence of a favorable pressure gradient (accelerating flow).

I. INTRODUCTION

Flows on a concave surface suffer centrifugal forces which have a destabilizing effect and lead to the creation of counter-rotative vortices. First observation of this phenomenon goes back to Clauser and Clauser[1] and then to the work of Bippes[2]. Görtler[3] gave the theoretical background for the instability and his name to the vortices. Because of its implications in aeronautics this instability has been widely studied. This instability is characterized by a single number, the Görtler number G based on the boundary layer thickness δ, the freesteam velocity U, the kinematic viscosity ν and the radius of curvature R: $G = U\delta/\nu \, (\delta/R)^{1/2}$. Classically, the instability is supposed to take place for G of order .5 the vortices been visible after $G \sim 5.6$. Following Görtler's analysis, the linear theory predicts the most unstable mode presented on the classical picture 1. It must be emphasized that the main effect of the vortices is the spanwise deformation of the boundary layer. Low speed streaks are formed by small upward motion from the wall.

Different facilities give different neutral curves and critical wave numbers. This particularity of the system is usually attributed to different reasons:
-The sensibility of the theoretical[3-4] neutral curve to the precise boundary profile.
-The non-parallel effects[5-7] which, following various studies, either cause a "$k = 0$" instability, or make the neutral curve problem ill-posed[8].

Recently Huerre[9] introduced the idea that the bad selectivity, together with the dependence of results to the facility, originate in the convective[10] nature of the instability. An unstable system is convective if the linear impulse response, the Green's function, decays in the laboratory frame. This criterion is linear and, on an intuitive basis, it allows us to discriminate between open flows with propagation of information (waves) in both directions (absolute instability) and open flows with propagation of information only downstream (convective instability).

An absolute system presents its own intrinsic response, whereas a convective system reacts

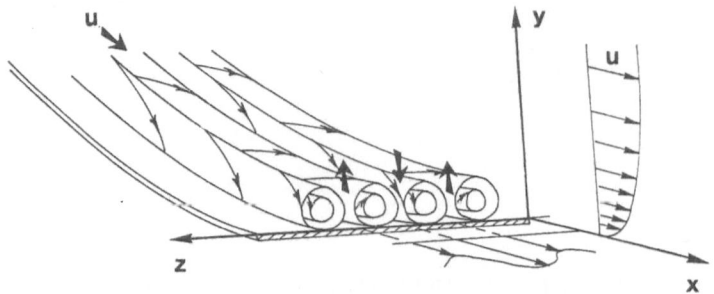

Fig.1. Schematic Görtler vortices. (following ref 3)

to the extremal perturbations. Paradoxically[11] the absolute systems are easier to simulate numerically than convective ones. For the latter if you do not specify the perturbation the numerics will do it for you, resulting in spurious results[12]. New theoretical results[13] on the asymptotic suction profile, indicate that the nonlinear terms are destabilizing, resulting in a subcritical bifurcation for the onset of the Görtler vortices. We extend the convective and absolute theory[14] to nonlinear systems. The same classes of behavior exist between:

-non-linear intrinsic systems showing hysteresis in the subcritical regimes called absolute systems;
-non-linear spatial amplifier not showing hysteresis systems called convective systems.

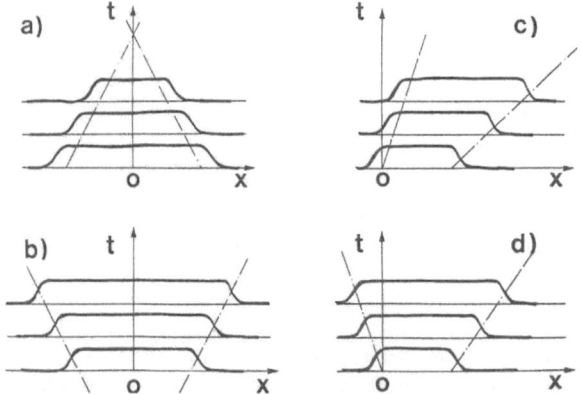

Fig.2. Evolution of a "droplet" of bifurcated state a) stable case and b) unstable case with parity; c) convectively unstable case and d) absolutely unstable case (the parity is broken by experimental conditions).

The general definition of non-linear absolute or convective instability may be given as follows:

-the basic state of a system is stable if, for all initial perturbations of finite extent and amplitude, the system relaxes to the basic state everywhere in any moving frame;
-a system is unstable if it is not stable in the above sense;
. the instability is convective if for all initial perturbations of finite extent the system relaxes

to the basic state everywhere in the laboratory frame;
it is absolute if there exists an initial condition of finite extent and a location where the system does not relax to the basic state.

In the supercritical case, for a range of the control parameter μ, two distinct homogeneous stable states coexist. Classically[6] the non-linear stability is determined by considering whether a sufficiently large "droplet" of bifurcated states surrounded by the basic state expands or shrinks (Figure 2a, b). A similar situation occurs in a first order phase transition. For systems with $x \to -x$ symmetry, the velocity $v_f(\mu)$ of a front between the basic state extending to $-\infty$ and the bifurcating state to $+\infty$ determines the stability: stable for $v_f(\mu) > 0$, unstable for $v_f(\mu) < 0$. At the critical value μ_M where the basic state becomes metastable v_f equals zero.

For systems where the parity is broken by assumption, (e.g., because the laboratory frame is specified), this consideration extends naturally to non-linear absolute or convective instability. The non-linearly unstable case is split in two: it is convective if the expanding droplet has fronts moving in the same direction, absolute if the droplet expands in both directions. For a right advected system, the sequence is presented on Figure 2 c, d.

In the convective case the system behaves like a spatial nonlinear amplifier as demonstrated in reference [14]. The response at x_0 to a steady forcing at the origin for the subcritical pitchfork bifurcation is presented on figure 3.

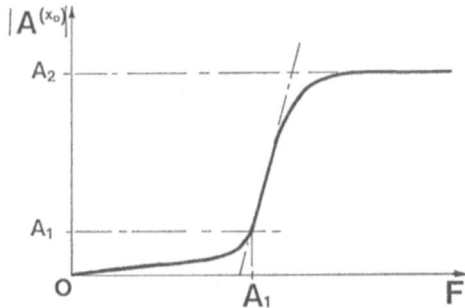

Fig.3. Response at x_0 to a forcing in amplitude at $x = 0$ in the convective case.

The numerical simulation of Deissler [16] illustrates the nonlinear convective case using the modified Ginzburg Landau equation (GL). It shows how turbulent slugs and spots may be viewed as a selective noise amplification by a subcritical system with an advection term. The system is linearly stable but a finite amplitude perturbation introduced at the origin is advected away without attenuation. When the artificial forcing by noise at the origin is turned off, the system relaxes to zero.

II. EXPERIMENTAL SET UP

II.1 The facility

All experiments were run in the large water channel[17] of the French meteorological research center. It measures 1 m deep, 3 m wide and 22 m long. The velocities used range from 5 cm/s to 50 cm/s. The concave convex model was set 15 m from the last grid regulating the flow. At such a distance all the small scale turbulent structures have disappeared. The turbulence level integrated over scales smaller than 20 cm is better than 0. 05%. (order of the electrical noise in the measurement apparatus). This makes the flow laminar at potential scales for the instability ($\sim$ 1.5 cm from previous experiments). The large scale (of order 50 cm) and the velocity inhomogeneity out of the boundary layer (of order 10 cm at 150 cm for u=10 cm/s) represent a 0. 7% perturbation.

II.2 The concave-convex model

We study the model defined and studied by Peerhossaini *et al.*[18] (fig 4). According to computation and measurements made by Peerhossaini *et al.* in a wind tunnel, the pressure is nearly constant downstream of the stagnation line. Only at the far end of the concave section ($36cm$-$46cm$) does a strong negative pressure gradient develop. The model is set in flow, out of the floor boundary layer ($20cm$ from the floor) and the concave part oriented upwards. The blocing effect of the flow is 23%. To compensate this and kill the pressure gradient it is possible to add an upper part: a counter-profile. The normal distance to the profile is then constant, equal to $15cm$. In this second configuration only visualization is possible. Special devices will be used in the future to allow measurement in this configuration. Two flaps are designed to act on the upstream stagnation point. We run the experiment with a stagnation point slightly on top of the elliptic leading cylinder, as sketched on figure 4, to avoid instability and separation bubbles.

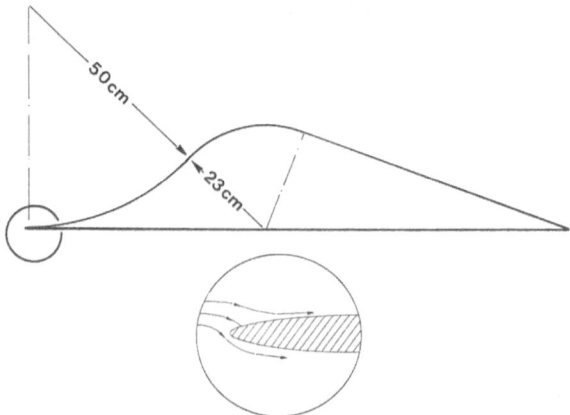

Fig.4. Concave-convex model.

II.3 Measurements

Measurements are obtained using a classical hot film fixed on an x-y-z computerized displacement device. This allows us to perform fixed measurements of two velocity components. We were abble to explore the velocity field by a slow motion of the probe, without a long integration on account of the small turbulence level. All the measurements presented here, have been carried on with probes bent at right angles set horizontally, which kept us from getting close to the wall. Straight probes will replace them in future experiments.

II.4 Visualization

We use the fluorescence induced by the argon laser technique. The fluorescine dye is emitted by a porous strip of $80cm$ by $1.5cm$ placed at the beginning of the concave section. The emission is very uniform, creating a totally negligible upward velocity, smaller than $0.001cm/s$. The lasers of $1W$ or $5W$ are tuned on the blue ray. A mirror vibrating at $1200\,Hz$ creates a laser sheet of uniform brightness. The laser plane normal to the flow and to the concave part intersects the fluorescine surface normally. Sequences of pictures taken by a camera are recorded by a VCR and then analysed on a computer. It is also possible to create a large laser cone with a small angle ($10°$) at the top, using a small glass cylinder with a nearly tangential incidence. The cone set tangent to the fluorescent surface is a powerful device to detect any small perturbation of the flow. The low brightness of this technique, when used for a $1m^2$ field, keeps us from showing pictures here. The first technique, when used with a field of $10x10cm^2$ gives us after computer processing, the same very low level of detection as the second.

III. EXPERIMENTAL RESULTS

III.1 The natural instability

III.1a) Geometry without counter-profile

We measure the velocity profiles at x= 10, 20 and 30 *cm* from the head of the concave part, with U=10 *cm*. A good agreement with Blasius theory is obtained. Because of the negative pressure gradient the flow is slightly accelerated: 8% between 10 and 20 *cm* and 13% between 20 and 30 *cm*. At the same streamwise location we measure the total (u, v) field in the (y, z) plane. We perform separated visualization of the flow for the full range of velocity between 5 and 50 *cm/s*. The corresponding maximum Görtler number accessible is of order 20 (G is computed classically with $\delta = \sqrt{\nu x/U}$. The surprising result is that neither streamwise nor upward motion of the dye are observed despite the strong over-criticality. Bippes[2,19] reported, under equivalent conditions (R=50 *cm*, U=7. 5 *cm/s* with water), appearance of vortices at $G = 5.6$.

We have mainly documented U=10 *cm/s*. At the end of the concave part G equales 11.7 and no streamwise perturbation corresponding to a Görtler vortex appears. We observe that any large enough perturbation (as a small air bubble) triggers the emergence of downstream longitudinal Görtler type vortices associated with an upward motion of the dye and a slow speed streak.

III.1b) Geometry with counter-profile

We run the same experiments in the second geometry. Adding the transparent counter-profile suppresses the favorable pressure gradient. We only use the visualization because of the impossibility of introducing a probe. At the same mean velocity of 10 *cm/s*, a striation of the fluorescent dye sheet is visible after 15 *cm* (G=7. 8). Figure 5 presents a cross section at x=30 *cm* of the fluorescent sheet. The five small bumps marked by arrows are the signature of the Görtler vortices. They measure about 0. 1 *cm* height. The corresponding mean wave lenght is about 1. 5 *cm* and corresponds to the Bippes[2] observations.

III.2 The forced instability

As we get no vortex in the accelerating flow and small vortices in the constant speed flow,

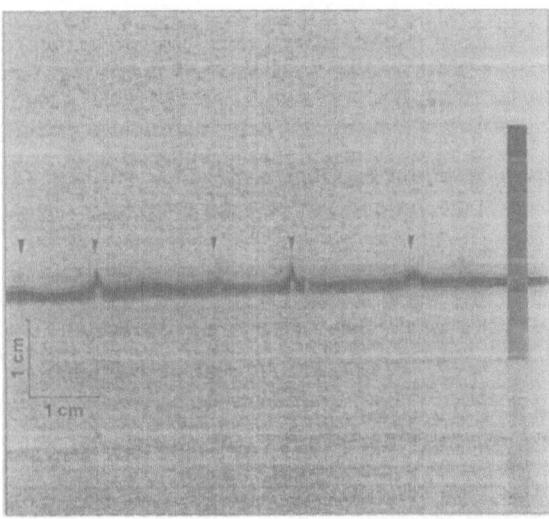

Fig.5. Natural Görtler vortices for the flow with a counter-profile.

we decide to observe the response of the system to various forcings localized in space. We realize forcings with a small jet and a small ribblet (needle) variable in intensity and in height. We report here mainly the first type of forcing, the second type being discussed in [14].

III.2a) Forcing means

We force the flow at the head of the concave section in front of the porous strip in the central part of the profile. The jet comes from a small circular hole of diameter $d = 0.1 cm$ in the wall of a rectangular cavity, fed with fluid by a porous cylinder. The porous cylinder insures a nice control and regularity of the jet but keeps us from doing short impulsive jets. The jet is operated by the difference of hydrostatic pressure measured by the elevation between the driving reservoir and the water surface. The operating vertical velocity of the jet, u_j, ranges between 0 and about $100\ cm/s$. The geometry of a jet in a cross flow on a flat plate is well documented experimentally, numerically and theoretically (see for example ref [20, 21, 22] for details and references.). The velocity ratios $R_j = u_j/U$ vary from 0 to 10 in the 10 cm/s mean flow.

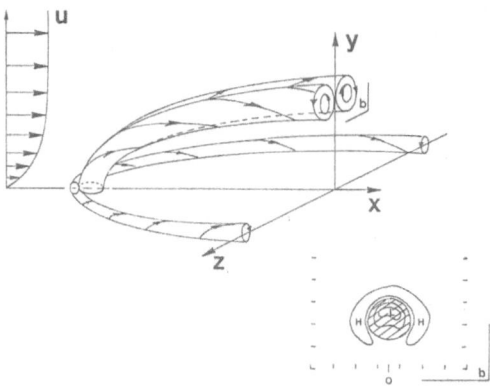

Fig.6. Forcing geometry : jet in a crossflow at $R_j = 4$, a)general view, b)streamwise velocity at $x = 11.5d$, $L = 0.8U$, $H = 1.05U$ from [21].

The flow due to the jet is schematized in figure 6. A weak horseshoe vortex is created at the foot of the jet by the lifting of the boundary layer. The two parts of this vortex stay in the boundary layer, away from each other, and decay slowly downstream. The core of the jet gives rise to a vortex pair stronger than the horseshoe vortex and with an opposite rotation sign. It comes from the wrapping of the streamwise vorticity at the side of the initial jet. Its height and intensity depend on the velocity ratio R_j. Following [21], it goes up to 3 jet diameter for $R_j = 2$, 7d for $R_j = 4$ and more than 15d for $R_j = 8$. This vortex pair resembles a Görtler vortex pair but is of very small size and energy. The vortex slow-speed core (fig 6) mesures about 3 diameters (0. 3 cm here) at its maximum.

In absence of ring instability this vortex pair diffuses slowly downstream. In our case, after bending the perturbation energy equals 1.4 $10^{-7}W$ for $R_j = 4$ and $10^{-6}W$ for $R_j = 8$, roughly proportional to R_j. We conclude that the jet forcing acts by seeding the flow with a very small and weak vortex pair variable in height as R_j (0.3 cm for $R_j = 2$) and in energy as R_j ($10^{-6}W$ for $R_j = 8$). The horseshoe is totally neglected :
-it rotates in the opposite direction compared to the observed Görtler vortex;
-it is one order of magnitude smaller than the vortex pair. The pressure gradient observed in one of our geometries causes only a 8 % strengthening at x=20 cm and 21% at x=30 cm of the streamwise velocity, which in the absence of the Görtler instability have no important effects. Even close to the maximum jet speed ($R_j \sim 8$), the streamwise velocity deficit, when the core reaches a 1.0 cm size, is smaller than 2% of the mean velocity²1 (0.1 cm/s here).

28

III.2b) Effect of the forcing
Results from visualization

We observe that the jet forcing is very efficient but in very different ways in the different geometries. As we may only use the visualization when the counter-profile is present, the effect of the jet in the two configurations will be compared with the laser-induced fluorescence technique, as presented in figure 7. In each picture the laser plane cuts the fluorescent manifold at $x = 30\ cm$. Considering the difference in the vision angle, the two pictures 7b) and 7d) correspond to the same height of the dye pic. We already notice that the two jet intensities R_j, which drive the instability largely differ.

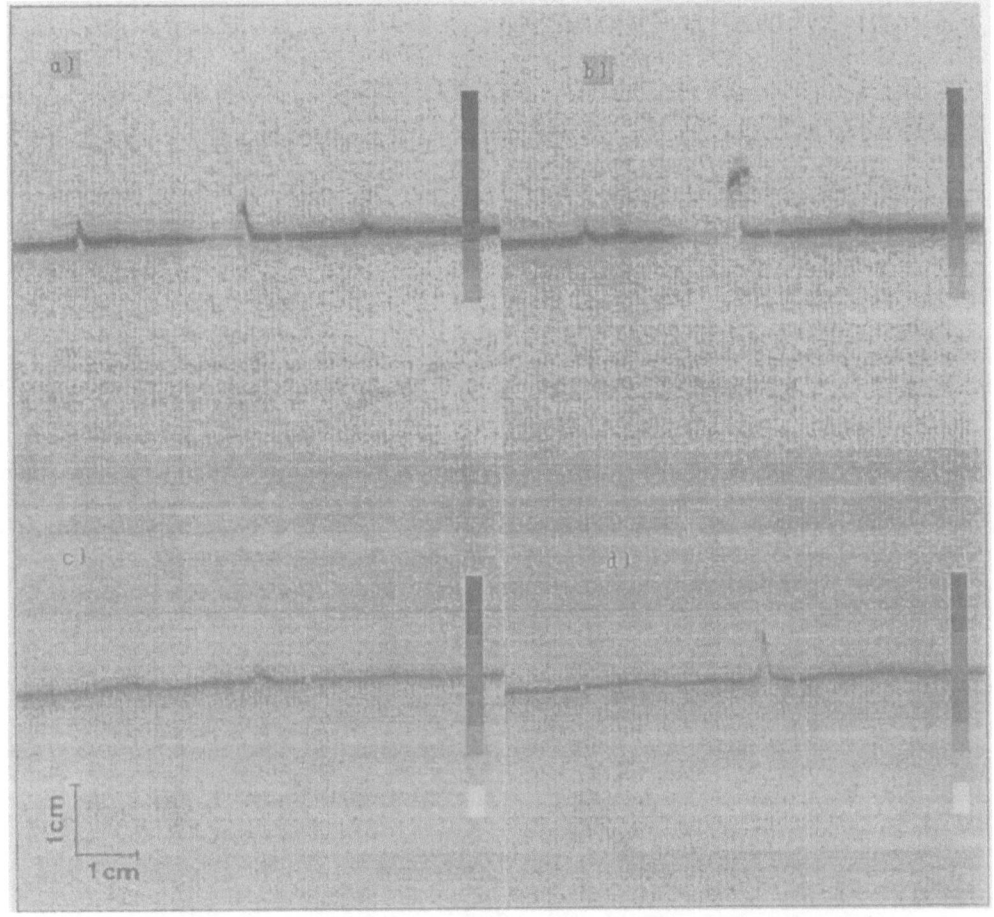

Fig.7. Picture of the forced Görtler pair, a), b) with the counter-profile, c), d) without the counter-profile; a) $R_j = 1.15$, b) and c) $R_j = 3.5$, d) $R_j = 10$.

Figure 8 presents the visual height of the Görtler pair as a function of R_j for both geometries. With the counter-profile, no pressure gradient acts on the flow, and the forced Görtler pair takes over the natural one in a linear and continuous maner at $R_j \sim 0.35$. Whereas without counter-profile, a favorable pressure gradient is present, and a ten times faster jet is required to create a Görtler pair: $R_{jc} \sim 3.5$. Past this threshold the growth is approximately linear in forcing with a smaller slope than in the counter-profile geometry. In this case, where large R_j are accessible, a beginning of saturation is observed at an height of $1.2\ cm$ corresponding to the top of the boundary layer measured by a velocity probe (5δ for a Blasius profile).

For both experiments no hysteresis is detectable and the same Görtler vortex height is measured when increasing or decreasing the forcing. Surprisingly the acceleration of the flow does not only decrease the instability but also delays it with a large threshold effect.

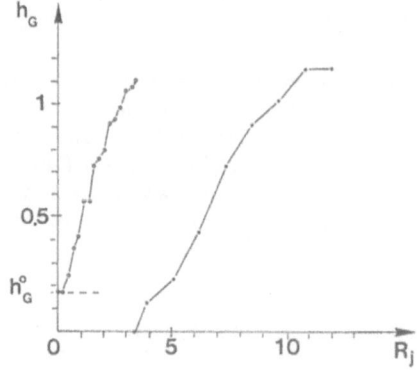

Fig.8. Response to the forcing : visual height h_G (in cm) of the Görtler pair at x=30cm versus R_j for the open geometry, o; for the counter-profile set up, ▪ .

Results from measurements

In the open geometry visual results have been compared to hot wire measurement and the threshold confirmed. As the Görtler pair was observed steady we measure the total (u, v) velocity field in (y, z) planes by slow lateral motion of the probe at velocity 0.357 cm s^{-1}. The low turbulence ratio allows us to measure the longitudinal velocity very precisely. Figure 9 presents the longitudinal velocity field at x=20 cm for $R_j = 9$. On 9a) and 9b) modulation of the boundary profile by the Görtler pair is visible and on 9c) the natural profile without forcing has been subtracted. We observe that the pair is associated to a concentrated low speed streak surrounded by extended and weak high speed regions. The pair does not appear symmetric, certainly because of an amplification of a slight dissymmetry of the forcing.

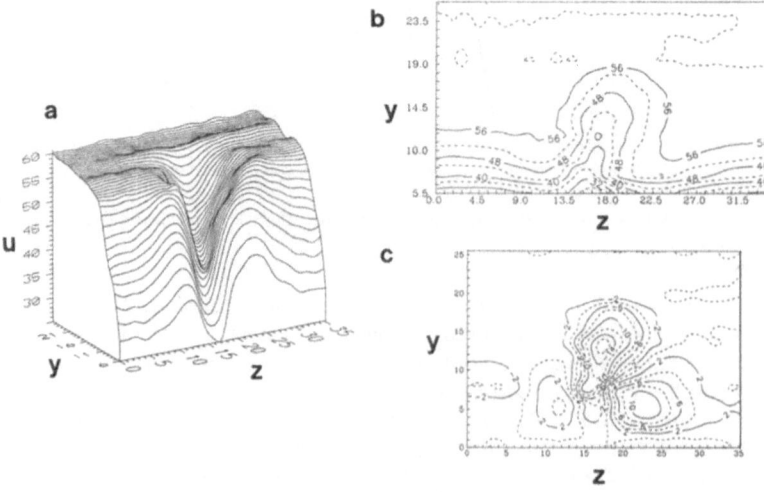

Fig.9. a) Three dimensional representation of the longitudinal velocity field in the (y, z) plane at x=20cm. for $R_j = 9$ b) corresponding velocity field. c) corresponding velocity perturbation field showing the low speed core of the Görtler pair.

IV. CONCLUSION

We confirm that Görtler instability is very sensitive to external perturbations[23-25] and belongs to the convective instability class. Then spatial instability theory can be applied and the system behaves as a spatial amplifier. We test two geometries with and without a counter-profile. In the first case natural Görtler vortices are present and weak forcing by a small jet gives a linear response. Surprisingly, in the second case, no natural Görtler vortex was detectable. The system was insensitive to the jet forcing up to a threshold R_{jc} and then the response was proportional to $(R_j - R_{jc})$. This behavior corresponds exactly to the one we predict for subcritical convective instability. Thus we claim that Görtler instability in presence of a pressure gradient is subcritical and convective in nature. With no pressure gradient, we just observe a convective system, as the turbulence level is high enough to already create vortices. A recent theoretical analysis by Park and Huerre[5] demonstrates for a special model (the asymptotic suction profile with zero pressure gradient) that the Görtler instability is convective and subcritical. The same restrictions hold as in the experiments: changes or inhomogeneity in the velocity profile affect the computed non-linear terms in an unknown manner as the pressure gradient does in the experiment.

We propose to implement a regular spanwise forcing in the future because the criticality may be different for different wavenumbers. Suction will be also tested in order to simplify the forcing geometry. Effects of the pressure gradient on non-linear terms are also to be tested by deforming the counter-profile.

We wish to thank V. BALAJI, V. HAKIM, P. HUERRE, H. PEERHOSSAINI and J. E. WESFREID for comments and contributions. The experiment was made possible by the team SPEA and we wish to thank for their kind help, enthusiasm and efficiency: B. BEAUDOIN, J. C. BOULAY, A. BUTET, S. LASSUS-PIGAT, C. NICLOT, M. NICLOT and H. SCHAFFNER.

REFERENCES

[1] M. CLAUSER and F. CLAUSER, The effect of curvature on the transition from laminar to turbulent boundary layer; Nat. Advisory Committee for Aeronautics, *Techn. Note N° 613*, (1937).

[2] H. BIPPES and H. GÖRTLER, *Acta Mech.* 14, 251 (1972).

[3] H. GÖRTLER, Uber eine dreidimensionale Instabilität laminarer Grenzschichten an konkaven Wänden, *J. Rat. Mech. Anal.* 4, pp 271-321, (1955).

[4] G. HAMMERLIN, Uber das Eigenwertproblem der dreidimensionalen Instabilität laminarer Grenzschichten an konkaven Wänden, *J. Rat. Mech. Anal.* Vol. 4, pp. 279-321 (1955).

[5] A. M. O. SMITH, On the growth of Taylor-Görtler vortices along highly concave walls, *Quart. Appl. Math.* Vol. 13 n° 3, pp 233-262 (1955).

[6] M. J. FLORYAN, W. S. SARIC, Stability of Görtler vortices in boundary layers, *AIAA Journal* Vol. 20 n° 3, AIAA-19-1497R (1982).

[7] S. A. RAGAB, A. H. NAYFEH, Effect of pressure gradients on Görtler instability, *Fluids 2 Plasma Dynamics Conference*, AIAA-80-1377 (1980).

[8] P. HALL, The linear development of Görtler vortices in growing boundary layers, *J. Fluid Mech.* 130, pp 41-58 (1983).

[9] P. HUERRE, On the absolute:convective nature of primary and secondary instabilities. *In Propagation in Systems Far from Equilibrium*, ed. J. E. Weisfreid, H. R. Brand, P. Manneville, G. Albinet, N. Boccara, Berlin: Springer-Verlag pp 340-53 (1988).

[10] P. HUERRE and P. A. MONKEWITZ, local and global instabilities in spatially developing flows, *Annu. Rev. Fluid Mech.* 22, 473 (1990).

[11] J. M. CHOMAZ, P. HUERRE, L. G. REDEKOPP. Effect of nonlinearity and forcing on global modes, *New Trends in Nonlinear Dyn. and Pattern-forming Phenom.*, ed P. Coulet, P. Huerrre, New York/London: Plenum (1990).

[12] J. C. BUELL, P. HUERRE, Inflow/outflow boundary conditions and global dynamics of spatial mixing layers, *Proc. NASA Ames/Stanford Cent. Turbul. Res. Summer Program. Rep. n° CTR-S88*, pp 19-27 (1988).

[13] D. S. PARK and P. HUERRE, *preprint* (1990).

[14] J. M. CHOMAZ, Generalization of absolute and convective instability to non-linear systems,*preprint* (1990).

[15] E. M. LIFSHITZ and L. P. PITAEVSKII, *Physical kinetics*, Pergamon, Landau (1981).

[16] R. J. DEISSLER, Noise-sustained structure, intermittency, and the Ginsburg-Landau equation,*J. Stat. Phys.* 40:371-95 (1985).

[17] M. PERRIER, A. BUTET, Veine hydraulique du CNRM fonctionnement en écoulement neutre,*note de trav. EERM n°216* (1988).

[18] H. PEERHOSSAINI, H. BIPPES and D. STEINBACH,*preprint* (1990).

[19] H. BIPPES, Experimental study of the laminar-turbulent transition on a concave wall in a parallel flow,*NASA* TM-75243 (1978).

[20] J. ANDREOPOLOUS, W. RODI, Experimental investigation of jets in a cross- flow,*J. Fluid Mech.* 138, 93-127 (1984).

[21] R. I. SYKES, W. S. LEWELLEN and S. F. PARKER. On the vorticity dynamics of a turbulent jet in a crossflow,*J. Fluid Mech.* 168, 393-413 (1986).

[22] D. S. NEEDHAM, N. RILEY and J. H. B. SMITH, A jet in crossflow, *J. Fluid Rech.* 188, 159-184 (1988).

[23] V. KOTTKE, Taylor-Görtler vortices and their effect on heat and mass transfer, *proc. of 8th Int. Heat and transfer Conference*, San Francisco (1986).

[24] M. J. FLORYAN, W. S. SARIC, Stability of Görtler vortices in boundary layers,*AIAA Journal* Vol. 20 n° 3 AIAA-19-1497R (1982).

[25] I. TANI, Y. AIHARA, Görtler vortices and boundary layer transition, *ZAMP* Vol. 20, pp609-618 (1969).

DYNAMICS OF LARGE SCALE STRUCTURES IN TURBULENT TWO DIMENSIONAL WAKES

D.G. Christakis,[*] and D.D. Papailiou

Laboratory of Applied Thermodynamics
University of Patras, Patras, Greece

INTRODUCTION

The generation and dynamics of large scale vortices in turbulent shear flows, as well as their participation in the structure of turbulence and the transport processes associated with it, have been the subject of investigation for the past few decades (1-10).

The present experimental work was initiated with the purpose of addressing certain questions regarding the origin and evolution of large scale vortical structures in a turbulent wake, as they interact with its mean shear flow field. The main question under investigation has been that of turbulence memory of initial flow conditions, due to the vortical structures formed at the starting stage of flow development. However, the obtained information revealed some interesting aspects regarding the dynamical behaviour of these structures, pertaining to phenomena such as vortex generation, decay and pairing of vortices, which are also discussed in the present work.

The role of large scale vortices as carriers of turbulence memory of initial flow conditions in turbulent wakes is dominently related to the development of the flow towards a state of self preservation and similarity. This question has been addressed by a number of investigators (11-14) and presently there are two different views discussed in the following.

In a number of publications Townsend (1,11-13) has developed his "equilibrium hypothesis" stating that the general structure of turbulence in all self preserving flows is the same. Based on experimental evidence (15,16,17) he also postulated that the eddy form in all unidirectional turbulent shear flows is nearly similar while existing principal differences such as the variation of the entrainment constant, are attributed to "spesial" eddies, characteristic of individual flows. These vortical structures, according to Townsend, originate in the stretching of unorganized turbulence by the mean shear flow and, in self-preserving flows, they attain a condition of "moving equilibrium" resulting from a balance between growth due to stretching and dissipation caused by the action of smaller eddies.

Bevilaqua and Lykoydis (13) made the remark that the "equilibrium hypothesis" implies that the structure of these flows shows no memory of initial conditions, their development depending on the net force on the fluid but not the details of how the force was applied. This, according to the authors, is not consistent with the results of a number of experiments

---

[*] Present Address: Technical University of Grete

demonstrating, that vortices formed at the initial stage of flow development persist into the region of self preservation (5,14,18). To answer the above questions they performed a series of experiments in which the turbulent wakes of a sphere and a porous disk, with approximatelly the same drag but without exhibiting vortex shedding, were compared. They found that owing to the initial condition differences, two distinct self-preserving states with differences in the large vortex structure evolved. They concluded that, contrary to the implication of Townsend's hypothesis, there exist a turbulence memory caused by the persisting vortices shed at the origin of the flow and suggested a distinction between self-preserved and fully developed flow. They also advanced the concept of the existence of a hiererchy of successive states of self-preservation. Subsequent experiments conducted by Gray and Sheldon (14) in two-dimensional wakes verified some of the above conclusions.

The mentioned experiments cover a downstream distance of approximately 100 diameters and the principal remaining question is what happens far downstream, that is, does the turbulence memory fade in the far wake or it continues to persist? The conducted experiments described in this work, offer a qualitative answer to the above question as well as some insight regarding the dynamics of large scale structures, especially in the far wake.

THE EXPERIMENTS

The conducted experiments consist of comparing the structure of the turbulent two-dimensional wakes formed behind a solid cylinder and a porous plate exhibiting the same drag. Cylinders of diameter 15mm and 25mm and a length of approcimatelly 800mm were used. The porous plate, consisting of a number of equally spaced parallel wires, had a porosity equal to 0.523.

The two bodies were mounted on a moving cart and towed in a water tank of dimensions 7mx2mx2m with side glass windows allowing flow visualization studies to be conducted (fig. 1). The moving cart rode on bearings attached to its undersides, over a track of two parallel polished steel rods placed horizontally on the top of the water tank. The motion of the cart was controlled by a system consisting of a cable and a number of pulleys properly connected with a D.C. motor of adjustable rpm. All moving parts and the rods were carefully aligned, to provide for smooth and continous motion of the cart. The velocity of the body was estimated by measuring the rate of change of the resistance of a thin wire stretched along the tank's wall, remaining in contact with the sharp edge of a stylous attached to the moving cart. The system proved very sensitive to small disturbances in the motion and was therefore used to check its smoothness.

The drag and lift forces on the bodies were measured by means of the device shown in fig. 1, consisting of a number of straingauges properly attachend to the body's support and connected to a Wheatstone bridge.

Visualization of the flow was achieved by using polysterene particles of diameter less than 70 and of almost neutral buoyancy. To obtain the flow pattern as it developed behind the moving bodies, consequtive photographs were taken by employing a motor-drive camera, for more than 1000 diameters and for Reynolds numbers between 10^3 and $3x10^3$.

DISCUSSION OF THE EXPERIMENTAL RESULTS

The visualized flow patterns of the near wake behind the cylinder and the porous plate, are shown for comparison in the obtained photographs (plates 1 to 6). From their inspection it becomes evident that the flow structure in the two cases develops differently. For approximatelly the first ten diameters, no vortices are formed behind the plate, while the corresponding flow region behind the cylinder is dominated by the shed vortices. This, results to an almost complete lack of entrainment in the wake of the plate, contrary to the substantial broadening occuring in the wake of the cylinder.

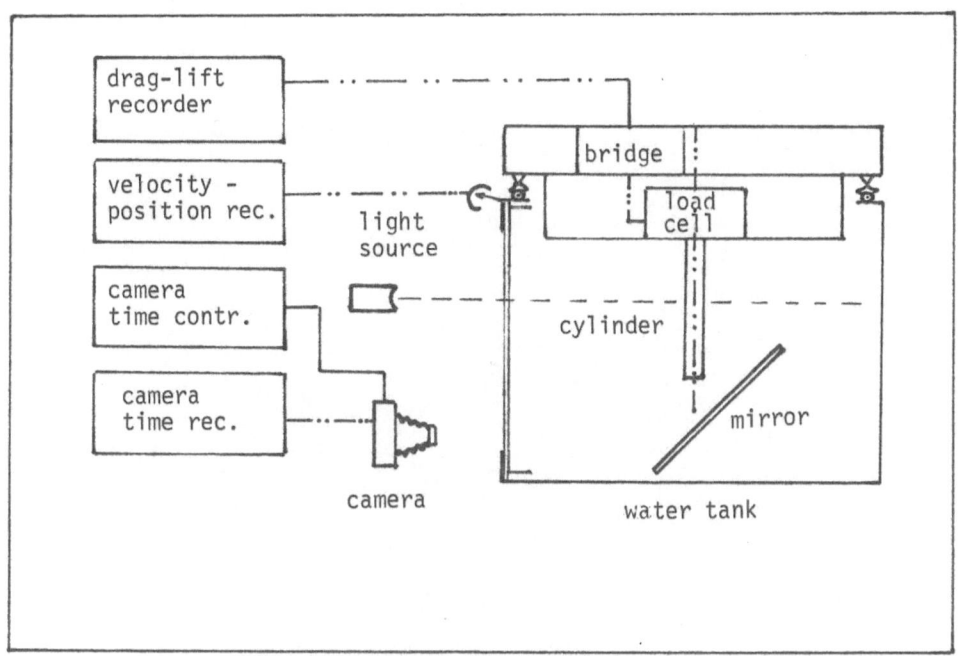

Figure 1. Experimental set-up

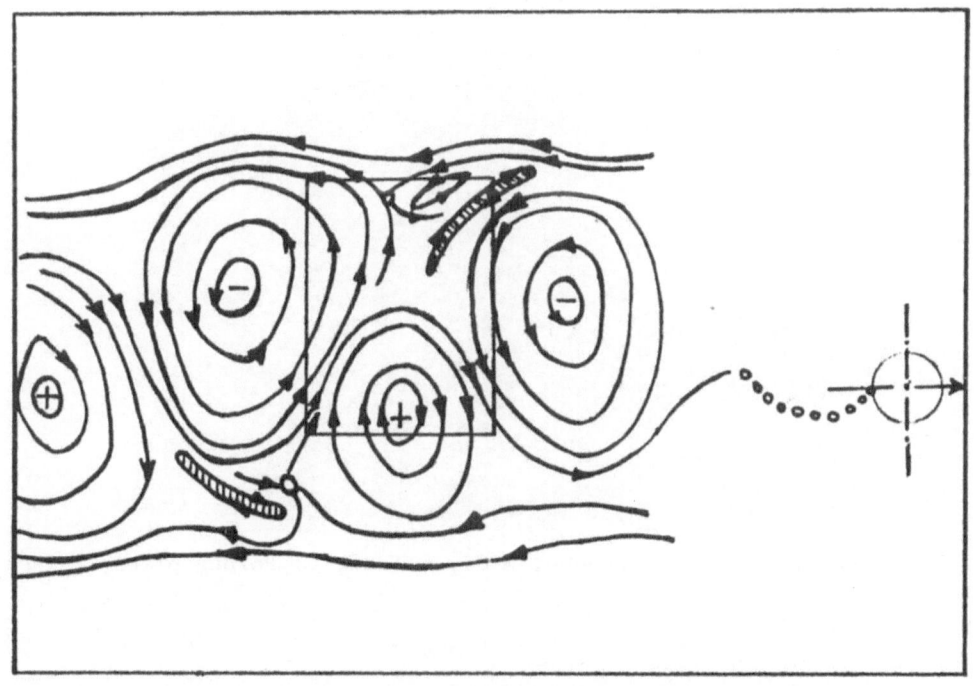

Figure 2. The topology of the far wake region

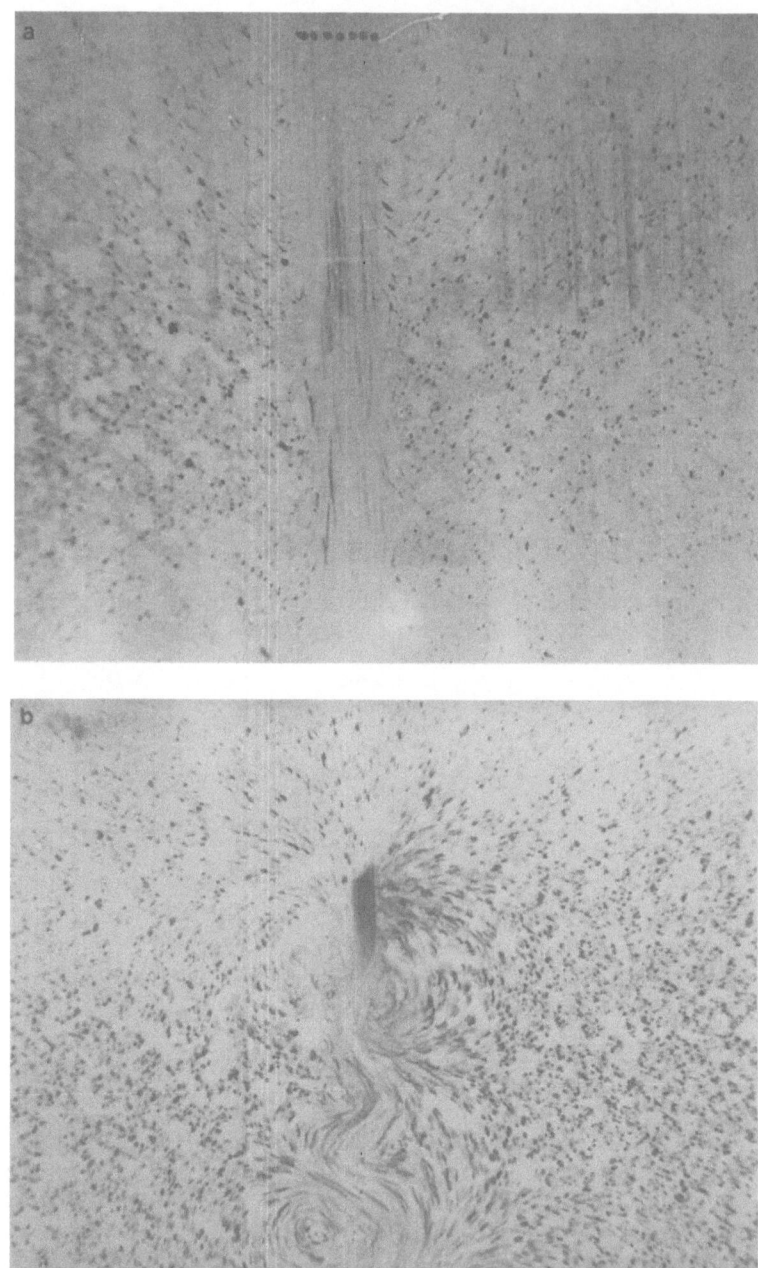

Plate 1. a) Cylinder, Re=3400, x=6d, t=1sec, b) Porous plate, t=1sec

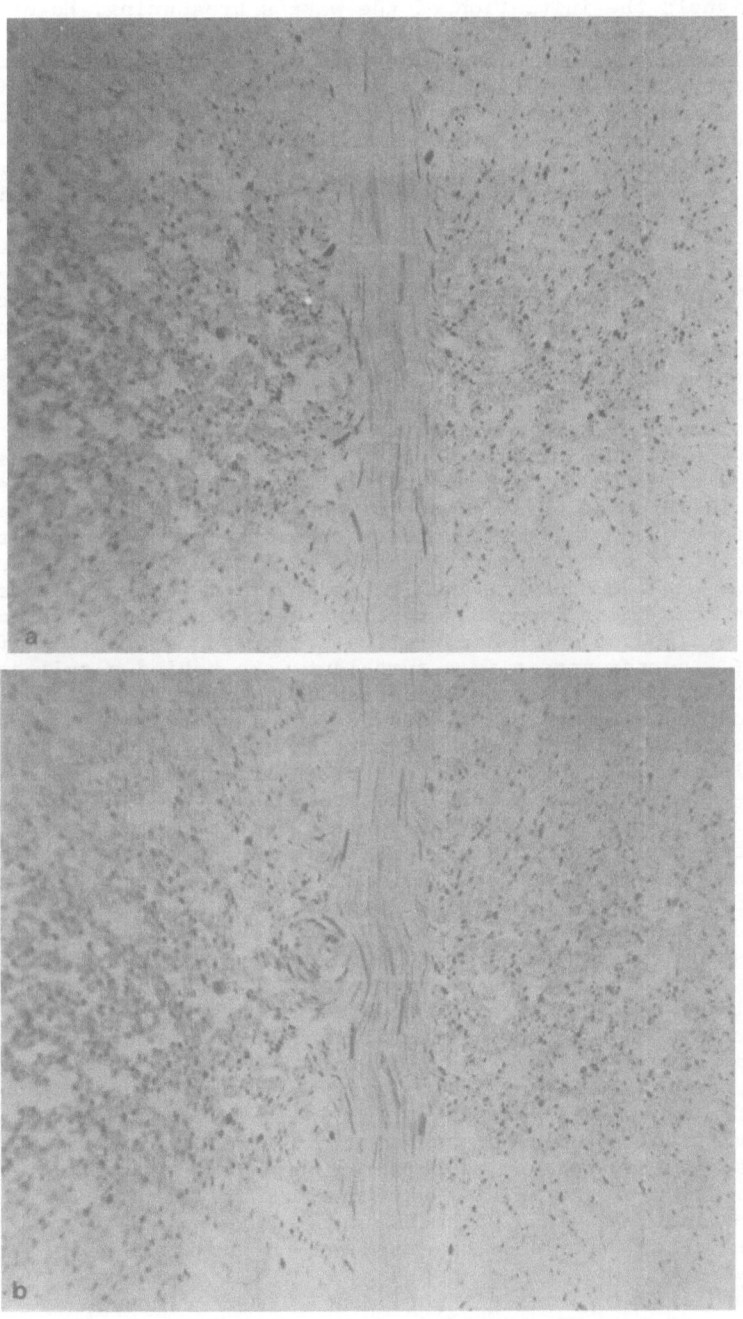

Plate 2. Porous plate, a) t=2sec, b) t=3sec

At a distance of about ten diameters, a sporadic formation of vortices occurs at the edges of the core of the plate's wake, apparently as a result of the development of flow instabilities. The appearence of vortices, with a geometry bearing no resemblance to the periodic formation of a v. Karman street, signals the initiation of the wake's broadening. However, entrainment occurs at a considerably reduced rate as compared to that corresponding to the cylinder's wake. This can be seen in the flow pattern of the two wakes depicted in a series of photographs, covering distances from about 50 to 300 diameters (plates 3 to 6). It is also evident, that after a distance of approximatelly two to three hundred diameters, the above described differences between the two wakes disappear. They both appear visually identical, not allowing any inference to be made regarding their origin and initial conditions.

Inspection of a large number of photographs of the cylinder's wake have shown that, in the range of low Reynolds number under study, vortices shed at the origin of the wake were able to survive for distances over two hundred diameters that is, comparable to the distance needed for the two wakes to achieve a state of visually similar development. Unfortunately, experimental restrictions limiting the view field of the camera, prohibited the tracing of identified vortices further downstream.

Visualization experiments describing similar observations have been reported by previous investigators (3,5,18,19). Also, reference should be made to the experiments by Townsend (15) in a turbulent wake behind a cylinder, in which an array of hot wires was used. In these experiments, the existence of periodic flow patterns consisting of groups of four or five consequtive vortices were detected at 180 cylinder diameters downstream. Although no explanation was given regarding their origin, it is plausible to surmise that these groups of vortices were surviving parts of the vortex street formed behind the cylinder. The preceding description of the development of a vortex street in a turbulent wake behind a cylinder could also fit with the observation made by Teneda (19). Finally, it should be added, that vortex generation probably due to flow instability, has been also observed in this region of the far wake.

Notwithstanding the limitations of the presented experimental results due to their qualitative nature, the presented evidence leads to the following conclusions regarding the development of the two wakes and the question of turbulence memory of initial conditions caused by the presence of vortices shed at the origin of the wake.

The conducted experiments verified the persistence of vortices shed behind the cylinder for a distance of a few hundred diameters downstream. Their presence constitutes a turbulence memory, as claimed by previous investigators, which persist not only for the reported distance of one hundred diameters, but for a few hundred diameters that is, as long as these vortices remain intact, therefore influencing the structure of the turbulent flow field. Beyond this distance, their destruction results to the fading of turbulence memory of initial conditions as the visually identical structure of the two wakes, indicates. It appears however, that the growth-decay cycle of vortical structures, suggested by Townsend in his equilibrium hypothesis, is also present in this region while it dominates the flow after the obliteration of the initial vortices. Finally, apart from the decisive role of the initially shed vortices in the development of the wake towards self preservation, they also dominate the entrainment processes causing its broadening.

In the following, certain phenomena will be discussed pertaining to the dynamics of large scale vortices in a turbulent shear field occuring in both wakes. Particular interest presents the obtained evidence of the existence of different modes of vortex generation in a turbulent wake which is discussed in relation to existing views.

Events of vortex pairing are frequently found depicted in the obtained photograhs of the flow in both wakes as characteristically shown in the two

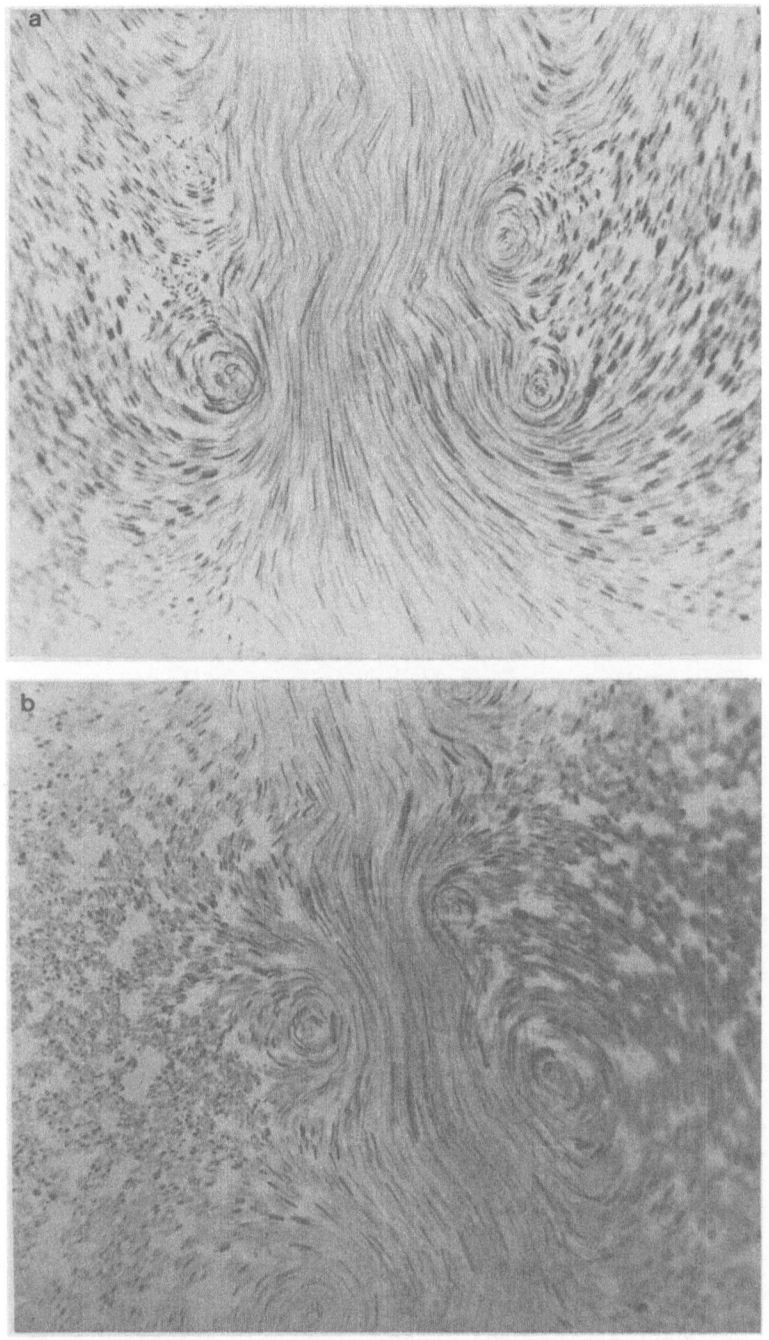

Plate 3. a) Cylinder, x=114d, t=19sec, b) Porous plate, t=19sec

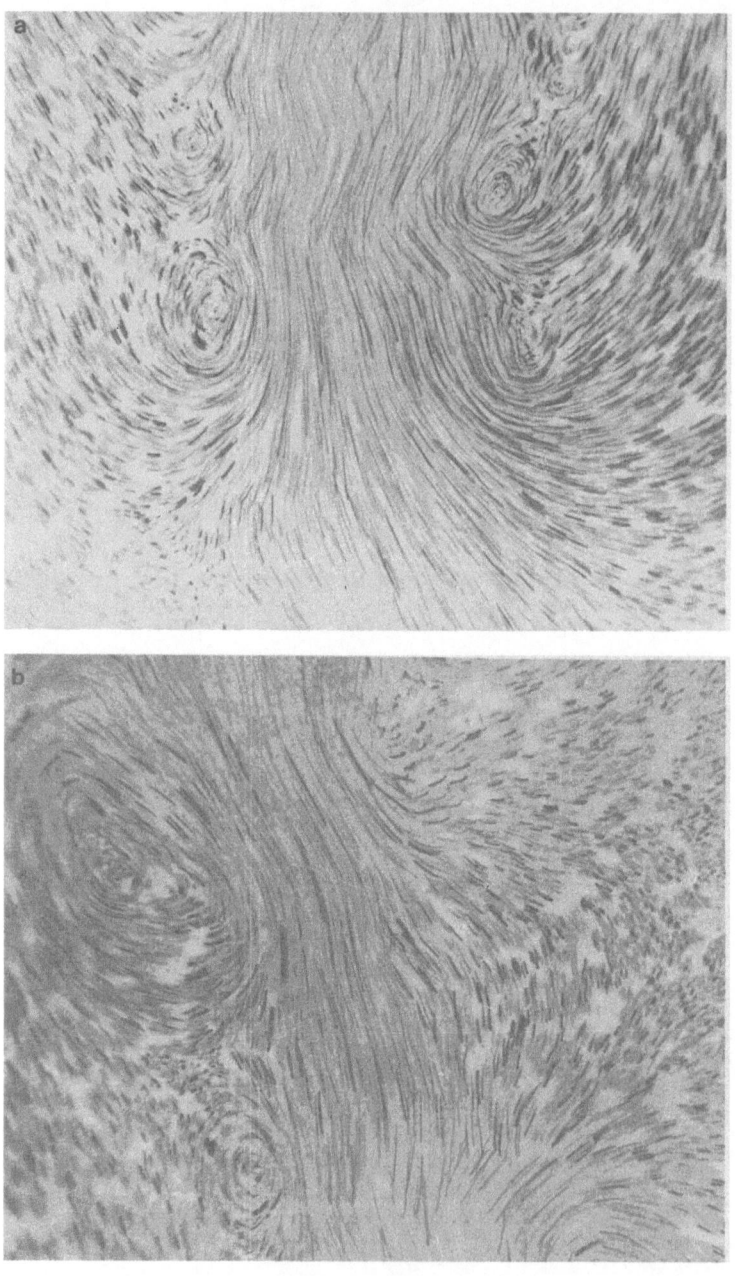

Plate 4. a) Cylinder, x=150d, t=25sec, b) Porous plate, t=25sec

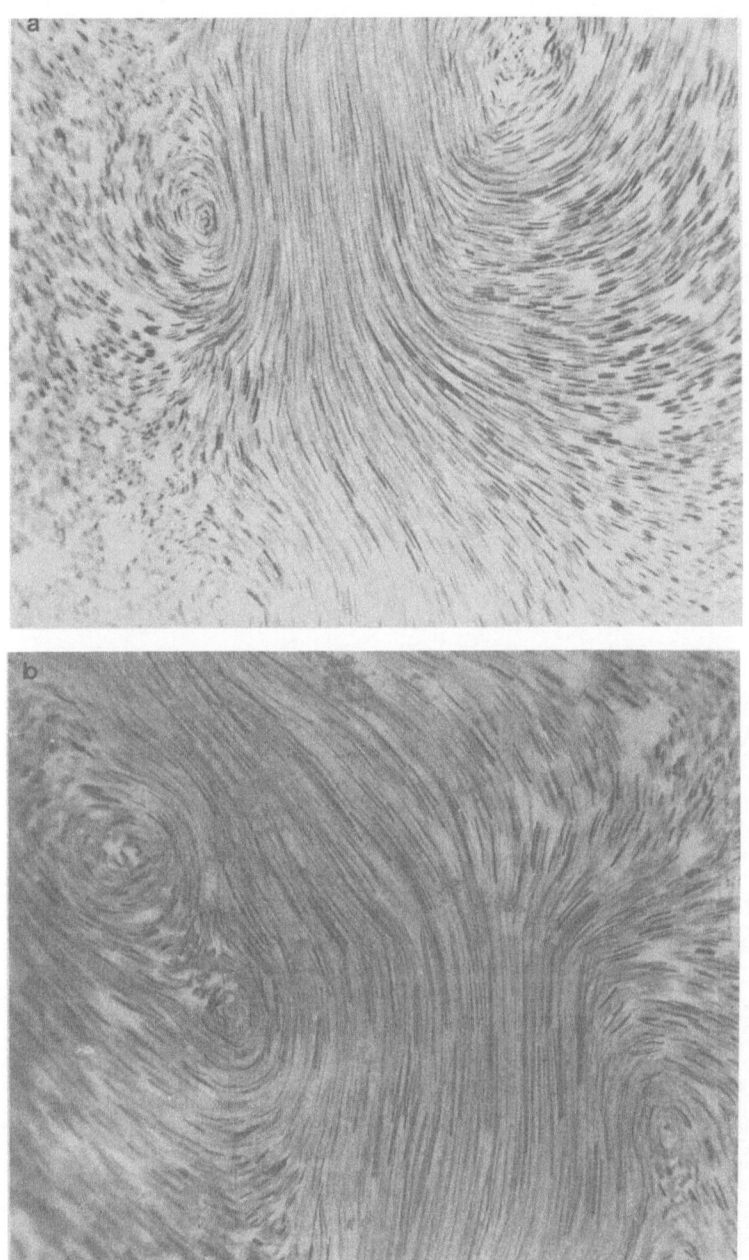

Plate 5. a) Cylinder, x=300d, t=50sec, b) Porous plate, t=50sec

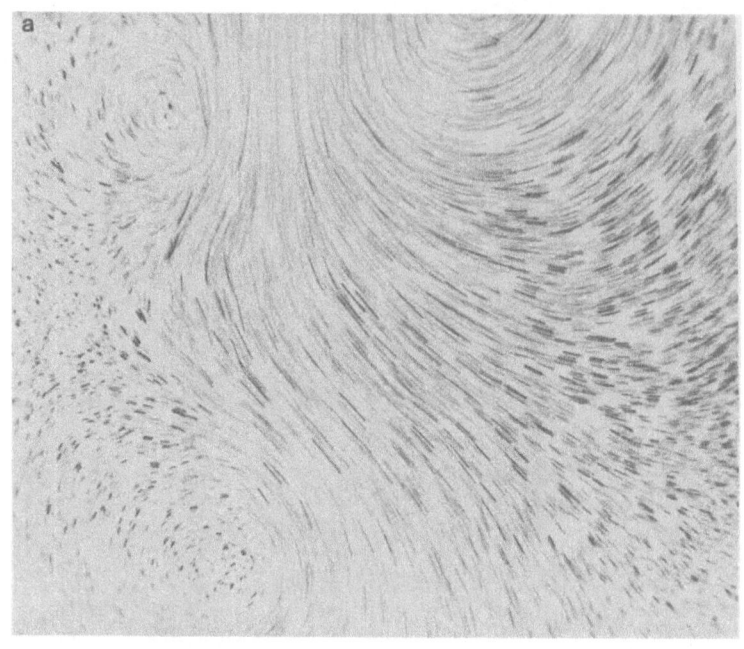

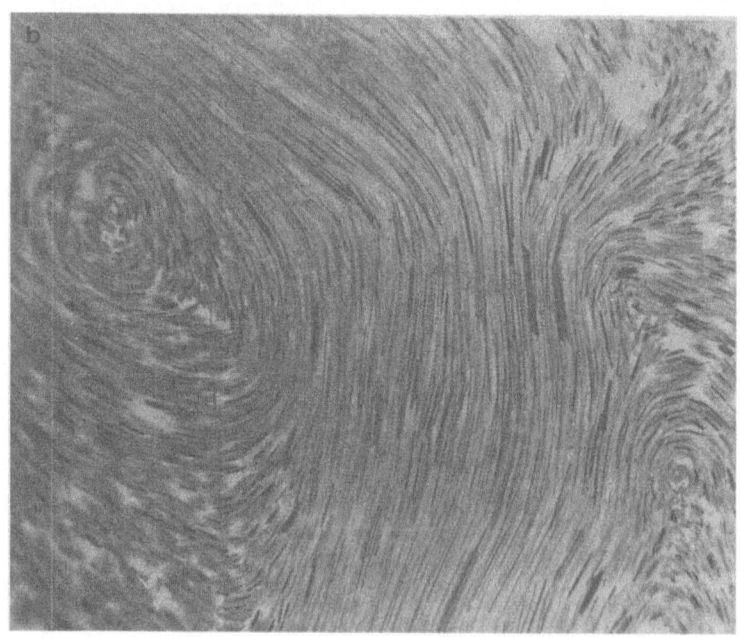

Plate 6. a) Cylinder, x=540d, t=90sec, b) Porous plate, t=90sec

42

sequences of exposures, corresponding to the case of a cylinder (plates 3a,4a,5a) and that of a porous plate (plates 3b,4b,5b,6b). In general, a pairing process appears to take place in three successive phases that is, an initial phase of gradual approach of the participating vortices, followed by the merging of their outer parts into a common layer of rotating fluid enveloping them (plate 5b). Finally, during the last phase the process of pairing is completed, resulting to the emergence of a single large vortex structure. In several instances, the merging of two vortices has been observed to take the form of a "vortex dipole" as shown in plate 7. Similar phenomena have been reported to occur behind bodies in oscillating flows (20,21), a pair of bluff bodies (29) and in magnetofluid-mechanic wake flow behind a cylinder (23).

A most interesting process of vortex generation has been observed to occur in the present experiments and is discussed in the following. It is shown clearly in a sequence of exposures obtained at distances beyond 7000 cylinder diameters (plates 8 to 12). To depict the flow pattern at this late stage of the wake's development, it was necessary to continue taking photographs after the cylinder stoped that is, after it reached the end wall of the tank. Since the flow motion in the wake was extremelly slow at these far distances, it was required to increase gradually the exposure time up to 60 seconds, in order to allow the particles to trace a detectable path.

The described experimental conditions pose the question of possible alterations in the structure of the wake, caused by the early termination of the cylinder's motion. No apparent reason for such influence could be found therefore, the effects of the restricting walls on the flow were only considered. It was decided to terminate the experiments when the growth of the wake reached the one third of the tank's width.

The sequence of events depicted in the obtained photographs is discussed in the following. It should be noted that the part of the wake controlled by the camera is indicated in figure 2, where the topology of the flow field is shown.

The study of the photographs reveals that initially, an instability develops in the flow field between two like-sign vortices, in the vicinity of the sagmatic point, along the line of maximum stress which is also a region of maximum dissipation. Subsequently, this instability grows to a series of elementary vortices undergoing successive merging processes as clearly shown in plates 9 to 11. The final result of this elaborate procedure, is the formation of a new large vortex structure shown in plate 12. The successive phases of the generation of the new vortex are also shown schematically in the figures 3-6. Finally, it should be noted that, contrary to the strongly irregular flow motion observed in the photographs of the wake at distances closer to the cylinder, caused by the turbulence action, those corresponding to the far wake region under discussion show a regular flow pattern. This probably indicates that a laminarization process is taking place during this phase of the wake's development.

As discussed in the following some insight can be gained, regarding the nature of the involved physical processes, by considering existing information. From the inspection of the obtained photographs it becomes evident, that the new structure is the final product of the development of an instability initiated within a local shear layer forming between two original vortices, in the vicinity of a saddle point. The mathematical description of this instability is probably different than that of a Kelvin-Helmholtz type, since both the geometry of the local flow field and the physical processes involved are strongly non-linear. In this context, viscosity dominated vortex generation processes have been reported in the works by Santangelo et al (24) and by Dritscell (25) in which, the results of some numerical simulation models are described and the instability producing mechanism is discussed. It should be noted, that the discussed findings in the mentioned sources including the present experiments, are in agreement with the experiments of Hussain and Hayakawa (26) and Antonia

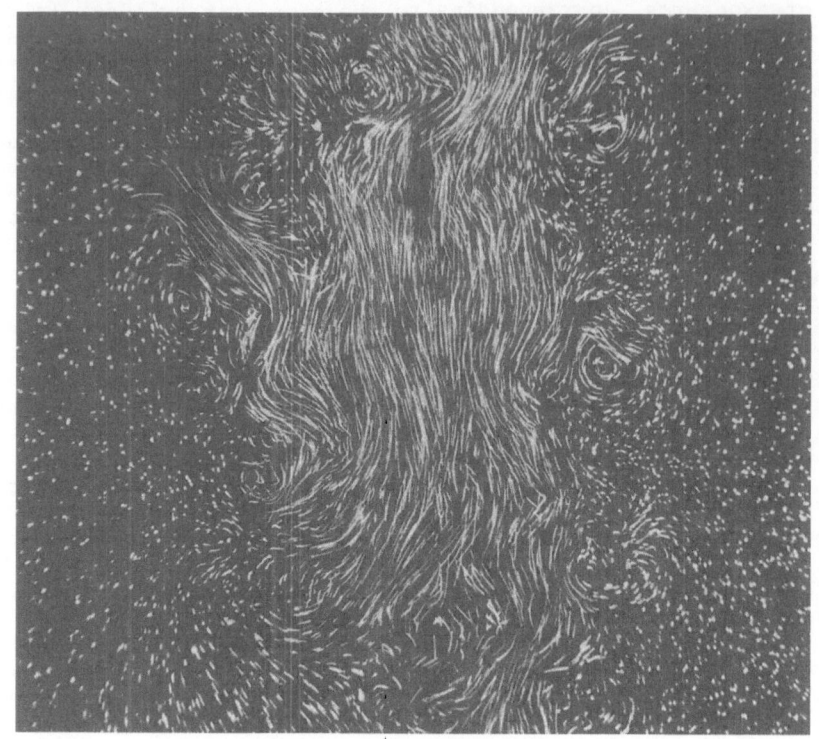

Plate 7. Formation of a "vortex dipole"

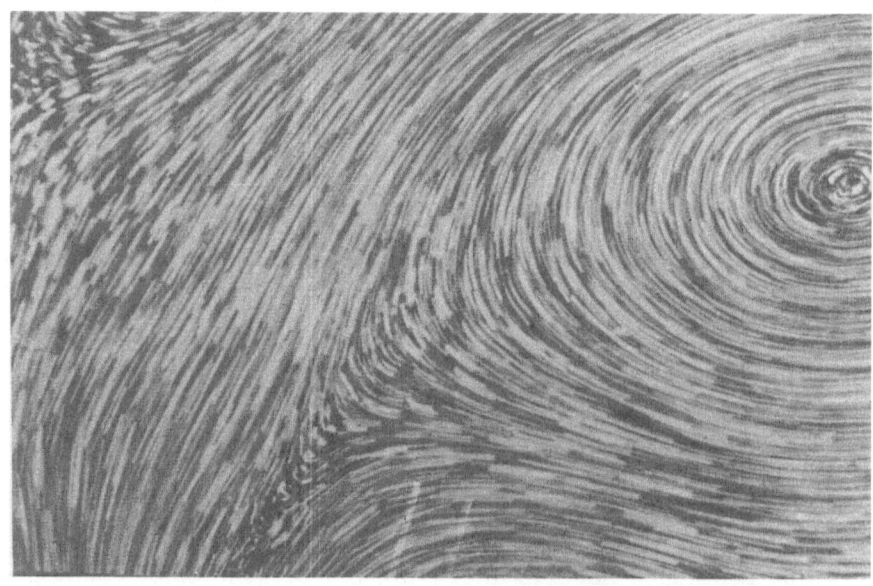

Plate 8. The local shear layer, Re=6540, x=7250d

44

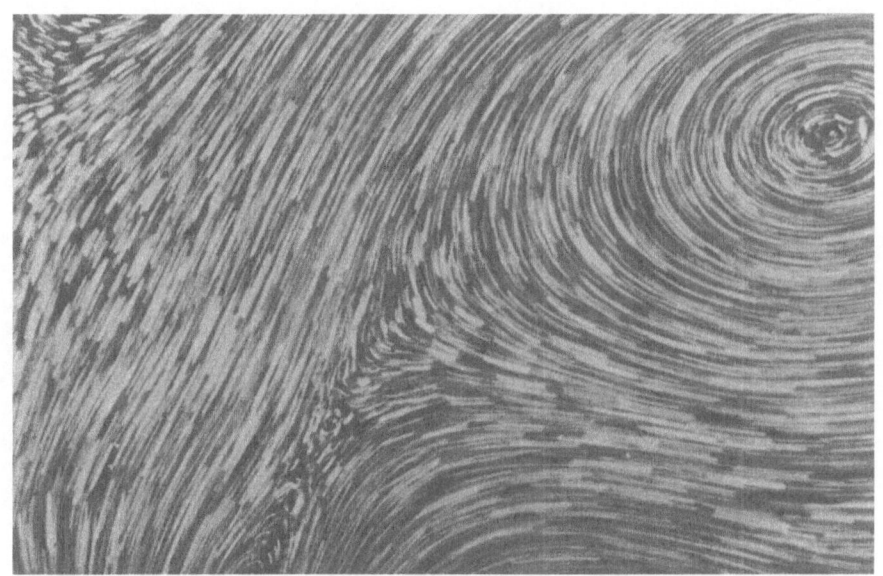

Plate 9. Instability, and elementary vortex formation, x=7400d

Plate 10. Successive vortex merging

Plate 11. Formation of the new vortex structure, x=7600d

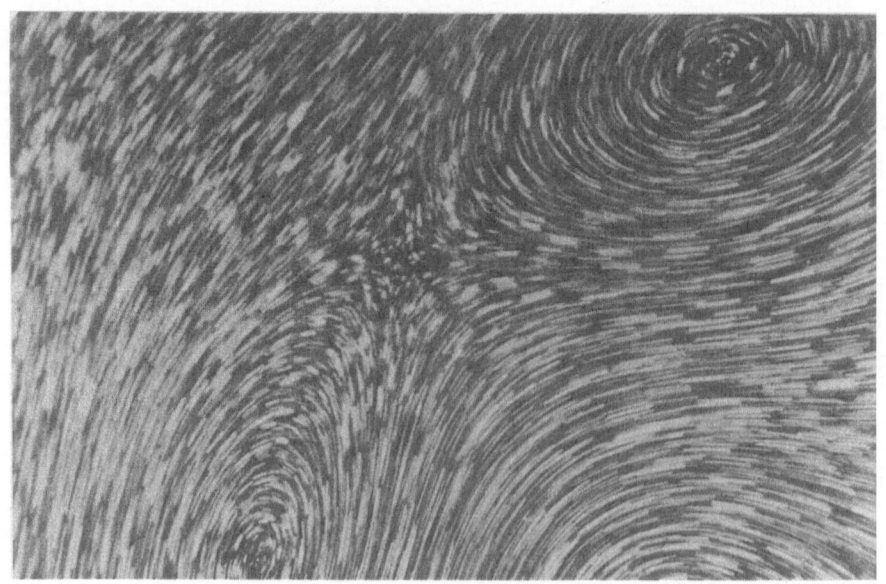

Plate 12. Final stage of vortex formation, x=7840d

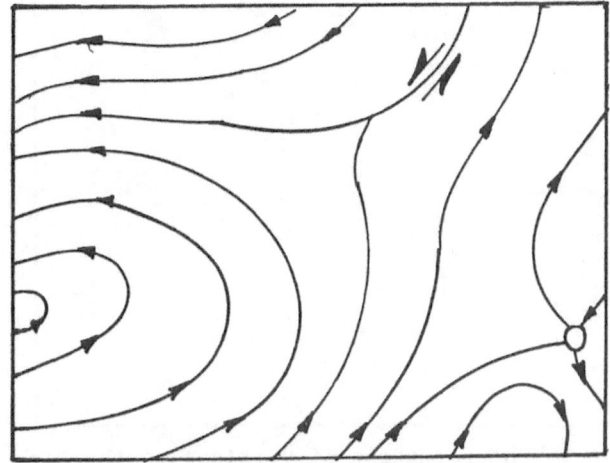

Figure 3.The local shear layer.

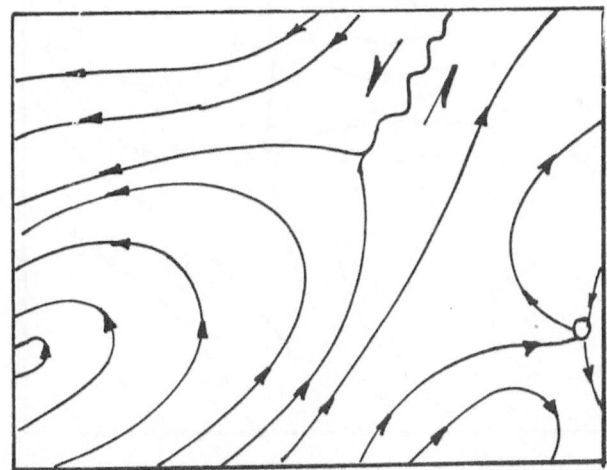

Figure 4.The development of the instability.

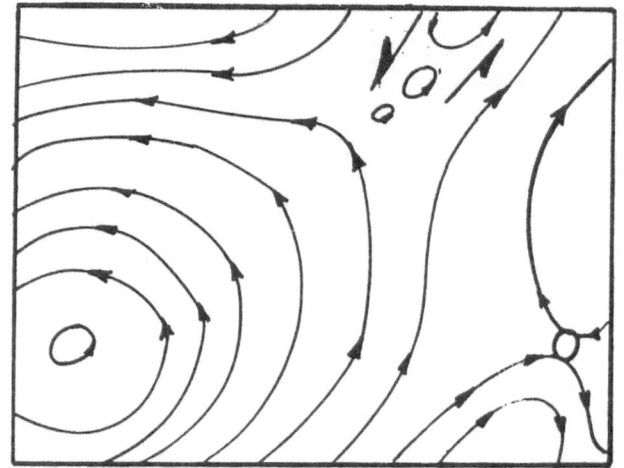

Figure 5. The formation of vortices on the high dissipation area.

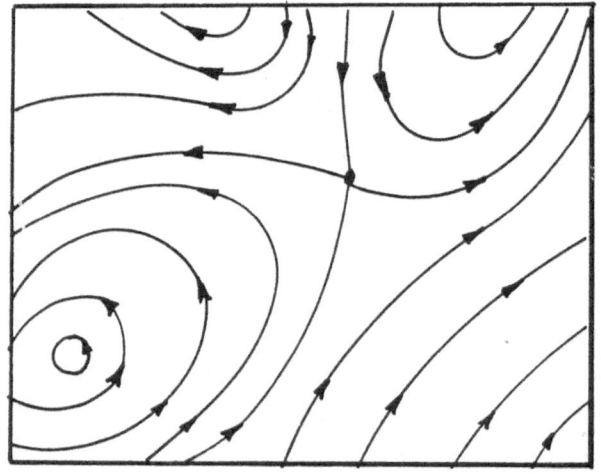

Figure 6.Successive vortex merging—the formation of a new large scale structure.

et al (27) on turbulent wakes behind circular cylinders in which the importance of stretching in the vicinity of the saddle point in production and transport processes is recognized. Finally, the generation of a sequence of elementary vortices in the shear layer behind a bluff body and its subsequent transport to form a larger vortex structure is a mechanism known for the last few decades, as it has been described in the experiments by Pierce (28).

In conclusion, a new mode of large scale vortex structure generation in the far wake not previously observed, is described in the present work. However, the presented experimental evidence has raised a number of basic questions regarding the nature of the described vortex formation mechanism as well as its significance in shear turbulence dynamics. More specifically, such questions are those pertaining to the nature of the observed laminarization process, the repeatability of the phenomenon in the wake, the role of stretching and viscosity in its generation and the spatial extend of the vortex structure, which should be the subject of further future clarification.

References

1. A. A. Townsend "Structure of turbulent shear flow" Cambridge University Press, 1956.
2. H. L. Grant, J. Fluid Mech., Vol. 4, p. 149, 1958.
3. G. L. Brown, A. Roshko, J. Fluid Mech. vol. 64, p. 775, 1974.
4. S. J. Kline, W. C. Reynolds, F. A. Schraub, P. W. Runstadler, J. Fluid Mech., vol. 30, p. 741, 1967.
5. D. D. Papailiou, P. S. Lykoudis, J. Fluid Mech., vol. 62, p. 11, 1974.
6. S. C. Crow, F. H. Champane, J. Fluid Mech., vol. 48, p. 547, 1971.
7. J. Laufer., J. Applied Mech., vol. 50, p. 1079, 1983.
8. B. Cantwell, Ann. Rev. Fluid Mech., vol. 13, p. 457, 1981.
9. H. Aref, E. D. Siggia, J. Fluid Mech., vol. 100, p. 705, 1980.
10. A. K. M. F. Hussain, J. Fluid Mech., vol. 173, p. 303, 1986.
11. A. A. Townsend, J. Fluid Mech., vol. 41, p. 13, 1970.
12. A. A. Townsed "Structure of turbulent shear flow", 2nd Edn. Cambridge University Press, 1976.
13. P. M. Bevilaqua, P. S. Lykoudis, J. Fluid Mech., vol. 89, p. 589, 1978.
14. D. D. Gray, G. F. Sheldon, AIAA J. vol. 20, p. 150, 1981.
15. A. A. Townsend, J. Fluid Mech., vol. 95, p. 515, 1979.
16. J. F. Keefer, J. Fluid Mech., vol. 22, p. 135, 1965.
17. P. Bradshaw, D. H. Ferriss, R. F. Johnson, J. Fluid Mech., vol. 19, p. 591, 1964.
18. P. M. Bevilaqua, P. S. Lykoudis, AIAA J., vol. 9, p. 1657, 1971.
19. S. Taneda, J. Phys. Soc. Japan, vol. 14, p. 843, 1959.
20. J. M. R. Graham, J. Fluid Mech., vol. 97, p. 331, 1980.
21. M. J. Downie, P. W. Bearman, J. M. R. Graham, J. Fluid Mech., vol. 189, p. 243, 1988.
22. C. H. K. Williamson, J. Fluid Mech., vol. 159, p. 1, 1985.
23. D. D. Papailiou, Progress in Astronautics and Aeronautics Series, Amer. Inst. Aeron. and Astron. Inc. vol. 100, 1985.
24. P. Santangelo, R. Benzi, B. Legras, Physics of Fluids, A!, (6) June 1989.
25. D. G. Dritschel, J. Fluid Mech., vol. 194, p. 511, 1988.
26. M. Hayakawa, A. K. M. F. Hussain, "Advances in Turbulence", Eds G. Compte - Bellot, J. Mathieu, Springer-Verlag, p. 416, 1986.
27. R. A. Antonia, L. W. B. Browne, D. K. Bisset, "Advances in Turbulence", Eds G. Compte - Bellot, J. Mathieu, Springer-Verlag, p. 337, 1986.
28. D. Pierce, J. Fluid Mech., vol. 11, p. 460, 1961.

VORTEX SPLITTING AND SHEDDING CELL INTERACTIONS IN WAKES
BEHIND LINEARLY TAPERED CYLINDERS AT LOW REYNOLDS NUMBERS

Charles W. Van Atta* and Paul S. Piccirillo

Dept. of Applied Mechanics and Engineering Sciences
(*) also Scripps Institution of Oceanography
University of California, San Diego
La Jolla, California 92093 U.S.A.

INTRODUCTION

The large scale or "global" vorticity structure of two-dimensional, non-tapered, bluff bodies at moderate and large Reynolds numbers is dominated by an essentially two-dimensional geometrical structure. However, when the body geometry is three-dimensional and varies in the spanwise direction, three-dimensional vortex dynamics and geometry play an essential role in determining the vortex topology of the global wake structure.

There have been few studies of structure in these cases, even for the simplest geometry of a uniformly tapered cylinder or cone. Plane wakes behind a flat plate subjected to periodic spanwise perturbations have been studied experimentally and numerically by Meiburg & Lasheras (1988).

The present paper summarises results of experimental and theoretical work in progress at UCSD and NASA Ames to examine bluff body wake structure in low Reynolds number wakes of tapered cylinders and cones. One may conjecture that similar large scale structure will be found in turbulent wakes of bluff bodies at much larger Reynolds numbers.

The earliest work on vortex shedding from tapered cylinders and cones was done by Gaster (1969, 1971). Gaster discovered a low frequency modulation which gave hot-wire wake velocity signals a two-frequency quasi-periodic appearance. Van Atta & Piccirillo (1990) found that for cylinders with taper ratios in the range 32:1-13:1 the spectrum of the velocity contained a series of peaks at multiples of the modulation frequency. The amplitude of each peak depended on spanwise location z along the cylinder, with lower frequencies being more excited for larger diameter $d(z)$ in a continuous transition as the hot-wire was moved in the spanwise direction.

Our ongoing experimental work (Piccirillo (1990) and Piccirillo & Van Atta (1991)) is directed at determining the cause of the low frequency modulation and corresponding multiple peaks in the spectrum, especially the connection between properties of the hot-wire signals and the geometry of the vortex wake structure. Numerical simulations of the flow are currently being carried out by Jesperson & Levit (1991) at NASA Ames.

CURRENT EXPERIMENTS

The experiments of Piccirillo (1990) and Piccirillo & Van Atta (1991) studied the connection between the wake vortex interactions, (i.e. mutual induction, splitting or bifurcation of vortices), wake velocity spectra and relative phase variation with z, and the dynamical systems properties as measured from time series of velocity at fixed points in the wake.

The experiments were conducted in the low turbulence wind tunnel in the Department of Applied Mechanics and Engineering Sciences at UCSD. The experimental cylinder had a taper ratio of 100 to 1. The Reynolds number based on diameter ranged from 90 at the narrow end to 145 at the wide end of the cylinder. Both flow visualisation (smoke wire) and hot-wire anemometry were used to study the flow.

Hot-wire measurements of the streamwise velocity showed that the flow was cellular in nature, consisting of several vortex shedding cells of individually constant but distinct dominant shedding frequency spaced along the axis of the cylinder, similar to the results of Gaster (1971). The dominant frequencies of adjacent cells differed by a constant frequency difference equal to the modulation frequency. Time series of the streamwise velocity and its power spectrum were similar to those found for cone wakes by Van Atta & Piccirillo (1990). Within each cell there is periodic, oblique vortex shedding, which is increasingly modulated as one moves towards the cell boundary.

Through smoke-wire flow visualisation, the cause of the modulation frequency was discovered to be vortex splitting occurring at the cell boundaries, in which a vortex in the lower frequency cell splits into two vortices in the upper frequency cell. Spatially, this process is complex, involving the bending of vortex lines in the boundary regions in order to accommodate the two cells. This behaviour is similar to that found at the end cells in flow behind a non-tapered cylinder by Eisenlohr & Eckelmann (1989) and Williamson (1989)

End conditions were found to have little effect on the global geometry and other quantitative features of the flow, a perhaps surprising result considering the sensitivity of non-tapered cylinder wakes to end conditions. The imposition of a deliberate symmetry-breaking in the form of a taper appears to stabilise the flow by removing the degeneracy of the two-dimensional case, e.g. the degeneracy connected with the equal probability of both positive and negative shedding angles with respect to the cylinder axis.

In the cell boundary region, the flow appears to be 3-frequency periodic, due to interaction between the cells which cause the vortex splitting. Comparison with other tapered cylinder results suggests that the ratio of the modulation frequency to the shedding frequency is an increasing function of Reynolds number and a decreasing function of the taper ratio. A geometric scaling for the cell size is also suggested by the data.

In the far wake, the interaction between the modulation and shedding frequencies causes the wake to collapse to a chaotic state within fifty diameters downstream of the cylinder. This is in sharp contrast to the far wake of a straight cylinder, described by Cimbala, Nagib, & Roshko (1988), for which the wake decays in an orderly fashion in the first 120 diameters behind the cylinder, and only a few discrete new frequencies, subharmonics of the shedding frequency, appear in the flow.

NUMERICAL SIMULATIONS

Jesperson & Levit (1991) have computed the unsteady three-dimensional low Reynolds number flow past a tapered cylinder using the compressible Navier- Stokes equations in generalised curvilinear coordinates and strong conservation law form. The ratio of span to diameter at mid-span was 32 to 1, and the ratio of tip diameter to base diameter was 0.6774. The local Reynolds number ranged from 100 at the tip to 145 at the base. They used second-order central space differencing for the fluxes and an implicit approximate factorization method for the time advance. Their computations and associated flow visualisations were done on the Connection Machine. The complete time series for the crossflow velocity component was saved at several spanwise locations to simulate a hot-wire probe. Time series of the "hot-wire " data were modulated similar to the experimental data, and spectral analysis of the numerical simulation data produced spectra very similar to the experimentally measured spectra. Computed flow visualisations showed that the mismatch between the vortex shedding frequencies at different spanwise locations resulted in vortex "dislocation" events similar to the vortex splitting observed in our laboratory experiments.

FINAL COMMENTS

Although we have gained a good picture of what causes the modulation frequency and the vortex interactions associated with it, we do not as yet have a satisfactory understanding of the mechanism producing it. This is undoubtedly associated with base flows along the cylinder and might be related to start-up conditions. The physics governing the selection mechanism for the location of individual cell boundaries is not yet understood, and one does not know why particular shedding frequencies are selected, and not others. Further experiments and numerical simulations on flow past cylinders and cones with different taper ratios would be useful to help answer these and other related questions.

ACKNOWLEDGMENTS

We thank Dr. Dennis C. Jesperson and Dr. Creon Levit of the NASA Ames Research Center for their continuing stimulating scientific interaction with us.

REFERENCES

Cimbala, J., Nagib, H., and Roshko, A. 1988 Large structures in the far wakes of two dimensional bluff bodies, *J. Fluid Mech.*, 190: 265-298

Eisenlohr, H. and Eckelmann, H. 1989 Vortex splitting and its consequences in the vortex street wake of cylinders at low Reynolds number, *Phys. Fluids,* A 1: 189- 192

Gaster, M., 1969, Vortex Shedding from Slender Cones at Low Reynolds Numbers, *J. Fluid Mech.*, 38: 565-576

Gaster, M., 1971, Vortex Shedding from Circular Cylinders at Low Reynolds Numbers, *J. Fluid Mech.*, 46: 749-756

Jesperson, D. C. and Levit, C. 1991 Numerical Simulation of Flow Past a Tapered Cylinder, Abstract for AIAA 29th Aerospace Sciences Meeting, Reno, Nevada, January, 1991

Meiburg, E. and Lasheras, J. C. 1988, Experimental and numerical investigation of the three-dimensional transition in plane wakes, *J. Fluid Mech.*, 190: 1-37

Piccirillo, P. S. and Van Atta, C. W., 1991, An experimental study of vortex shedding behind a linearly tapered cylinder at low Reynolds number, submitted to *J. Fluid Mech.*

Piccirillo, P. S., 1990, An Experimental study of Flow Behind a Linearly Tapered Cylinder at Low Reynolds Number, M.S. Thesis, Univ. of California, San Diego

C.W. Van Atta and P. Piccirillo, 1990, Topological Defects in Vortex Streets behind Tapered Circular Cylinders at Low Reynolds Numbers, *in:* "New Trends in Nonlinear Dynamics and Pattern Forming Phenomena: The Geometry of Nonequilibrium", NATO ASI Series B (Physics), eds. P. Huerre and P. Coullet, Plenum Press

C. W. Van Atta, 1990, Fluid Dynamical Chaos in Vortex Wakes, in: "The Ubiquity of Chaos ", ed. S. Krasner, American Association for the Advancement of Science

Williamson, C.H.K., 1989, Oblique and parallel modes of shedding in the wake of circular cylinders at low Reynolds numbers, *J. Fluid. Mech.*, 204: 523-542

SHEAR FLOWS:

CONTROL EXPERIMENTS

CONTROL OF TURBULENT SHEAR FLOWS

VIA STATIONARY BOUNDARY CONDITIONS

H. E. Fiedler, D. Hilberg

Hermann-Föttinger-Institut
Technische Universität Berlin
Berlin, FRG

Abstract

Present day's knowledge about origin, dynamics and topology of coherent structures provides a basis for its efficient and specific management and control. Earlier attempts on controlling turbulent flows - observed and reported already in last century - remained incompletely understood and therefore failed to provide a "recipe" as to a general (and thus generally and technologically applicable) principle of the procedure.

In the first part of this paper a short survey on possibilities of turbulence control by stationare boundary conditions is given. Next, the special case of the turbulent mixing layer in a narrow slit - a flow which develops a distinct periodicity - will be discussed in some aspects. Visualization and frequency spectra are presented to illuminate and explain the basic phenomenon, where features of feedback as well as of absolute instability were found.

1. Introduction

To explore the possibilities of flow control has for long time inspired and challenged the engineer as much as the scientist, and many classical examples might be quoted. The topic has lately received increased attention both from technological demands - as e.g. improvement of combustion processes or suppression of aerodynamic noise - and on the other hand as an outcome and side benefit of the study of coherent structures, the stability of which appears to provide the most efficient handle for turbulence control.

Turbulent flows can be effectively controlled via steady boundary conditions, which essentially alter or modify the stability characteristics of the large scale coherent structures.

2. Passive Control = control by steady boundary conditions

Two characteristic examples for this kind of flow control reported previously by one of the authors are the mixing layer in a favourable pressure gradient and the distorted mixing layer (Paschereit et al (1989), Fiedler et al (1990)).

Often a distinction into "active- " and "passive-" control is made in the literature, which is based on the source of the energy needed to activate the control mechanism. A controlling method is considered passive, when by means of certain circumstances, e.g. boundary conditions, feedback loops or mechanical devices, the controlling energy is drawn directly from the flow to be controlled. Passive methods are often of technical simplicity and practicability, whereas active control serves most conveniently (being itself easily controllable) for a systematic and scientific study of turbulent phenomena.

Flow stability - and in particular the stability of the coherent structures - is obviously the key to control. It is responsible for such mechanisms as the laminar/turbulent transition, and for the origin (the creation) and to some extent the decay of structures by creation of smaller structures causing instability of the former. We therefore distinguish between primary and secondary instability, which may manifest in various modes. While the basic mechanism leading to growth of perturbations may be one of many, a major and more general distinction in stability behaviour, of particular importance for controllability is provided by the temporal and spatial growth behaviour of the initial disturbances. Major distinctions as particularly disczussed by Huerre & Monkewitz (1985, 1990) are between absolute- and convective- and between local- and global instability.

There are basic differences between those four conditions with respect to their control behaviour: Convectively unstable flows have the characteristic of a frequency selective, non-linear amplifier. Accordingly they are sensitive to external forcing at a given frequency, while the amplifier characteristics themselves may be modified by steady boundary conditions.

Absolutely unstable flows, which have more of an oscillator characteristic, are on the other hand comparatively resistant against external (periodic) forcing (self- excited flows). Flows of this kind may be effectively influenced by changes of the flow in the unstable regimes, as was e.g. demonstrated by Strykowski & Sreenivasan (1989) in the near wake behind a cylinder.

Locally unstable flows may - by modifications of their boundary conditions - become globally unstable. Global instabilities are often indicative of a feedback situation.

Finally there are "mixed" situations, where a flow has locally limited regions of absolute instability. Those may lead to self excited global response in a way similar to that of feedback mechanisms.

Temporal development of the perturbation frequency amplitude is similarly found in absolutely and in globally unstable flows, however, transient times in true feedback cases are apparently shorter and the flow is more easily influenced by external forcing.

Static ("passive") control is achieved by

- modification of structural stability by stationary boundary conditions, e.g. pressure gradients, e.g. transverse by curvature, or longitudinal (Fiedler et al (1990),
- geometric distorsion (continuous variation of the flow cross-section in downstream direction - Paschereit et al (1989)),
  side- and bottom walls (proximity of walls - Hilberg & Fiedler (1989), Michalke (1990));
- by introducing or causing structural instability and disruption of structures either by directly introducing three- dimensional vorticity at the trailing edge (upstream boundary conditions), or by selecting a trailing edge geometry to trigger undulations of the dominant structures. In both cases the production of streamwise vorticity is enhanced, which destabilizes the large structures and at the same time increases small scale mixing (e.g. lobed- or other non-circular nozzles and specially shaped trailing edges) . Disruption or destruction of large scale structures is also possible directly

by screens or gauzes in the flow. This possibility has been examined – among others – by Yajnik & Acharya (1988), Nieberle (1985) and Wygnanski et al (1979).

- by entrainment: Direct influence is executed by the Coanda effect, i.e. when the entrainment is obstructed by the proximity of adjacent walls (not to be confused with the stability modification due to wall proximity as noted above).
- Indirect influence is provided by background turbulence according to its intensity and scale. Related is the aspect of
- fluid characteristics, which in most applications is introduced via entrainment. Here we think of density or viscosity inhomogeneity (temperature, species), and of additives (polymers).

The overall efficiency of static control methods is comparable to that of dynamic methods, since in both cases the coherent structures are controlled and modified. The spatial extent of control depends on the method applied: side-wall boundary conditions control most of the exposed flow. Entrainment and fluid characteristics provide comparable (overall) effects. Upstream boundary conditions have a more limited, local effect. Often a combination of methods is successfully employed (as e.g. the periodic forcing of a three dimensional jet or a distorted mixing layer. In this way it is often possible to increase the effect considerably. In the following we shall discuss a special case which displays various characteristics as discussed above.

2. The Laterally Constrained Plane One-Stream Mixing Layer
2.1 Experimental Arrangements

The investigation of this phenomenon is done in two test sections of identical proportions yet of different dimensions, scaled as 1 : 5. Figure 1 shows the general layout and major dimensions. The test section consists of two side walls with adjustable distance B. There is a bottom wall and an open top. The height of the side walls H is twice the height of the nozzle. The larger test section is fitted with a loudspeaker arrangement at the trailing edge for artificial excitation. Inserts were used for special experiments and test – e.g. the variation of the height h, injection and suction at the face of the trailing edge. The basic flow is supplied from a speed controlled blower via a plenum chamber fitted with screens etc, to provide a smooth flow with low turbulence level.

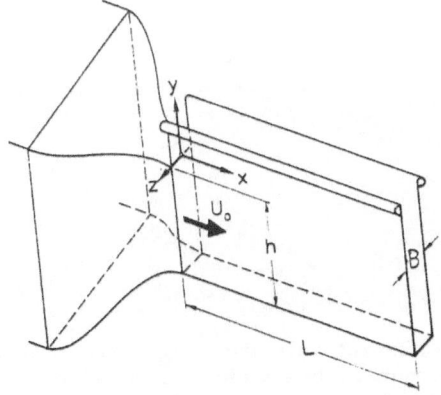

Fig. 1 . general layout and major dimensions

2.2 The Basic Phenomenon

It was observed (first results were reported by Hilberg & Fiedler (1989)), that at reduction of the ratio B/L below, say, 1/10, a single frequency peak emerges in the turbulence spectrum, which increases at further reduction of B/L. Maximum relative amplitues of order $A = u'_{eff.per} /U_0 = 0.08$ are found for B/L ≤ 1/40, at a frequency according to a Strouhal number $S_F = f L/U_0 \approx 0.56$. which characterizes a feedback system. Fig. 2 shows spectra measured in the B/L = 1/40 (narrow case) test section for four different flow velocities, together with the smooth spectrum for B/L = 1/4 for comparison. Paraffin smoke Streak-line-visualizations of the two different flow conditions elucidate the phenomenon.

While, however, these visualizations suggest a considerable increase in the mean spread of the turbulent regime as compared to the "wide" case, this is not found in the mean velocity distributions, nor in the turbulent intensity profiles, which show little difference between the two cases. A possible explanation might be intermittent occurrence of the flow periodicity.

a

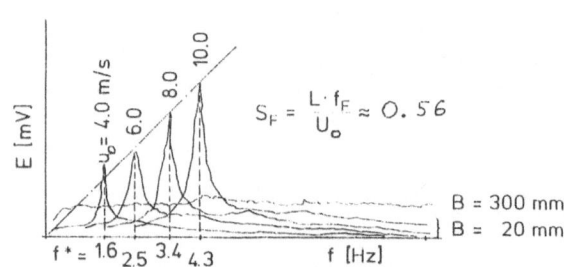

b

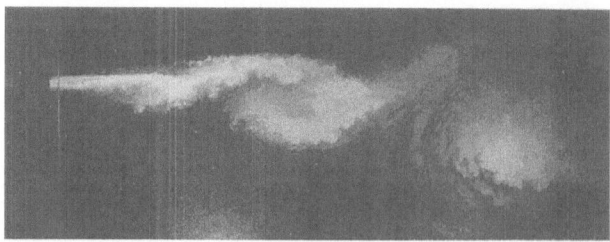

c

Fig. 2 a-c   spectra of different velocities U_0 (narrow case) - for comparision typical spectrum of the wide case is also outlined (a)
streaklines in the narrow case, U_0 = 4.2 m/sec (b)
streaklines in the wide case, U_0 = 8.0 m/sec (c)

2.3 Investigation of the phenomenon

The major part of the investigation was dedicated to the explanation of the phenomenon described. Feedback via pressure waves from the trailing edges of the sidewalls was identified to be the primary cause for the periodicity. Verification was obtained from a comparison with artificial feedback of a combined hotwire signal from two symmetrical positions in the flow at $x = L$ and $y = 0$ and various spacings b, which was fed into a loudspeaker positioned near the upstream trailing edge of the flow. The principle of his experiment and its main result is shown in figure 3 a-c: Similar flow behaviour was found for small B and for large B yet small b. On the basis of this observation the following picture emerges: In the narrow case the principal structures are essentially two-dimensional and therefore he pressure signals emitted at the sidewall trailing edges during the exit of the structure, which provide the feedback, are well correlated. At larger distances those two feedback signals are – as a consequence of three-dimensional distorsion of the structures (secondary instability etc.) – increasingly uncorrelated and their combined feedback signal at the upstream trailing edge is unable to maintain periodicity. Periodicity becomes perceptible for $B/\Theta_L \approx 3.5$, reaching saturation at $B/\Theta_L \approx 1$, where Θ_L = momentum thickness of the mixing layer at $x = L$. The overall content of the periodic motion on the total fluctuation energy is approximately 25% at saturation.

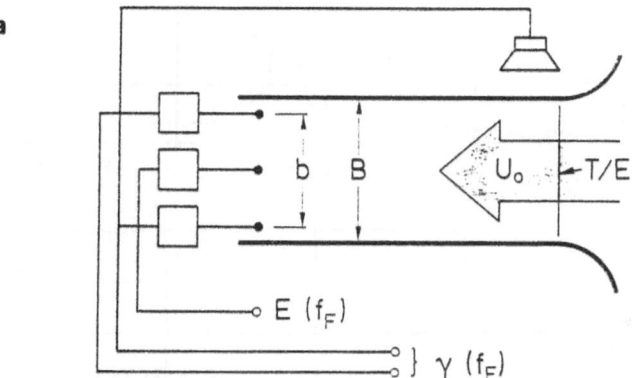

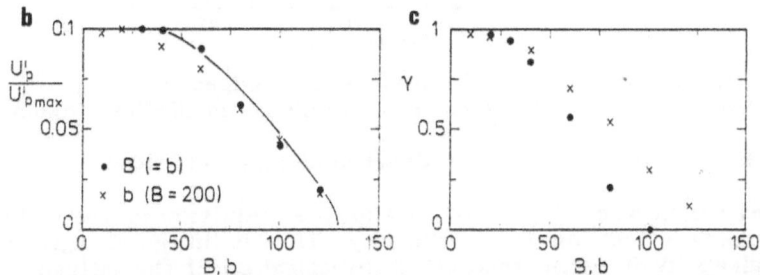

Fig. 3 a-c: the signals of two hw-probes of width b is added and triggers the loudspeaker at the trailing edge (a). Variation of b is similar to variation of B in development of u'$_{eff,per}$ (b) and of the coherence function γ (c)

Dimotakis (1976) has already - on the basis of his correlations - conjectured about the existence of a feedback frequency as a facility effect. It was, however, not substantiated owing to the dimensions of his test section. Only during the final preparation of this manuscript an inverstigation by Buell & Huerre (1988) became known to the authors, in which global resonances of the kind observed in this experiment are assumed.

2.4 Control Characteristics

If feedback from the leading edges is the cause for the periodicity observed, then its modification should control the flow behaviour. Indeed rendering the leading edges "soft" to absorb the pressure signal proves to be amazingly successful. Experiments were done with various soft paper extensions and with geometrical modifications. Best results with almost full suppession of periodicity was achieved with gauze- strips of width B fitted over the whole height of both trailing edges (figure 4 a,b).

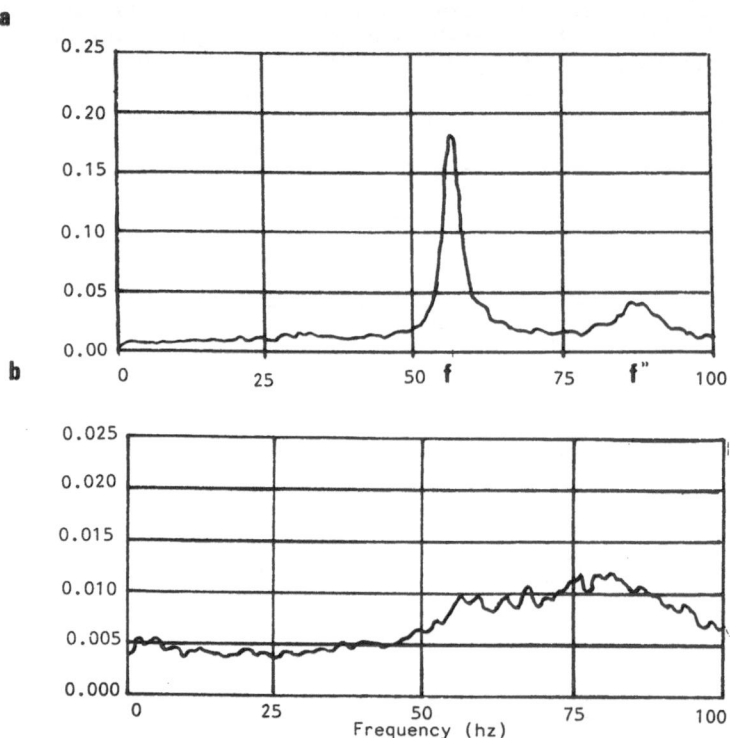

Fig. 4 a, b. spectra measured with sharp leading edges (a), with soft leading edges (b) - amplitude is 10-times expanded

Other conditions investigated were also found to be effective:

(1)   Raising the bottom of the test section (i.e. reduction of the ratio h/L) has a damping effect on the periodicity. This is in good agreement with Michalke's (1990) recent theoretical investigation of the influence of a wall in the vicinity of a parallel flow with infexion point un the latter's stability.

(2)   In addition to its self-excitation this flow shows considerable receptivity for periodic forcing at the trailing edge - a phenomenon which would not be found in a flow with periodicity driven purely by (locally limited) absolute instability.

Thus the natural (feedback-) frequency can be enhanced externally. This is found to be considerably more efficient if not the natural frequency but its harmonic is excited (a case of subharmonic resonance). The feedback frequency is suppressed by excitation at some incompatible frequency in the range between its harmonic and the subharmonic. This method was used to investigate the temporal (transient) excitation- and decay behaviour of the basic frequency, which displayed characteristic behaviours akin to absolute instability as can be described by the Landau equation (see below). A behaviour of this kind was hitherto not observed in plane mixing layer flow.

(3) Blowing and suction at the face of the trailing edge is also of considerable effect on the periodic flow structure, where increase is found when suction is applied and vice versa. This observation points at the possibility of a locally limited region of absolute instability near the trailing edge.

These findings suggest a more complex flow situation than previously assumed. A number of special tests were done which are so far not yet conclusive, showing, however, some interesting aspects.

2.5 Some Special Points

Closer inspection of the velocity spectra shows another frequency $f^* \approx 1.7 f_F$ (fig. 4 a). which is apparently not independent of yet with regard to its phase de-coupled from the feedback frequency.

The period of a wave of frequency f (or a multiple n thereof) travelling from the trailing edge to the position where the feedback signal originates is equal to the time delay L/U_c of the wave in the flow plus the time delay of the feedback signal L/a (L = length of test section, U_c = convection (phase-) velocity and a = velocity of sound). Thus

$L/U_c + L/a = n/f$. For $U_c \ll a$ and (for $U_c = f(x)$ - as is the case here)

$L/U_c = \int dx/U_c (x)$.

Then we have a feedback Strouhal number $S_F = f/\int dx/U_c = 1 (=n) \approx 1$.

The second peak in the spectrum at f^* follows a Strouhal number $S^* = f^* L/U_0 \approx 1$, suggesting a standing wave. The ratio f_F/f^* is constant, the the ratio of the two amplitudes depends on geometry as well as on the particular position in the flow.

The transient behaviour of the periodic signal is described by the non-dimensional Landau equation:

$$dA^{*2}/dt = aA^{*2} - bA^{*4},$$

where $A^* = A/U_0$ and $t^* = t U_0/L$.

From measurements we find $a = 2(\Delta\log(A^*)/\Delta t^*) = 1.77$, and

$$b = a/A_{max}^{*2} = 6.37$$

Those values are - expectedly - equal over most of the flow region and different from the values known from true absolute instability characteristics. The value of b was, however, found to oscillate in what appeared to be a random manner. This observation and the peculiar way of its oscillation gave rise to its analysis as a possible candidate for a low dimensional dynamical system (Chaos Dimension): For this end we applied the algortithm suggested by Theiler (1986) to the envelope of the amplitudes of the u'-fluctuation constituent at feedback frequency. The result obtained so far, shown in figure 5 a-c, is not conclusive. It suggests, however, the existence of a dimension of the order 10 (Eggers (1989).

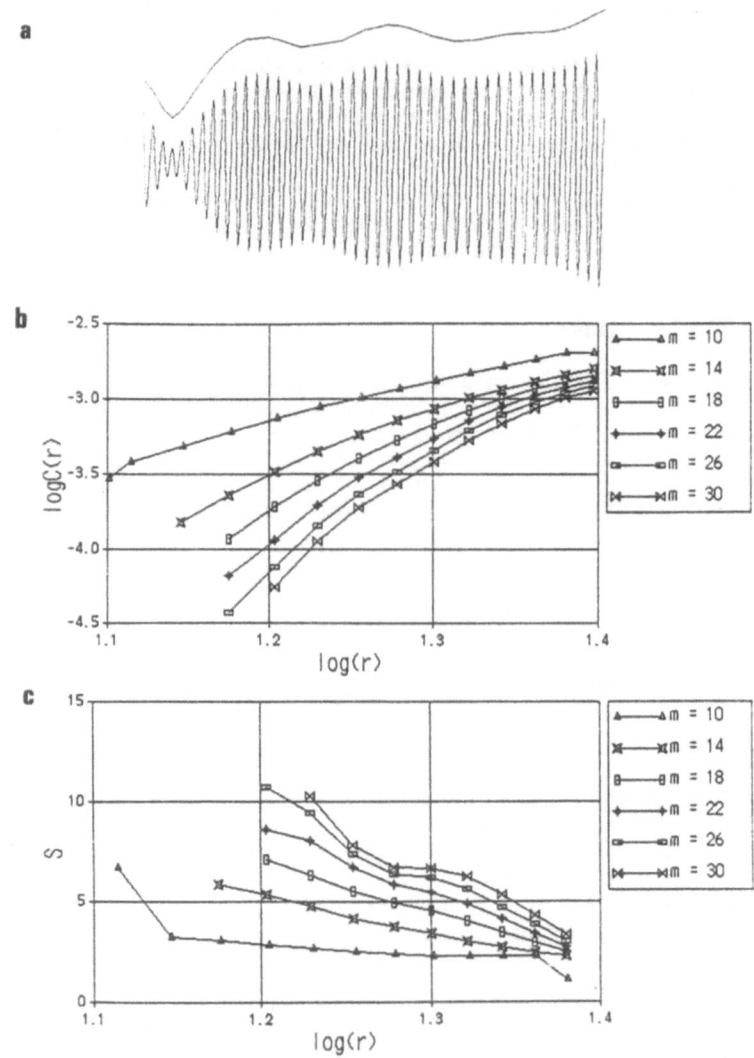

Fig. 5 a-c. calculated dimensions (b) and their gradients (c) of the envelope of the amplitudes of the u'$_{eff, per.}$- fluctuations (a)

To provide a cleaner situation, some of the experiments were repeated in a test section provided with an upstream fetch, which has a true channel flow in the feedback region. The obvious yet unexpected effect was a considerable reduction of the feedback effect, probably as a consequence of the higher turbulence level in this flow and the ensuing randomization of the feedback signal (figure 6).

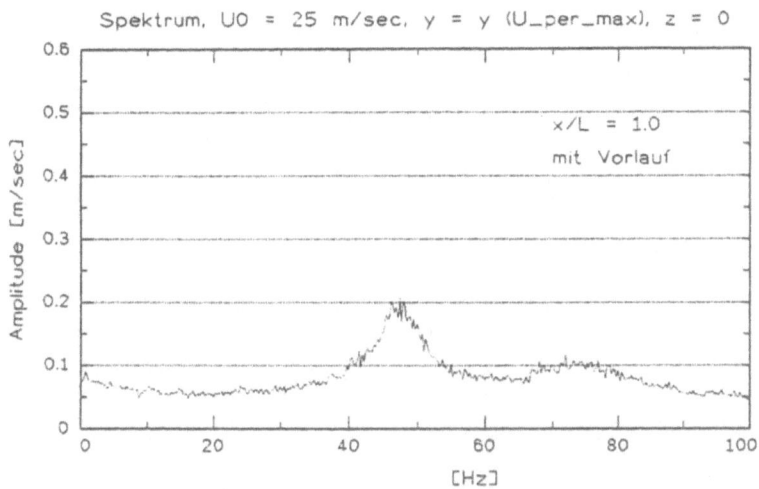

Fig. 6. spectrum shows influence of upstream fetch (compare with fig. 4 a)

6. Acknowledgements

This investigation is supported by Deutsche Forschungsgemeinschaft (DFG). Some of the measurements reported were done by W. Lazek.

7. References

Buell, J.C., Huerre, P.: Inflow/outflow boundary conditions and global dynamics of spatial mixing layers. Proc. NASA Ames/Stanford Cent. Turbul. Res. Summer Program. Rep. No. CTR-S88, pp. 19-27

Dimotakis, P.E., Brown, G.L.: The mixing layer at high Reynoldsnumber: large-structure dynamics and entrainment. JFM, Vol. 78, pp. 535-560, 1976

Eggers, J.: Untersuchung zur Dynamik des turbulenten Übergangs in einer Scher-schicht. Diplomarbeit; HFI / TUB 1989

Fiedler, H.E., Kim, J.-H., Köpp, N.: The accelerated mixing layer in a tailored pressure gradient. To appear in Journal of Mechanics

Hilberg, D., Fiedler, H.E.: The spanwise confined turbulent mixing layer. In Advances in Turbulence (eds. H.-H. Fernholz and H. E. Fiedler), Springer 1989

Huerre, P., Monkewitz, P.A.: Local and global instabilities in spatially developing flows. Ann. Rev. Fluid Mech. 1990. 22: 473-537

Nieberle, R.: Entwicklung einer Methode der Mustererkennung zur Analyse kohärenter Strukturen und ihre Anwendung im turbulenten Freistrahl, Diss. TU-Berlin, 1985

Paschereit, O., Schüttpelz, M., Fiedler, H.E.: The mixing layer between non-parallel walls. In Advances in Turbulence (eds.: H.-H. Fernholz and H. E. Fiedler), Springer 1989

Theiler, J.: Spurious dimension from correlation algorithms applied to limited time series data. Physical Review A, 34, 3, 1986, p. 2427-2432

Wygnaski, I., Oster, D., Fiedler, H., Dziomba, B.: On the perserverance of quasi two-dimensional eddy-structure in a turbulent mixing layer, J. Fluid Mech. 93, 2, 1979, pp. 325-335

Yajnik, K.S., Archarya, M.: Non-equilibrium Effects in a Turbulent Boundary Layer due to the Destruction of Large Eddies. In Structure and Mechanisms of Turbulence I (ed. H. Fiedler), Lectures Notes in Physics 75, pp. 249-260, Springer Verlag, Berlin, 1977

THE EFFECTS OF EXTERNAL EXCITATION ON THE REYNOLDS-

AVERAGED QUANTITIES IN A TURBULENT WALL JET

Y. Katz, E. Horev, and I. Wygnanski

Department of Aerospace and Mechanical Engineering
The University of Arizona
Tucson, AZ 85721

ABSTRACT

The effects of external, two-dimensional excitation on the plane turbulent wall jet were investigated experimentally. Measurements of the streamwise component of velocity were made throughout the flow field for a variety of Reynolds numbers, imposed frequencies, and amplitudes. The present data were always compared to the results generated in the absence of external excitation. Although the bulk of the unexcited flow is self-similar, it depends on the momentum flux at the nozzle and on the viscosity and density of the fluid and not on the width of the nozzle, which was commonly used to reduce the data. These similarity scales were used to check the consistency of the skin friction measurements, which are otherwise determined with considerable difficulty. It was shown that external excitation has no appreciable effect on the rate of spread of the jet or on the decay of its maximum velocity. In fact, the mean velocity distribution did not appear to be altered by the external excitation in any obvious manner. The flow near the surface, however (i.e., for $0 < Y^+ < 100$), was profoundly different from the unforced flow, indicating a reduction in wall stress exceeding at times 30%. The production of turbulent energy near the surface was also reduced, lowering the intensities of the velocity fluctuations. It was also shown that the inviscid, logarithmic portion of the "law of the wall" cannot be valid in either the forced or the unforced flows, because the Reynolds stress decreases rapidly beyond the distance at which the viscous stress becomes vanishingly small. This casts a doubt on the existence of a "constant stress layer" in the wall jet.

INTRODUCTION

The plane, turbulent wall jet, evolving over a flat surface in the absence of an external stream, is a generic flow governed by the boundary layer equations. Although this flow was extensively investigated over the years because of its many engineering applications (see the review articles of Launder and Rodi, 1981, 1983), it is still poorly understood. There is general agreement that the mean velocity in the wall jet is self-similar, but the parameters scaling it are controversial, in spite of the massive statistical data available in the literature. The complexity of this flow stems from the fact that its outer part resembles a free jet while its inner part resembles a turbulent boundary layer. In fact, most models attempting to predict the average characteristics of the turbulent wall jet superimpose a free jet on top of a boundary layer by using strip integrations and matching the most obvious boundary conditions.

The rate of spread of the wall jet and the decay of its maximum velocity in the direction of streaming appear to be dependent on the Reynolds number (Tailland and

Mathieu, 1967). Since such a dependence was not observed in a free jet at comparable Reynolds numbers, one cannot help but wonder as to its origin. It is convenient to attribute this dependence to the presence of the solid surface, but the latter contributes very little to the total momentum loss. In fact, Narasimha et al. (1973) suggested that the traditional scaling of the relevant distances in the wall jet by the characteristic dimension of the nozzle might be erroneous. They proposed to scale the streamwise evolution of the flow by the momentum flux and the viscosity of the fluid, but their suggestion was either ignored or discarded on the grounds of poor two-dimensionality of their data (Launder and Rodi, 1981).

Experimental investigations of large coherent structures in turbulent shear flows bypassed the wall jet, concentrating either on wall-bounded flows like a boundary layer or a channel (e.g., Willmarth, 1975a, 1975b) or on free shear flows like the mixing layer, the wake, and the jet (e.g., Ho and Huerre, 1984; Wygnanski and Petersen, 1987). In the latter category of flows, the large coherent structures were identified as the predominant instability modes and were quantitatively analyzed in this context. However, the large coherent structures in wall-bounded flows are much more complex and more difficult to identify than those in free shear flows. Many forms of such structures were observed visually (Kline et al., 1967), but no consensus was reached as to their origin and their precise association with the enhancement of the skin friction or with the rate of growth of the boundary layer.

Free shear flows are inviscidly unstable, while boundary layers in the absence of an adverse pressure gradient are not. There is enough evidence (Katz et al., 1989) to suggest that a boundary layer on the verge of separation resembles a free mixing layer and responds to external stimuli, like a mixing layer. The receptivity of this boundary layer is attributed to an inviscid instability which may dwarf, in this case, any other form of instability and generate large, predominantly two-dimensional, coherent structures. The generic wall jet is inviscidly unstable in its outer region and may thus possess large coherent structures characteristic of a plane turbulent jet. The similarity between the outer part of the wall jet and the free jet will be explored in this context, and the relevance of the solid surface to the evolution of the large coherent structures will be assessed. From this point of view, the boundary layer in a strong adverse pressure gradient may be regarded as a wake evolving in the vicinity of a solid surface. In both flows the significance of the outer region is accentuated while the no-slip conditions at the solid surface are maintained.

In the present report we shall describe the major effects of a harmonic, two-dimensional forcing on the structure of a turbulent wall jet. However, the scaling laws governing the basic flow had to be identified and understood before subjecting the wall jet to external excitation and, thus, a major portion of this report is devoted to this issue. The experiments were carried out in air, on the simplest wall jet configuration in the absence of an external stream or surface curvature. A schematic diagram of the wall-jet facility which incorporates two excitation schemes is shown in Figure 1. The efflux velocity of the jet and the width of the nozzle were adjustable in addition to the frequency and amplitude of the imposed excitation. The flow was incompressible, and the Reynolds Numbers based on the efflux velocity and on the nozzle dimension varied between $3*10^3$ and $3*10^4$.

A single hot-wire anemometer was used to measure the streamwise component of velocity in the wall jet. The probe was mounted on a traversing mechanism controlled by the laboratory computer, which was also used to acquire the data and to store them for further processing. The flow near the wall was carefully examined in view of the scarcity of reliable data in that region.

RESULTS

The Unexcited Flow

The longitudinal velocity distribution in the unexcited wall jet was mapped seven times for a variety of slot widths and jet efflux velocities. Two of the seven measured

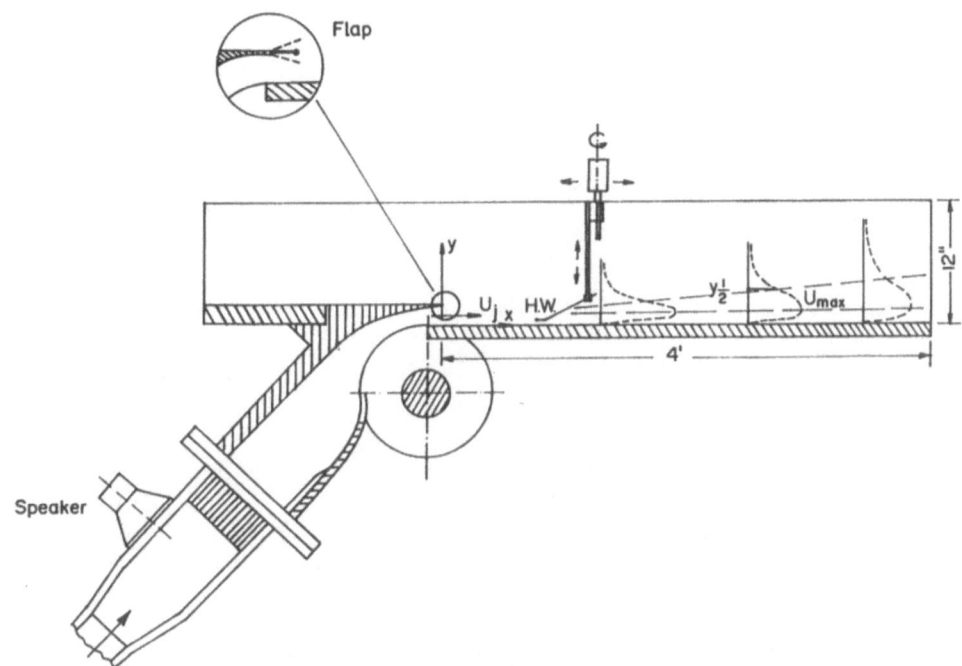

Fig. 1. A schematic drawing of the experimental facility.

sets of normalized mean velocity profiles are plotted in Figure 2. These two sets were chosen because the difference between them was the largest one observed. Each set represents a superposition of eight individual profiles measured at distances ranging from 30 to 140 slot widths from the nozzle for a given nozzle Reynolds number Re_j. ($Re_j = U_j b/\nu$, where b is the width of the nozzle and U_j is the jet efflux velocity.) The velocity at each X location was divided by its local maximum velocity U_m and the distance from the wall was divided by $Y_{m/2}$ (where $Y_{m/2}$ is the distance measured from the wall to the location at which the mean velocity decreases to 1/2 of its local maximum value in the outer part of the flow). The differences observed between the two sets of data may be attributed to the effect of Re_j. The nozzle Reynolds number might therefore have a slight effect on the dimensionless velocity profile at the outer part of the jet (i.e., at $Y/Y_{m/2} > 1.3$). However, the data taken in this region are not very reliable because of the possible presence of room drafts, which may affect the measurements as well as the entrained flow. In fact, most of the measurements taken in the absence of an external stream terminate at $Y/Y_{m/2} = 1.6$, and there is a fair amount of scatter reported in the literature for $Y/Y_{m/2} > 1.3$. The velocity profile measured at $Re_j = 19,000$ agrees very well with the velocity profile measured by Tailland and Mathieu (1967) at comparable Re_j. The latter is also plotted on Figure 2 for the sake of visual comparison. One may conclude that the mean velocity distribution in the wall jet is self-similar and almost independent of Reynolds number when it is normalized by the local length and velocity scales.

The decay of the maximum velocity in the jet with increasing distance from the nozzle is plotted in Figure 3. Traditionally, the ratio of $[U_j/U_m]^2$ is plotted versus X/b in order to accentuate the decay of the velocity scale, which is expected to vary approximately as $1/\sqrt{X}$. In this plot, the streamwise distance is measured from a virtual origin X_0 by requiring that the lines fitted through the data will converge to $U_m/U_j = 1$ at $X = X_0$. There is a somewhat irregular tendency of the virtual origin to move upstream with increasing Re_j. A notable shift in this origin occurred around $Re_j = 5000$, whereupon turbulence was first noticed near the solid surface in the plane of the nozzle. Near the upper lip of the nozzle, the flow is oscillatory as a result of the instability of the free mixing layer. The interaction between the flow oscillations near the wall and

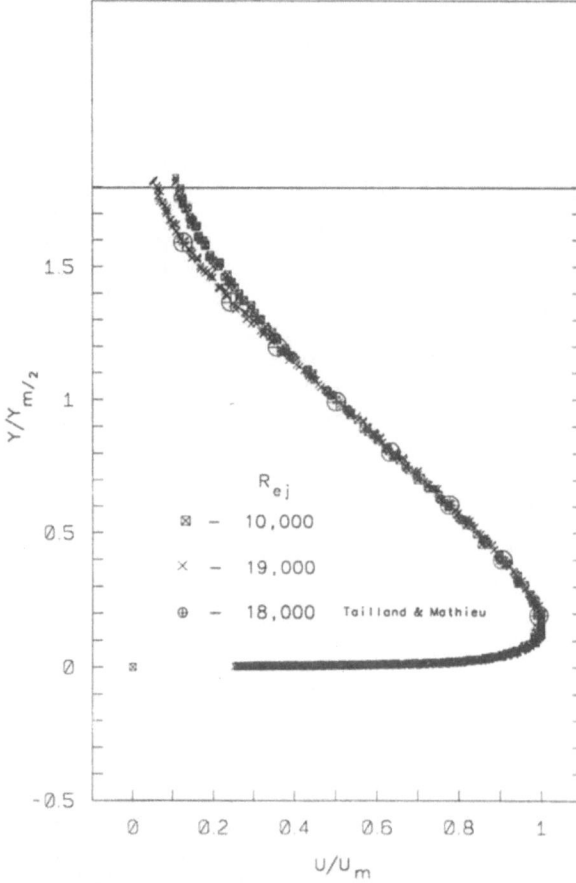

Fig. 2. A normalized mean velocity profile in the unforced wall jet.

near the outer lip of the nozzle might be responsible for the complex dependence of X_0 on Re_j. Straight lines represent fairly well the results plotted in Figure 3, but a power-law expression fits the data better when the exponent is approximately –0.47 rather than the implied –0.5. The corresponding exponent for the rate of spread of $Y_{m/2}$ with X is close to 0.88, which is in substantial agreement with the results of Narasimha et al. (1973), who suggested that this exponent should be 0.91.

The effect of Reynolds number on the decay of the maximum velocity and the rate of spread of the jet in the present investigation is significant. In fact, the slope of the lines drawn in Figure 3 changes by a factor of 2 in the range of Re_j considered. Narasimha et al. (1973) suggested that the fully developed wall jet should attain a local equilibrium that is independent of the detailed conditions at the nozzle. The sole parameter determining the evolution of an incompressible wall jet surrounded by an identical fluid should therefore be its initial, kinematic momentum flux, J. This is conceptually identical to the scaling laws proposed for jets by Newman (1961). Using Narasimha et al.'s suggestion one gets:

$$\frac{U_m \nu}{J} = F_1(\xi) ; \qquad \frac{Y_{m/2} J}{\nu^2} = F_2(\xi) ; \qquad \frac{\tau_w}{\rho}\left[\frac{\nu}{J}\right]^2 = F_3(\xi) \qquad (1)$$

where $\xi = [X \, J/\nu^2]$; $J = [U_j^2 \, b]$; ρ and ν are the density and kinematic viscosity of the fluid, respectively; τ_w is the shear stress at the wall; and X is measured from the nozzle.

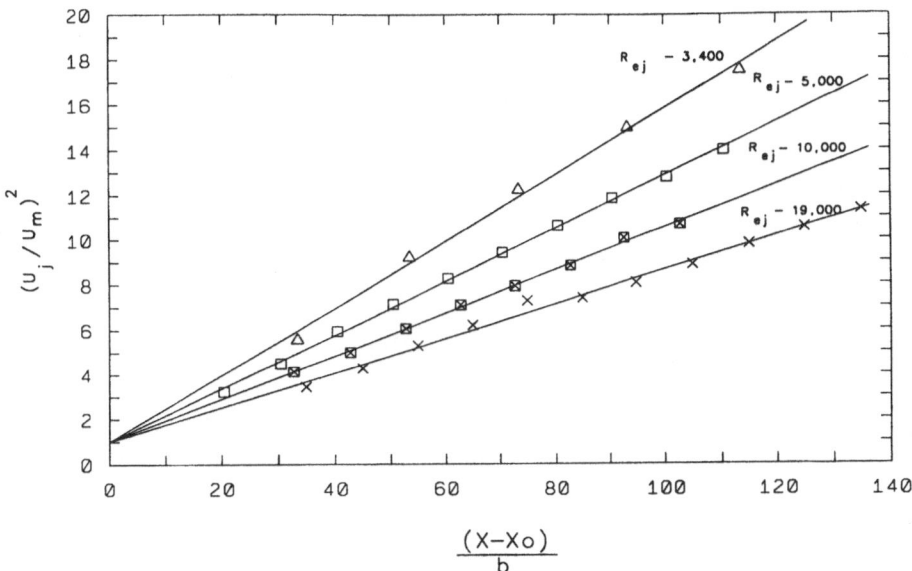

Fig. 3. The decay of the maximum velocity in the unforced wall jet with X/b.

The velocity and the length scales plotted in this fashion prove the validity of Narasimha et al.'s (1973) assumptions and indicate that the flow is indeed independent of Re_j provided the latter exceeds a threshold value of 5000 [Figure 4]. The various values of Re_j shown in Figure 4 were obtained by changing the slot-width between 2.5 and 7.5 mm or by changing the velocity between 10 and 57 meters/sec. The solid lines drawn through the high Re_j data can be expressed analytically by the following power laws:

$$\left[U_m \frac{\nu}{J}\right] = A_u \left[X \frac{J}{\nu^2}\right]^n \quad \text{and} \quad \left[Y_{m/2} \frac{J}{\nu^2}\right] = A_y \left[X \frac{J}{\nu^2}\right]^m \tag{2}$$

where A_u = 1.473 (n = -0.472) and A_y = 1.445 (m = 0.881). Even for $Re_j \leq 5000$, the exponent representing the decay of the velocity scale or the rate of growth of the length scale with ξ remains unaltered, only the constant coefficient changes. The results presented in figure 4 will be used in calculating τ_w and determining the consistency of the data.

The velocity distribution in the vicinity of the wall was replotted on a larger scale in Figure 5 for two values of Re_j. It appears that the normalized velocity profile near the surface is weakly dependent on Re_j (regardless of any shift in the virtual origin), particularly when the latter is lower than 5000. The self-similarity near the wall is thus limited to the variation of the velocity profile with X at a prescribed Re_j. Although the deviation from similarity is small enough as not to be observed in Figure 2, it is large enough to indicate that the inner portion of the wall jet might not be self-similar because of the presence of the wall.

It is observed that the maximum velocity in the jet occurs at $[Y/Y_{m/2}] \cong 0.15$, which agrees very well with most of the results reported in the literature (see Launder and Rodi, 1981, Table 1). Since these data were accumulated over a large range of Reynolds numbers and flow conditions [e.g., the results of Guitton and Newman (1977) on curved surfaces and the results of Irwin (1973) in the presence of a free stream and pressure gradient may also be included in this correlation], we may surmise that the ratio $[Y_m/Y_{m/2}]$ is insensitive to the the conditions affecting the flow. Consequently, one may consider $Y_{m/2}$ as the primary length by which the turbulent portion of the wall jet (i.e., the fraction of the flow which is dominated by Reynolds stresses) might be scaled. The

71

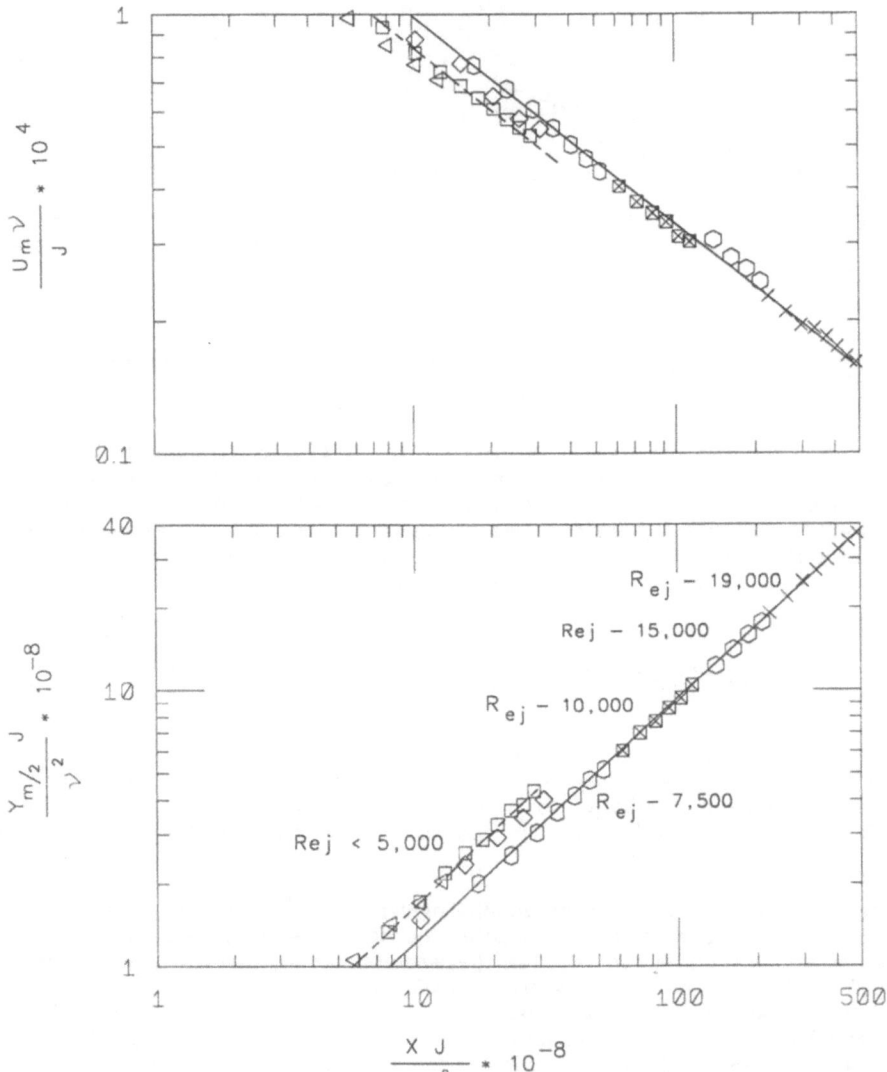

Fig. 4. The dependence of the velocity and length scales on ξ in the absence of forcing.

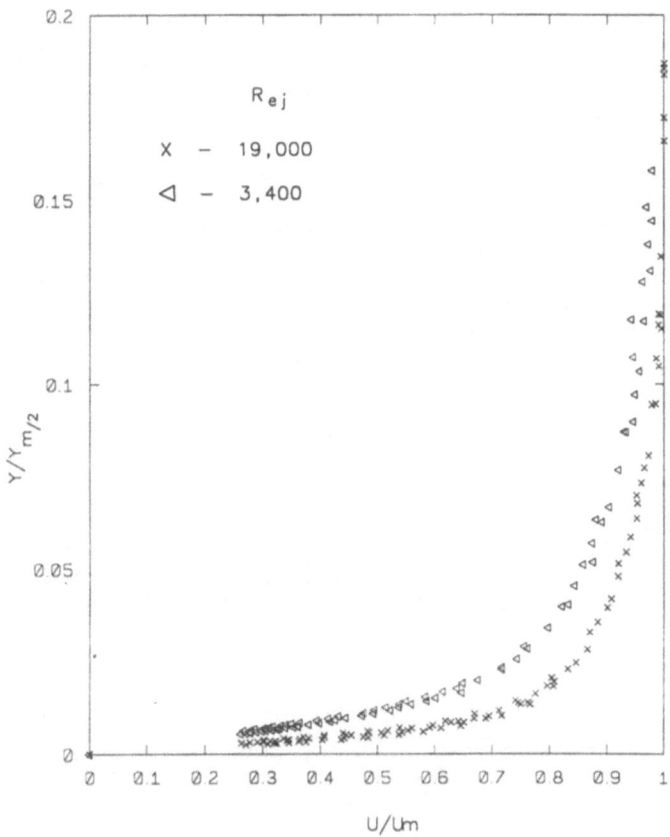

Fig. 5. The normalized, unforced mean velocity profile near the surface.

separate tabulation of the external width of the flow measured from the location of the maximum velocity $[Y_{m/2}-Y_m]$ (e.g., Tailland and Mathieu, 1967) seems to be superfluous.

The determination of wall shear stress in this flow is very difficult because of the thinness of the inner boundary layer that extends from the surface to the location at which the jet velocity has a local maximum (i.e., where $Y \le Y_m$). Since the wall friction has little or no influence on the spreading of the flow, attempts to correlate the wall stress with the depreciation of the momentum integral in the direction of streaming were not very successful (e.g., Schwarz and Cosart, 1961). The skin friction was either measured directly by using a small, floating drag-balance or indirectly by calibrated surface heat-transfer devices or impact probes like the Stanton or Preston tubes (Launder and Rodi, 1981; Ozarapoglu, 1973). Since these devices are usually calibrated in channel flows, they depend on the universality of the velocity profile and cannot extend in Y beyond the range in which the universal velocity distribution associated with the "law of the wall" applies.

The mean velocity gradient in the viscous sublayer, which has to be constant near the surface, can also be used to estimate the skin friction. This method, which was first used in the wall jet by Tailland and Mathieu (1967), was criticized by Launder and Rodi (1981) as being inaccurate because the estimates of C_f based on this technique "have produced values ranging from 20% to 35% below the consensus values of impact tube data." In the present experiment, the shear stress at the wall was estimated by using the momentum integral method, the mean velocity gradient in the viscous sublayer, and a Preston tube. The latter method was only used in those cases where it was established, by using the first two methods, that the mean velocity profile expressed in the "law of the wall" coordinates complies with the constants used for calibrating the tube.

The depreciation of the jet momentum in the direction of streaming for $Re_j = 5000$ is plotted in Figure 6. The data were normalized by the jet momentum measured at $X/b = 30$. The frictional losses estimated from $[dU/dY]_{wall}$ were checked against the losses estimated from the momentum integral equation between $X/b = 30$ and $X/b = 9$ 0. There is a reasonably good agreement between the two methods. An independent check was provided by using a calibrated Preston tube based on the design of Patel (1965) for an Re_j of 5000, at which the known universal constants of "law of the wall" based on the friction velocity, $U_\tau = \sqrt{\tau_w/\rho}$, measured by the other techniques were actually realized [i.e., $U/U_\tau = 5.5 \log((Y\ U_\tau)/\nu) + 5.5$]. The diameter of the Preston tube used in this experiment was 0.89 mm, corresponding to $14 < (U_\tau d/\nu) < 28$. One may therefore apply Patel's calibration curve as stated in his article. The result of these measurements confirmed that the momentum loss was estimated correctly by the other indirect means, thus justifying their use.

A plot of the dimensionless skin friction coefficient

$$C_f = \frac{\tau_w/\rho}{1/2U_m^2}$$

versus the local Reynolds number $(U_m Y_m)/\nu$ is given in Figure 7 and compared with the results of Tailland and Mathieu (1967). The agreement between these two sets of data is good. Since the present results are entirely self-consistent, they cast some doubt on most of the impact-tube data reported in the literature, which yield a much higher value of C_f.

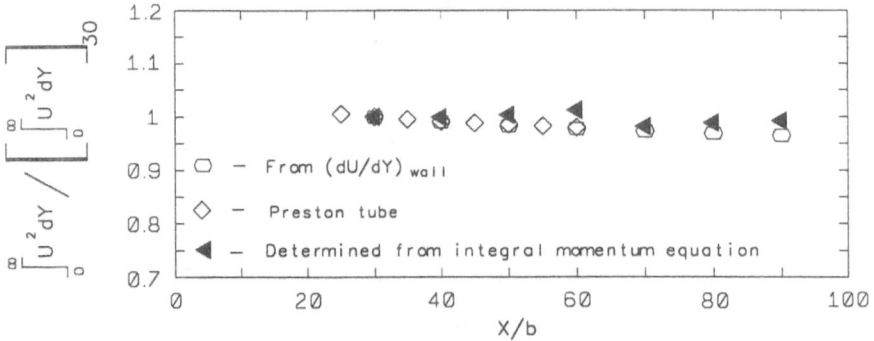

Fig. 6. The depreciation of the jet momentum with X/b in the absence of forcing.

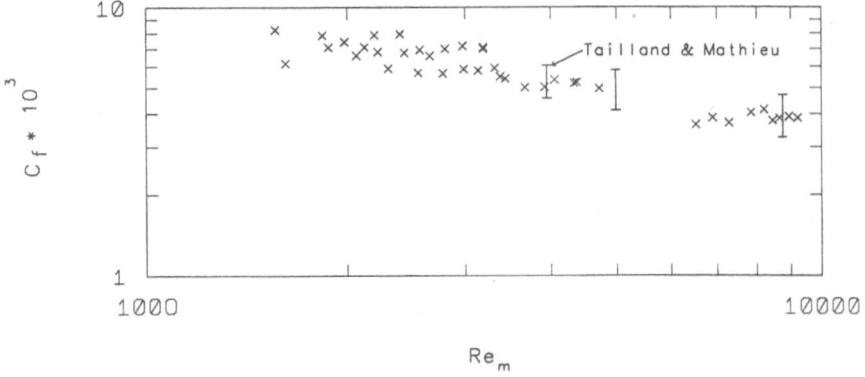

Fig. 7. The dependence of C_f on Re_m in the absence of forcing.

We have decided to use the slope of the mean velocity profile near the surface as the most reliable method for measuring the wall stress, knowing very well that the reliability of this method depends on the quality of the traversing mechanism used (a resolution of 0.01 mm might be required), the quality and size of the hot-wire probe, and the number of data points taken. Furthermore, heat loss from the wire to the surface may change the apparent slope of the velocity profile measured very close to the wall (Wills, 1962); consequently, such change in slope served as an indication for stopping the traverse. The data used for assessing the wall friction were taken at elevations exceeding 50 wire diameters above the surface and Reynolds numbers based on the wire diameter in excess of 0.7. The linear fit to the data extended in some cases to 100 wire diameters and Reynolds numbers of 1.5. For this range of variables, the heat loss to the wall is not significant in a laminar flow, let alone a turbulent one, whereupon the corrections suggested by Wills are approximately halved.

A plot of $[\tau_w/\rho \, (\nu/J)^2]$ versus ξ is presented in Figure 8, where all the data obtained through the use of the adopted method are plotted in the general similarity coordinates. Once again, the skin friction scales correctly with ξ and is independent of Re_j. The best fit to the data is given by

$$\frac{\tau_w}{\rho} \left[\frac{\nu}{J}\right]^2 = A_\tau \, \xi^k \tag{3}$$

where $A_\tau = 0.146$ and $k = -1.07$. A cross check on A_τ and k can be obtained by assuming that the mean velocity is self-similar, thus

$$\int_0^\infty U^2 \, dY = \mathcal{K} \, U_m^2 Y_{m/2}$$

where

$$\mathcal{K} = \int_0^\infty \left[\frac{U}{U_m}\right]^2 d\left[\frac{Y}{Y_{m/2}}\right] = 0.74$$

Then by using the momentum integral equation in conjunction with Eq. (2),

$$\frac{\tau_w}{\rho} = -\frac{d}{dx} \int_0^\infty U^2 \, dY = -\mathcal{K} \frac{d}{dx} [U_m^2 Y_{m/2}]$$

one gets the exponent k as

$$k = 2n + m - 1 = 2(-0.472) + 0.881 - 1 = -1.063$$

and the constant coefficient as

$$A_\tau = -\mathcal{K}(2n + m)A_u^2 A_y = 0.146$$

The agreement between the measured shear stress at the wall and the calculated shear stress on the basis of flow similarity is good and thus may provide an alternate indirect method of estimating wall friction in a two-dimensional wall jet.

The Effects of Forcing on the Reynolds-Averaged Quantities

External excitation of free shear flows, like the mixing layer, had a strong effect on a variety of time-averaged quantities like the mean velocity distribution, the rate of spread of the flow, the turbulent intensity, and the Reynolds stress. It is thus natural to

75

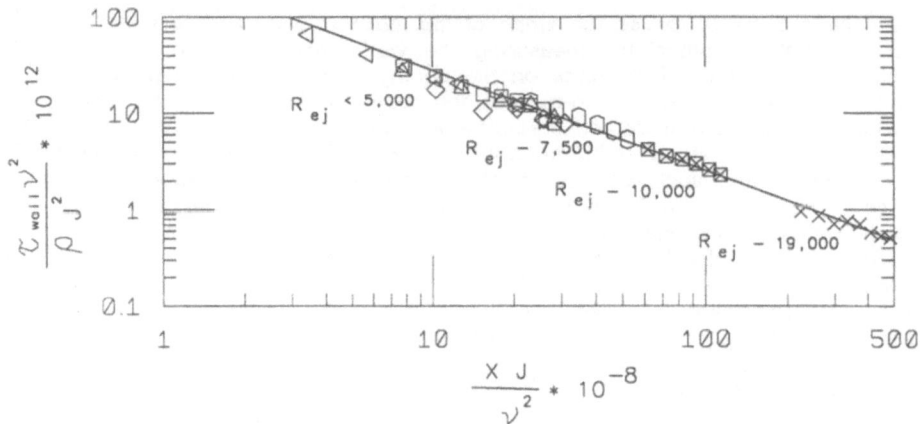

Fig. 8. The variation of wall stress with ξ in the absence of forcing.

start the present discussion of the experimental results by exploring parametrically the effects of forcing on the same quantities in the wall jet.

A variety of forced flows were considered in the present experiment. They differed in jet momentum; slot width; frequency and amplitude of the imposed oscillations; and, finally, the manner in which forcing was introduced to the flow. A change in each one of the above parameters required a concomitant traverse across the flow at various X locations. A sample of the data is shown in Figure 9. The velocity profile plotted on the left (Figure 9a) exhibits self-similarity with respect to $Y_{m/2}$ and U_m irrespective of the X location, while the profile plotted on the right (Figure 9b) proves the existence of self-similarity irrespective of the changes in all other parameters considered. Both sets of data are compared to the dimensionless, unforced, velocity profile shown in Figure 2 and replotted in Figure 9 as a solid curve. One may conclude that two-dimensional external excitation does not affect the normalized form of the velocity distribution within the range of parameters considered. Replotting the velocities adjacent to the wall on an expanded scale (Figure 9c) and comparing them to unforced velocity measurements at otherwise identical conditions indicate small but consistent differences between the two sets of data. These differences will be discussed later in some detail.

The rate of spread of the wall jet and the rate of decay of its maximum velocity are hardly affected by the external excitation. A more appropriate assessment of this statement may be made upon comparing the results by using the variables suggested by Narasimha et al. (1973) for the unforced wall jet. The solid lines drawn in Figure 10 represent the unforced data accumulated for $Re_j \geq 7500$ while the broken lines correspond to $Re_j \leq 5000$. The symbols plotted in Figure 10a and b represent data corresponding to various frequencies of excitation at nozzle Strouhal numbers ($St_j \equiv f*b/U_j$) varying between $3 \leq St_j\ 10^3 \leq 17$ (corresponding to $St \equiv f\nu^3/J^2$ of $0.4 \leq St\ 10^{15} \leq 444$) and amplitudes, based on the streamwise velocity perturbation near the nozzle, ranging from 2% to 20% of the efflux velocity at the nozzle. Although some consistent differences depending on the level of forcing and on the Strouhal number can be detected, they do not alter our initial conclusion about the lack of sensitivity of these mean-flow parameters to the external excitation. The cumulative effect of the threshold in Re_j occurring around $Re_j \cong 5000$ appears to be much more significant than that of the two-dimensional excitation. In the unforced case, this dependence on the threshold value in Re_j was essentially eliminated by accounting for the virtual origin of the flow which shifted downstream at higher Reynolds numbers. One should remember that the data are plotted on a compact logarithmic scale because of the "power-law" dependence of U_m and $Y_{m/2}$ on ξ. Dimensional analysis of the independent parameters in the externally excited wall jet suggests that variables other than ξ might affect the mean velocity distribution. Whenever the external excitation is harmonic, the broadening or distortion of mean flow will depend on the square of the local amplitude of the velocity

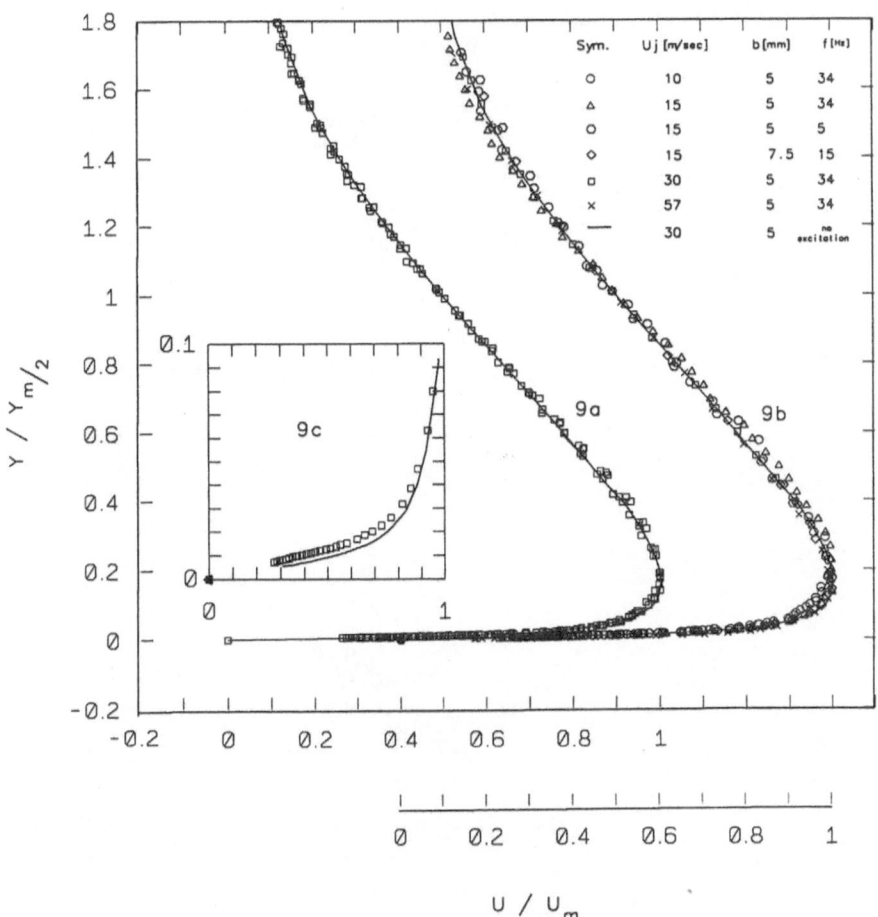

Fig. 9. The normalized velocity profile in the forced wall jet.

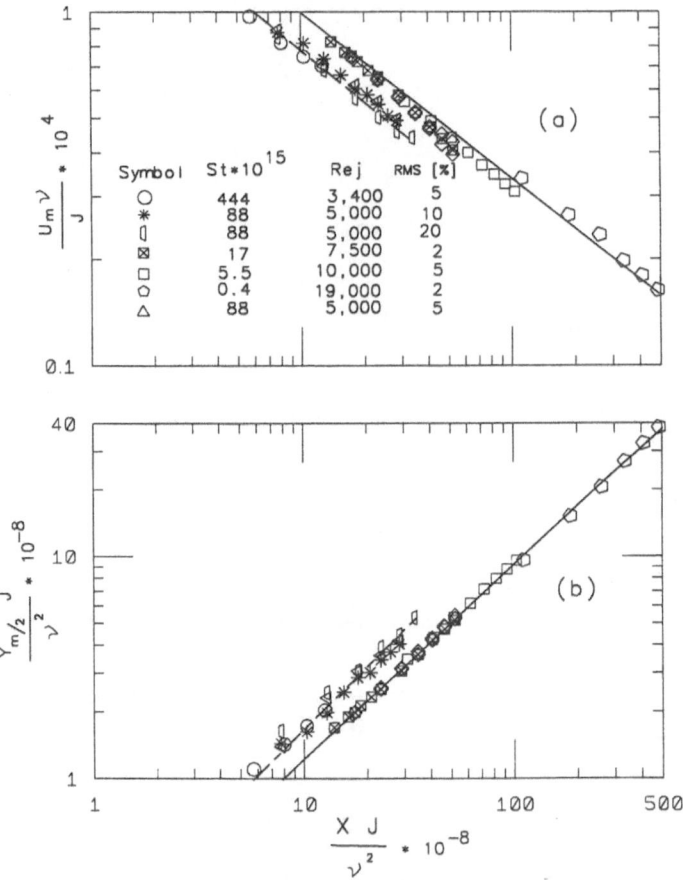

Fig. 10. The dependence of the velocity and length scales on ξ in the forced and unforced flows.

perturbations. In some cases, the finite amplitude might be introduced directly by the forcing mechanism, while in others, the flow itself might act as an amplifier.

We shall now examine the global effect of the excitation on the loss of momentum resulting from skin-friction drag. The velocities plotted in Figure 9c suggest that forcing the wall jet might have an effect on the drag. The velocity measurements plotted in Figure 11 are shown in dimensional form because this representation enables the reader to assess the quality and the amount of data acquired in the viscous sub-layer. There is no doubt that even a relatively low level of forcing (the data plotted correspond to an initial amplitude of 2%) results in a reduction of the wall shear stress, τ_w.

Plotting τ_w/ρ $(\nu/J)^2$ versus ξ and comparing the results with the skin friction measured in the absence of forcing indicates clearly the differences in τ_w in spite of the logarithmic scale chosen (Figure 12). At no instance did τ_w increase above its nominal unforced value; reductions in the skin friction of approximately 10% were prevalent at most frequencies corresponding to initial excitation amplitudes which were lower or equal to 5%. However, reductions in τ_w of approximately 40% were also recorded by forcing at much higher amplitudes corresponding to 10% or 20% of the efflux velocity at the nozzle. Alternately, forcing at a preselected frequency which is amplified by the flow may achieve the same drag reduction at a much lower input amplitude (Figure 12).

In attempting to sort out the independent contribution of frequency (i.e., $St \equiv f\nu^3/J^2$); amplitude and Reynolds number on the local τ_w, one may assume that the dimensionless $(\tau_w)/\rho$ $(\nu/J)^2$ depends on the distance from the nozzle $(XJ)/\nu^2$ and the local amplitude of the coherent and perhaps quasi two-dimensional perturbation present locally in the flow. Neglecting the effects of Re_j might be justified on the basis of the correlations derived for the unforced wall-jet, but by neglecting the effects of St one tacitly assumes that all two-dimensional coherent structures are similar regardless of their origin provided their local intensity is accounted for in some manner. This is a crude assumption which might correlate the distortion of the velocity profile near the surface and the Reynolds stress in the wall region with some (as yet unknown) universal coherent motion.

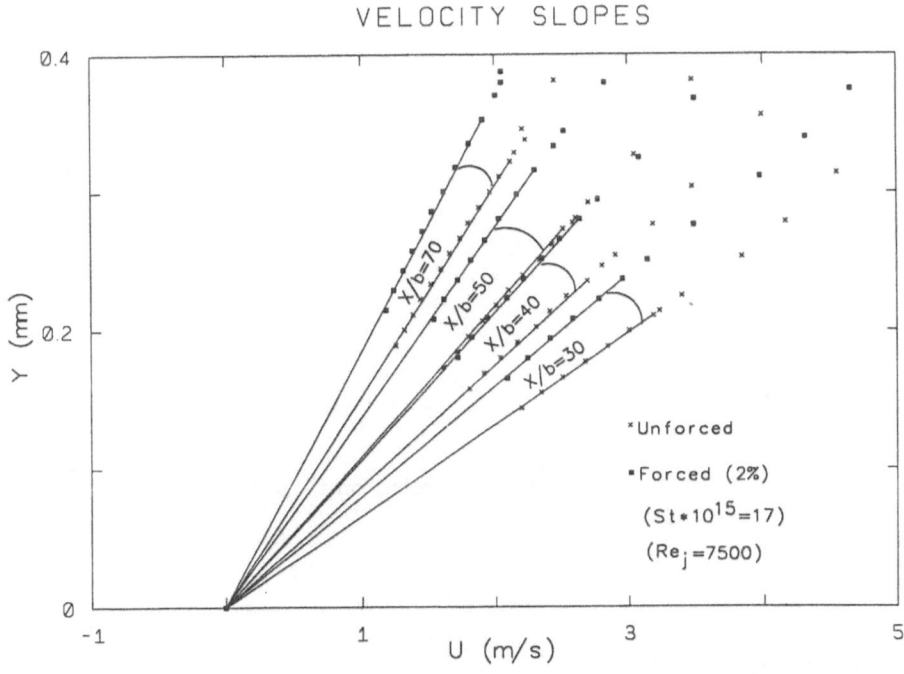

Fig. 11. The effect of forcing on the velocity gradient near the wall.

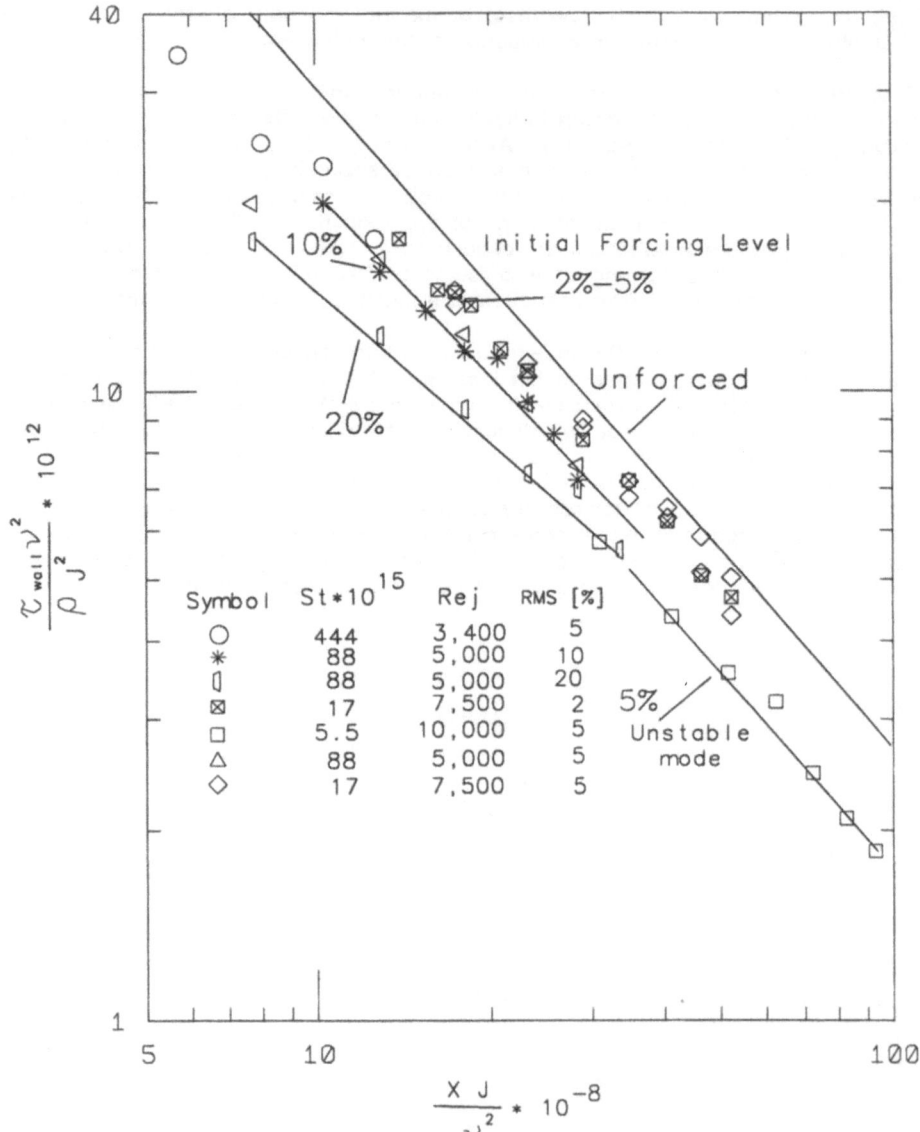

Fig. 12. The effects of forcing on the wall stress.

We computed the phase-locked, ensemble-averaged velocity signals and from them deduced the local r.m.s. levels of the coherent motion. These quantities were then integrated across the flow to provide a measure of the intensity of the coherent motion at a given X station:

$$\langle A \rangle = \frac{1}{U_j Y_{0.1}} \int_0^{Y_{0.1}} \sqrt{\frac{1}{T} \int_0^T [\langle U \rangle - \langle \bar{U} \rangle]^2 \, dt} \; dY$$

where

$\langle U \rangle$ – the phase-locked velocity at a given Y and $\langle \bar{U} \rangle = \frac{1}{T} \int_0^T \langle U \rangle \, dt$

$Y_{0.1}$ – the distance from the wall to the location at which $U/U_m = 0.1$ at the outer region

T – the period of the forcing signal

Some of the data sets shown in Figure 12 were calculated as functions of the computed local amplitudes and then replotted again with ξ as the abscissa and with the local, cross-flow-averaged r.m.s. values of the coherent motion as parameters (Figure 13a). One may deduce from this figure that a local coherent motion having an average amplitude of 0.3% may be responsible for a reduction in τ_w of approximately 15% while an additional increase in local amplitude by a factor of 10! (i.e., to 3% of U_j) resulted in an incremental reduction in $(\tau_w)/\rho \, (\nu/J)^2$ of an additional 15% only. This suggests that the reduction in the local skin friction is not linearly dependent on the strength of the coherent motion as defined by the single parameter $\langle A \rangle$.

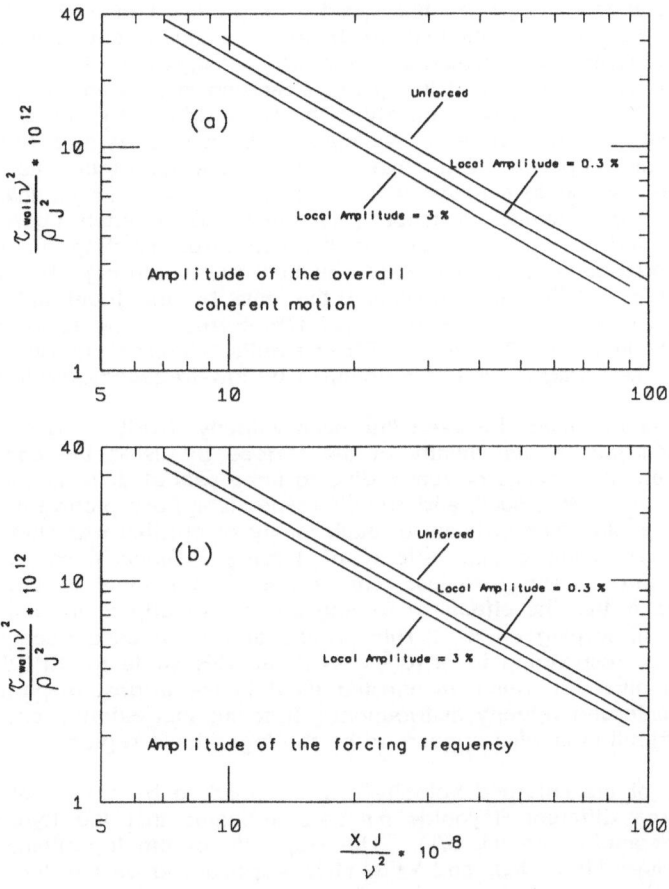

Fig. 13. The effect of the local amplitude of the coherent motion on the wall stress.

Fourier decomposition of the coherent signal prior to integration across the flow and a repetition of the procedure mentioned above correlates the dependence of $(\tau_w)/\rho \ (\nu/J)^2$ on the coherent amplitude of the motion at the forcing frequency and its various harmonics (Figure 13b). It appears that when the local amplitudes of the coherent oscillations at the forcing frequency are also 0.3%, the reduction in τ_w is also 15%, but a further increase in that amplitude to 3% resulted in even smaller reduction in τ_w than estimated for the overall coherent signal. One may conclude that most of the reduction in τ_w at low amplitudes of the coherent motion may be attributed to the forcing frequency.

Isolating the effect of frequency on τ_w is not a simple task, and there are not enough data available to provide a definitive correlation at this time. However, the available results suggest that the lowest possible frequency of forcing is most effective in reducing the local τ_w at some predetermined local mean flow conditions (i.e., at some prescribed $Y_{m/2}$ and U_m). This is because the low frequencies remain amplified over longer distances as a consequence of the divergence of the mean flow.

The fractional loss of momentum in the direction of streaming is equal to frictional losses integrated over a prescribed distance:

$$\frac{1}{J} \int_{X_0}^{X} \frac{\tau_w}{\rho} \, dX = \int_{\xi_0}^{\xi} \frac{\tau_w}{\rho} \left[\frac{\nu}{J}\right]^2 \, d\xi$$

Since the prediction of drag and its possible reduction by active means is of primary interest to the practicing engineer, this quantity is plotted versus the physical distance X/b (Figure 14) knowing full well that Re_j becomes an important factor in this plot. It was decided to begin the calculations from x/b = 30, where the wall jet was fully developed. Four cases are plotted in Figure 14 ranging in St from $444{*}10^{-15}$ to $5.5{*}10^{-15}$ and in Re_j from 3400 to 10^4, respectively. The amplitude of forcing was maintained at 5% throughout but the slot width was maintained at 5mm in parts a, b, and d while being increased to 7.5 mm in part c. This change in the slot width implies that the data shown in Figure 14c correspond to a physical distance which is 50% larger than in the rest of the plots. The dimensionless distance ξ is shown above each figure for reference purposes. The vertical distance between the curve representing the forced and the unforced data corresponds to actual drag reduction due to forcing. By comparing these differences at X/b = 100 and normalizing the data by the local unforced values of momentum loss, one gets a drag reduction of 17% corresponding to the case plotted in a; 19% in b; 22% in c; and 27% in d. These results reinforce our previous conclusion that the most efficient drag reduction is obtained by low-frequency forcing.

A comparison was made between the mean velocity distributions of the forced and the unforced wall jets in the vicinity of the surface by using the conventional "wall coordinates." Velocity profiles corresponding to three sets of data (i.e., at $St{*}10^{15}$ = 444, 88, and 5.5 and Re_j = 3400, 5000, and 10,000, respectively) are plotted in Figure 15 on a semi-logarithmic scale. The ordinate of each family of profiles was shifted to provide a clearer visual assessment of the effects of forcing. Since each of the measured velocities was rendered dimensionless with respect to the local friction velocity $U_\tau \equiv \sqrt{\tau_w/\rho}$, one expects that the effects of forcing on the velocity distribution will correlate with the effects of forcing on τ_w. This proved to be the case because the average reduction in τ_w corresponding to $St{*}10^{15}$ = 444 was 15% while for $St{*}10^{15}$ = 5.5, it was 34% (see also Figure 12), which manifested itself in the largest disparity between the forced and the unforced velocity distributions. It seems that external excitation modifies the velocity distribution in what appears to be the logarithmic region.

Examination of the unforced velocity profiles (marked by x symbols on Figure 15) measured at three different Reynolds numbers indicates that the logarithmic velocity distribution is dependent on $(U_j \ b/\nu)$. The slope, A, of the logarithmic profile (U^+ = A*log Y^+ + B where U^+ = U/U_τ and Y^+ = YU_τ/ν) appears to be the "universal constant" quoted in the literature and is equal to 5.5 while the additive constant B is strongly

82

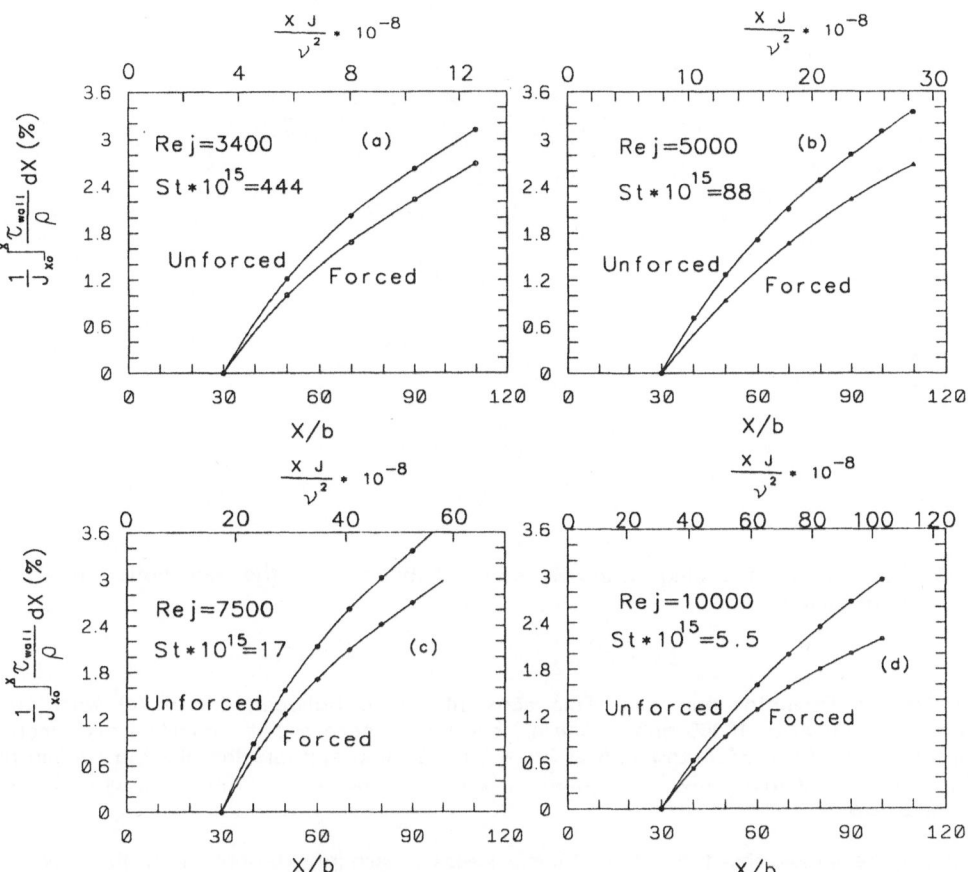

Fig. 14. The net drag reduction resulting from external excitation.

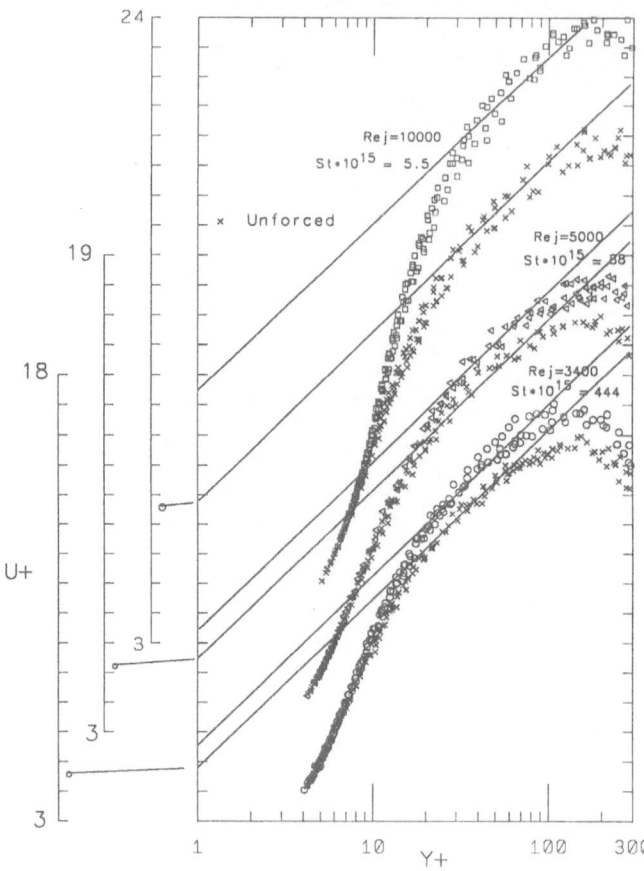

Fig. 15. The effects of forcing on the velocity distribution near the wall plotted in "wall" coordinates.

dependent on Reynolds number. This constant varied between 4.8 to 7.8 when Re_j changed from 3400 to 10000 only. External excitation increases the constant even more, making it 5.5 at Re_j = 3400 and 11.5 at Re_j = 10^4. It also appears that the logarithmic fit to the forced data may only be applied at larger values of Y^+ when compared to the unexcited flow.

It may be shown that the turbulent shear stress is strongly dependent on the distance from the surface even in the logarithmic region of the unexcited wall jet and, consequently, the existence of the logarithmic region appears to be fortuitous. We computed the stress distribution by integrating the momentum equation from the wall outward, requiring the entire stress to be viscous at the wall. We also computed the viscous stress from the mean velocity profile. The results of these computations are plotted in Figure 16 for Re_j = 3400 and 10000. The viscous stress vanishes at $((U_\tau Y)/\nu) > 30$ as anticipated from data accumulated in turbulent boundary layers. However, the total stress is not constant at $((U_\tau Y)/\nu) > 30$. In fact for Re_j = 34000, the total stress starts decreasing at $[YU_\tau/\nu] > 8$ while for the higher Re_j, it does so at $[YU_\tau/\nu] \cong 20$. Consequently the assumption of a constant stress layer does not apply to a fully developed, turbulent wall jet at these Reynolds numbers. In fact, the shear stress vanishes at the lower Reynolds number unforced flow at $((U_\tau Y)/\nu) \cong 70$ and at the higher Re_j at $((U_\tau Y)/\nu) \cong 100$. Thus, the basic assumption used in deriving the universal, logarithmic velocity distribution does not apply to the wall jet. It implies that the inertia terms increase as rapidly with Y in this flow as the viscous term diminishes, thus affecting the equilibrium to which we were so accustomed in the boundary layer.

84

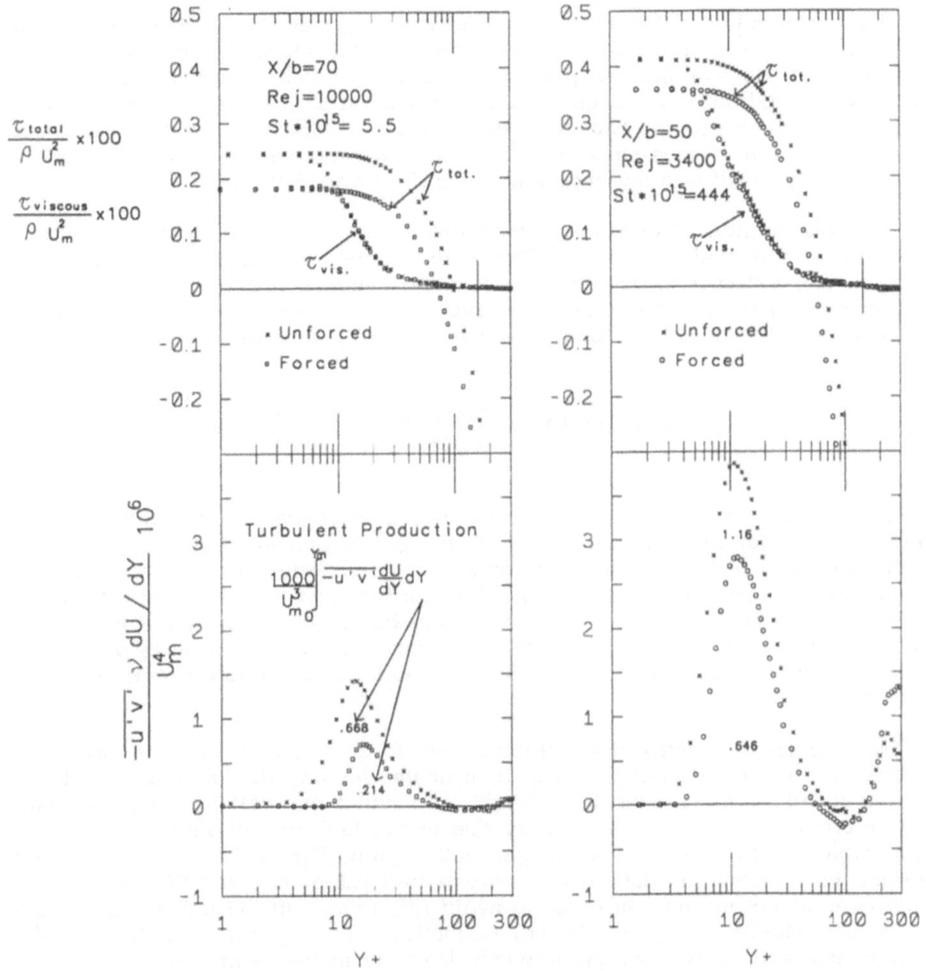

Fig. 16. The stress distribution and the production of turbulent energy near the surface for both forced and unforced wall jets.

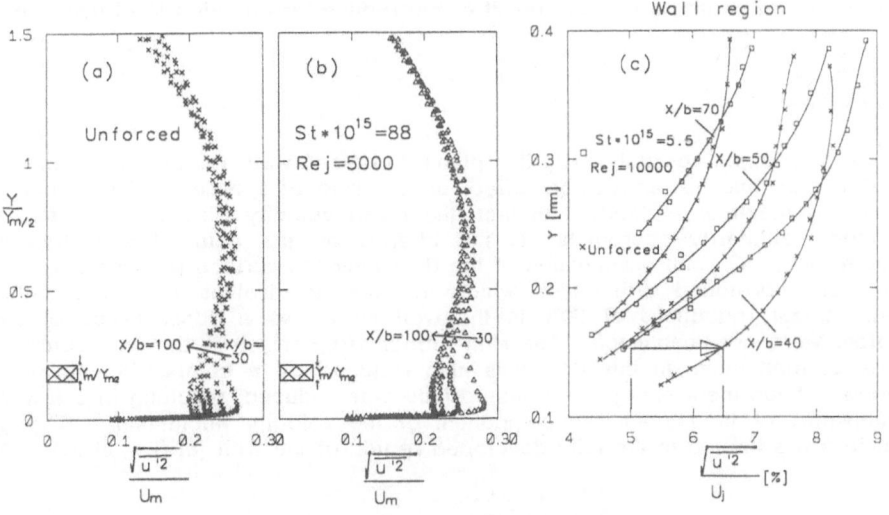

Fig. 17. The turbulence distribution.

The introduction of forcing not only affected τ_w but also reduced the extent of The constant stress layer (Figure 16). The total and the viscous stress distributions in the inner part of the wall jet are plotted at the top of this figure in order to assess the effects of excitation on these quantities. The wall stress, which is entirely viscous at the surface, is reduced by the two-dimensional forcing. However, for $Y^+ > 10$, the effect of excitation on the viscous stress vanishes since forcing ceases to have an effect on the mean velocity profiles in this region and therefore on $\nu(\partial U/\partial Y)$.

The distribution of turbulent energy production $-\overline{u'v'}\partial U/\partial Y$ with increasing distance from the surface is also plotted on Figure 16 for the two examples considered above. It is clear that external excitation reduces the Reynolds stress and with it the turbulent production. The integrated turbulent production in the inner region of the wall jet between the surface and the location at which $Y = Y_m$ is defined by

$$\text{PRODUCTION} = \frac{1000}{(U_m)^3} \int_0^{Y_m} \overline{-u'v'} \, \frac{\partial U}{\partial Y} \, dY$$

For $St = 5.5$ at $X/b = 70$, forcing reduced the value of the production integral from 0.667 to 0.214 while for $St = 444$ and $X/b = 50$, the production integral was reduced from 1.16 in the unforced case to 0.646. This is a significant effect which presumably alters the entire turbulent energy balance in this flow and not just the scales of the large coherent eddies. Since the mean velocity gradient vanishes at $Y^+ \cong 150$, the Y locations at which $\overline{u'v'} = 0$ and at which $\partial U/\partial Y = 0$ do not coincide, leading to a region of a weak but negative turbulent production just below the location at which $Y = Y_m$ (see also Kruka and Eskinazi, 1964).

No observable difference attributed to forcing could be seen in the level or distribution of the longitudinal component of the velocity fluctuations when the latter are plotted in traditional similarity coordinates spanning the wall jet (i.e., $0 \le Y/Y_{m/2} \le 1.5$; see Figure 17a and b). This is partly due to the lack of self-similarity in the unexcited flow, which is most noticeable in the wall region (Figure 17a). Consequently, one is obliged to compare the data at some prescribed values of ξ (or X/b) for the same efflux conditions at the nozzle. In order to avoid differences in the value and location of U_m, the results shown in Figure 17c are normalized also by the conditions at the nozzle. Forcing the wall jet reduces the turbulent intensity in the vicinity of the surface, and this reduction is strongly affected by St while all other parameters are maintained constant. The reduction in u'/U_j exceeds in some places 25%, as may be deduced from Figure 17c (see arrow comparing one data point at $X/b = 40$ and $St = 5.5$ which provides an example). Since the production of turbulence increases the level of u', the reduction in u' due to forcing might be a direct result of the reduced production level near the wall.

CONCLUSIONS

Two-dimensional excitation of the plane turbulent wall jet in the absence of an external stream has no appreciable effect on the rate of spread of the jet nor on the decay of its maximum velocity. In fact, the mean velocity distribution plotted in the conventional similarity coordinates is not altered by the external excitation in any obvious manner. Careful examination of the flow near the surface (i.e., for $0 < Y^+ < 100$) reveals some profound differences which manifest themselves in reducing the skin friction. Local reductions of 30% in the wall stress, as a consequence of such an excitation, were not uncommon. The skin-friction drag which is the only contributor to the loss of momentum in this flow was also reduced by a comparable amount. The production of turbulent energy near the surface was reduced, resulting in a lowering of the intensities of the streamwise component of the velocity fluctuations. The present conclusions are limited to the fully developed region of the wall jet (i.e., at $X/b > 30$).

ACKNOWLEDGEMENTS

The work was supported by AFOSR under contract number AFOSR-88-0176 and monitored by Dr. J. McMichael. The authors would also like to express their thanks to Dr. B. G. Newman for his help and advice.

REFERENCES

Guitton, D. E. and Newman, B. G., 1977, Self-preserving turbulent wall jets over convex surfaces, J. Fluid Mech., 81:155.

Ho, C. M. and Huerre, P., 1984, Perturbed free shear layers, Ann. Rev. Fluid Mech., 16:365.

Irwin, H. P. A. H., 1973, Measurements in a self-preserving plane wall jet in a positive pressure gradient, J. Fluid Mech., 61:33.

Katz, Y., Nishri, B., and Wygnanski, I., 1989, The delay of turbulent boundary-layer separation by oscillatory active control, AIAA Paper 89-0975.

Kline, S. J., Reynolds, W. C., Schraub, F. A., and Runstadler, P. W., 1967, The structure of turbulent boundary layers, J. Fluid Mech., 50:133.

Kruka, V. and Eskinazi, S., 1964, The wall-jet in a moving stream, J. Fluid Mech., 20:555.

Launder, B. E. and Rodi, W., 1981, The turbulent wall jets, Prog. Aerospace Sci., 19:81.

Launder, B. E. and Rodi, W., 1983, The turbulent wall jet--measurements and modeling, Ann. Rev. Fluid Mech., 15:429.

Narasimha, R., Narayan, K. Y., and Parthasarathy, S. P., 1973, Parametric analysis of turbulent wall jets in still air, Aeronautical J., 77:335.

Newman, B. G., 1961, The deflexion of plane jets by adjacent boundaries--Coanda effect, in: "Boundary Layer and Flow Control," Pergamon, London.

Ozarapoglu, V., 1973, Measurements in incompressible turbulent flows, Doctor of Sciences Thesis, Laval University. Quebec City.

Patel, V. C., 1965, Calibration of the Preston tube and limitations on its use in pressure gradients, J. Fluid Mech., 23:185.

Schwarz, W. H. and Cosart, W. P., 1961, The two-dimensional turbulent wall-jet, J. Fluid Mech., 10:481.

Tailland, A. and Mathieu, J., 1967, Jet parietal, J. de Mecanique, 6:1.

Willmarth, W. W., 1975a, Pressure fluctuations beneath turbulent boundary layers, Annual Review Fluid Mech., 7:13.

Willmarth, W. W., 1975b, Structure of turbulence in boundary layers, in: "Advances in Applied Mechanics," Academic Press, New York.

Wills, J. A. B., 1962, The correction of hot-wire readings for proximity to a solid boundary, J. Fluid Mech., 12:3.

Wygnanski, I. and Petersen, R. A., 1987, Coherent motion in excited free shear layers, AIAA J., 25:201.

CONTROL OF ORGANIZED STRUCTURES IN ROUND JETS

AT HIGH REYNOLDS NUMBERS

P. J.-D. Juvet and W. C. Reynolds

Department of Mechanical Engineering
Stanford University
Stanford, California 94305

1 INTRODUCTION

Researchers have studied jets for decades and have observed very early their sensitivity to sound. Dual-mode dual- frequency acoustic excitation can be applied to round air jets to alter dramatically their structure [3,4]. The axial excitation locks the formation of the shear-layer vortex rings on the excitation frequency and therefore controls the vortex spacing, while the helical excitation controls the eccentricity of the rings. The development of the jet depends on the ratio R_f of the axial excitation frequency f_a to the helical frequency f_h. When this ratio is a non-integer between 1.6 and 3.2 the jet is said to bloom [3] as the vortex rings are sent in all directions. When this ratio is equal to 2 the jet bifurcates into two distinct branches. As the axial excitation frequency is increased, the vortex spacing is reduced, leading to a stronger mutual interaction and jet spreading angle. The Strouhal number St_a of 0.55 based on the axial frequency, mean exit velocity and exit diameter was found to produce the largest spreading angles [3]. If the spacing is too small however, the rings tangle and the spreading effect disappears at once. The judicious combination of longitudinal and azimuthal excitation has been shown to enhance jet entrainment drastically [2].

1.1 Objectives

The control of the jet is obtained via the jet exit shear layer which acts as a flow amplifier for the acoustic disturbances. The maximum amplification is obtained for disturbances at the shear-layer natural frequency. This preferred frequency was shown to scale with the initial momentum thickness. Blooming and bifurcating jets are easily obtained at low Reynolds numbers for which the preferred frequency is near the optimum frequency for vortex interaction [3,4]. At high Reynolds numbers these phenomena can also be obtained provided that a sufficiently high forcing is applied. Earlier work [4] shows that this could be done in air with acoustic excitation up to Re of at least 100,000 and Mach number of 0.25. However, at these Reynolds numbers the thin initial shear layer is only weakly receptive to the relatively low applied frequencies required for blooming or bifurcation. The rings must then be organized by the axial excitation into larger vortices through collective interaction [1].

The present work tries to meet the challenge of increasing the receptivity of high Reynolds number jets and in particular very large jets which are of practical engineering interest. The idea is to study the enhancement of controllability through thickening of the nozzle boundary layer and hence the initial shear layer of the jet. This should reduce the frequency most rapidly amplified by the initial shear layer and make it easier to control the jet with weak forcing. Two approaches are explored; slight diffusion, which creates an adverse pressure gradient, and streamwise blowing near the jet exit. The response to the different shear-layer thickening and acoustic excitation is evaluated by measuring flow entrainment in a shrouded jet, such as might be found in jet ejectors or thrust augmenters. The results are compared with earlier results [2] obtained in the same jet facility by without shear-layer manipulation.

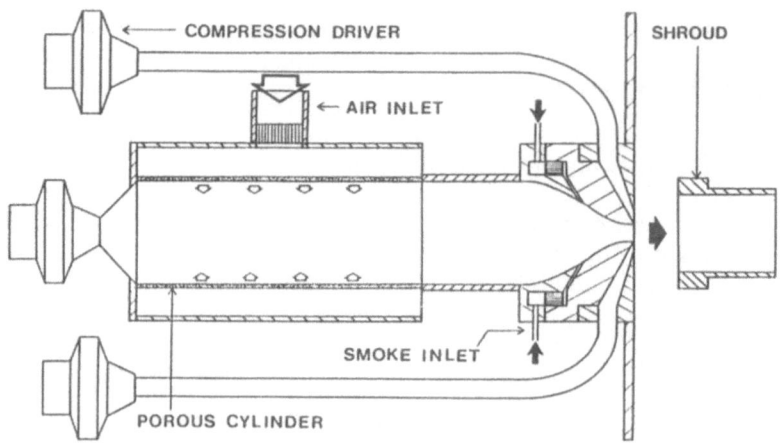

Figure 1. Schematic of the jet facility

2 EXPERIMENTAL FACILITY

2.1 Flow System

The apparatus shown schematically in Figure 1 has been described in details [4] in earlier work. Air, provided by a blower, flows into a plenum and through a nozzle with an exit diameter D of 20 mm. All the measurements presented in this paper were taken at a Re of 100,000 (based on the nozzle exit diameter D and exit mean velocity U_0) corresponding to a mean jet exit velocity of 75 m/s. The jet is surrounded by a shroud of approximately $4.5D$ in diameter and $5.7D$ in length. A gap is left between the jet exit plane and the shroud. Velocity profiles were measured at the shroud exit $8D$ from the jet exit plane with a hot-wire anemometer.

The different jet exit geometries used are sketched in Figure 2. The reference exit geometry is sketched in detail in Figure 2(a). This reference geometry was modified for shear-layer manipulation. Axisymmetric inserts were placed concentrically to the original jet exit geometry leaving a narrow annular gap (0.3 mm) between the outside of the main jet and the insert wall through which continuous blowing can be applied for jet control. Two inserts are used; a straight insert sketched in Figure 2(b) and a 8° conical diffuser insert shown in Figure 2(c).

2.2 Acoustic Excitation

The acoustic excitation can be applied in two modes at different frequencies with compression drivers. The axial excitation is provided by a driver located at the bottom of the plenum. The helical excitation is provided by four external drivers connected to the exit by steel tubes acting as wave guides. For the geometries modified for blowing the wave guides are positioned directly around the jet exit. The drivers are fed with sine waves 90° out of phase with respect to the adjacent one to create the helical excitation. For all the cases presented here, the axial Strouhal number St_a is 0.55 and the frequency ratio R_f is 2.3 for blooming. This corresponds to an axial excitation frequency of 2060 Hz and an helical excitation frequency of 896 Hz. The amplitude level of the axial excitation a' is quantified as the axial excitation velocity $u_{a'}$, measured at the jet exit centerline, as a fraction of the mean velocity at the jet exit U_0. The amplitude level of the helical excitation h' is quantified as the sound pressure amplitude p', measured in still air with a condenser microphone above the jet, as a fraction of the dynamic pressure at the nozzle exit $\frac{1}{2}\rho U_0^2$. The different cases of excitation amplitude used here are summarized in Table 2 and are also referred as natural, weak, modest and strong blooming cases.

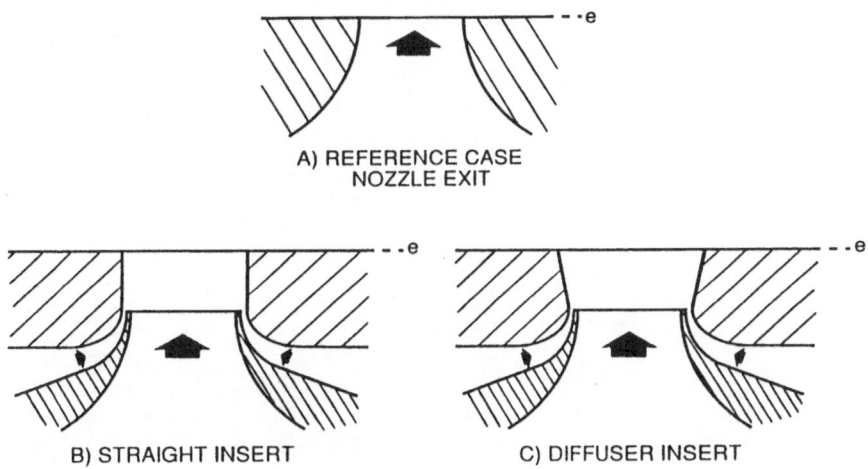

Figure 2. Schematic of the exit geometries

3 RESULTS AND DISCUSSION

Earlier experiments [2] were carried out using the reference geometry of Figure 2(a). Figure 3 shows instantaneous pictures of the jet at the shroud exit. The flow is visualized with smoke and a laser sheet. In the natural jet case (left) the smoke can only be seen in a limited region around the center of the shroud exit and does not seem to spread significantly. In the strong blooming case (right) the smoke fills the entire width of the shroud exit and spreads as it moves downstream. Clear vortices can be noticed at the exit of the shroud which demonstrates the strength and robustness of these large scale structures.

Later experiments were conducted with the modified geometries (b) and (c) of Figure 2. The first step was to study the effect of the different geometries and blowing rates on the initial shear-layer thickness. Figure 4 shows the mean velocity profiles of the shear layer slightly downstream of the exit plane ($x/D = 0.05$) for the reference case, the straight insert and the diffuser insert. The profiles for additional tangential blowing are also shown. Weak blowing refers to a blowing rate of 3.8% defined as the ratio of the secondary volume flow rate Q_b to the main jet flow rate Q_0. Strong blowing refers to a blowing rate of 6.9%. The slopes of the profiles are important since they are directly related to the vorticity thickness. The slopes for the straight insert profiles are close to those for the reference case. The effect of blowing is only marginal. In the diffuser case however, the slope for no-blowing is drastically reduced and weak blowing further reduces it, leading to an increased thickness. Table 1 presents these results quantitatively. The vorticity thickness δ_w, defined with the maximum profile velocity U_{max} and the profile maximum slope $(dU/dr)_{max}$, is normalized with respect to the main jet exit diameter D. In the no-blowing diffuser case the adverse pressure gradient imposed to the

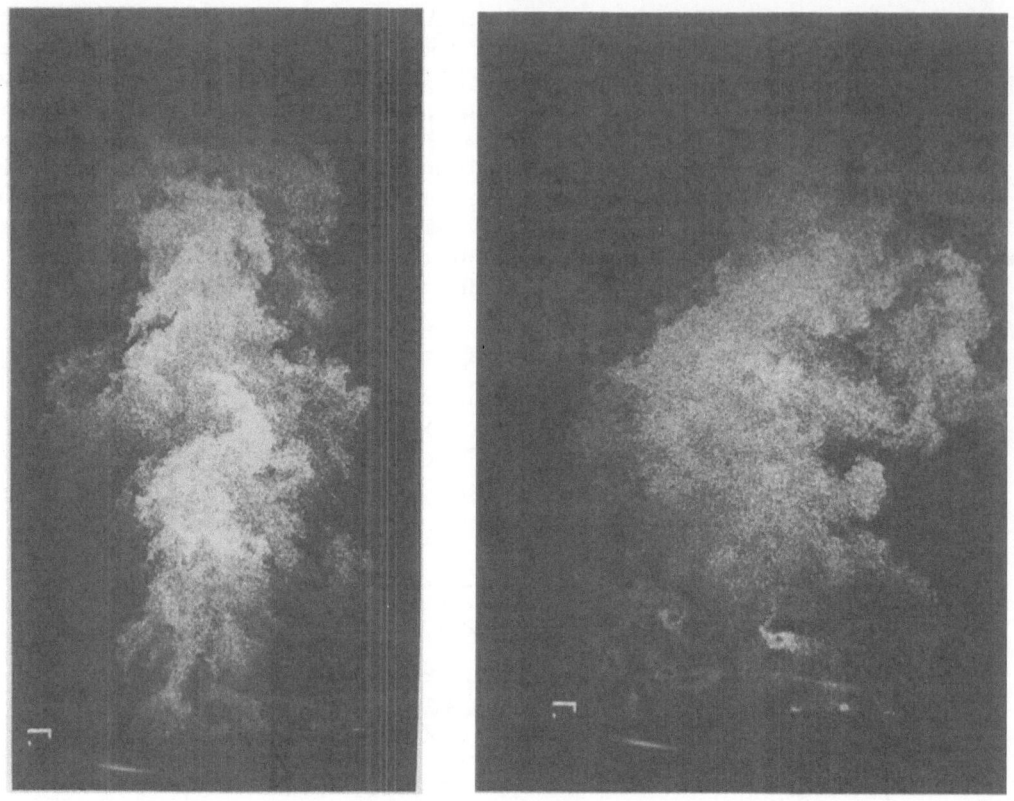

Figure 3. Photographs of a natural (left) and a strong blooming jet (right) at the shroud exit

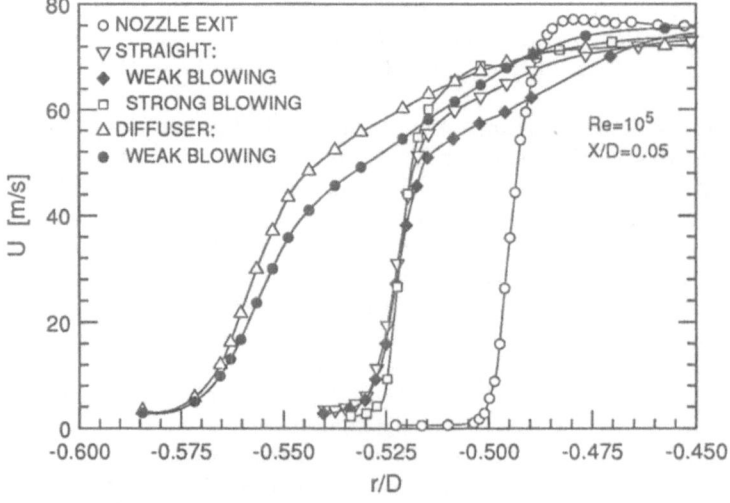

Figure 4. Shear-layer velocity profiles

Table 1. Shear-layer thickness for various cases

$$\delta_w = U_{max}/(dU/dr)_{max}$$

	δ_w/D	δ_w/D	δ_w/D
No Blowing	0.010	0.016	0.033
Weak Blowing	-	0.018	0.044
Strong Blowing	-	0.011	-
	Nozzle Exit	Straight Insert	Diffuser Insert

boundary layer allows a drastic thickening of the initial shear layer up to 3.3 times that of the reference case. The effect of blowing are much weaker. It is interesting to notice that a weak blowing increases the shear-layer thickness while a strong blowing brings it back to the reference value. The weak blowing case corresponds to a mean tangential velocity of about half that of the main jet and is therefore expected to have the strongest effect on the shear layer.

The second step was to study the effect of thickening on the jet response to acoustic excitation and in particular on the entrainment. Mean velocity profiles were measured at the exit of a shroud to estimate the entrainment rate for the different cases. The profiles for the diffuser insert are plotted in Figure 5. Blooming reduces the shroud centerline velocity and broadens the jet dramatically. The differences between the diffuser profiles and the straight insert profiles, not presented here, are minimal. The effect of blowing is small as one could have expected from the initial shear-layer profiles. The results are presented quantitatively in Table 2. The volume flow rate Q, calculated by integrating numerically the velocity profiles, is compared to the volume flow rate at the jet exit Q_e. The differences between the straight and diffuser insert are marginal as well as the effect of blowing. However, the entrainment rates obtained for these geometries are almost 20% larger than for the reference case (nozzle exit). Or in other words same entrainment rates can be obtained with weaker acoustic excitation amplitudes. For example the weak excitation in the diffuser case yields the same entrainment as the modest excitation for the reference case; this corresponds to a reduction in acoustic amplitudes of 2 for the axial excitation and 4 for the helical excitation.

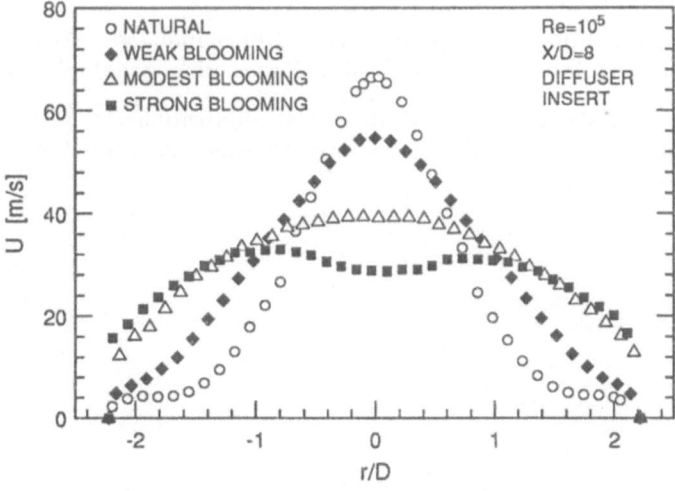

Figure 5. Velocity profiles at the shroud exit

Table 2. Entrainment rates for various cases

$$Q = 2\pi \int U r dr$$

	EXCITATION		ENTRAINMENT Q/Q_e		
	$a' = u'_a/U_0$	$h' = p'/(\frac{1}{2}\rho U_0^2)$	Nozzle	Straight	Diffuser
Exit	-	-	1	1	1
Natural	-	-	2.8	2.8	2.8
Weak Blooming	0.004	0.002	-	4.2	4.3
Modest Blooming:	0.008	0.009	4.4	5.9	5.7
Weak Blowing	0.008	0.009	-	6.2	5.6
Strong Blowing	0.008	0.009	-	6.5	6.1
Strong Blooming	0.028	0.018	5.4	5.8	5.7

4 CONCLUSIONS

The effects of dual-mode acoustic excitation on a shrouded jet are documented by hot-wire anemometry and flow visualization. Substantial broadening and increased entrainment are observed in the blooming jets. The modification of the exit geometry allows high levels of entrainment to be obtained with weaker acoustic excitation levels. The initial shear-layer thickness can be increased by tangential blowing and more efficiently by imposition of an adverse pressure gradient via a conical diffuser.

5 ACKNOWLEDGEMENTS

This work was sponsored by the Air Force Office of Scientific Research under Contract AF-F49620-86-K-0020.

References

[1] C. M. Ho and L. S. Huang. Subharmonics and vortex merging in mixing layers. *J. Fluid Mech.*, 119:443–473, 1982.

[2] P. J. Juvet and W. C. Reynolds. *Entrainment Control in an Acoustically Controlled Shrouded Jet*. Paper 89-0969, AIAA, 1989.

[3] Mario Lee and W. C. Reynolds. *Bifurcating and Blooming Jets*. Technical Report TF-22, Thermosciences Division, Dept. of Mechanical Engineering, Stanford University, 1985.

[4] D. E. Parekh, A. Leonard, and W. C. Reynolds. *Bifurcating Jets at High Reynolds Numbers*. Technical Report TF-35, Thermosciences Division, Dept. of Mechanical Engineering, Stanford University, 1988.

THREE-DIMENSIONAL STRUCTURE OF THE VORTICITY FIELD IN THE NEAR REGION OF LAMINAR, CO-FLOWING FORCED JETS

Juan C. Lasheras
Department of Applied Mechanics and Engineering Sciences
University of California, San Diego
La Jolla. CA 92093

Antonio Lecuona and Pedro Rodriguez
E.T.S.I. Aeronauticos
Universidad Politecnica de Madrid. 28040 Madrid. Spain

Abstract

We examine the vorticity dynamics in the near field of laminar ($Re=10^3$) co-flowing jets subjected to the single or combined effects of axial and azimuthal forcing. It is shown that the dynamic interaction of the three-dimensional vortex structure resulting from the growth of the two and three-dimensional instabilities may result into large changes in the entrainment and mixing characteristics of the jet. For each azimuthal forcing, and for a fixed velocity ratio between the inner and outer jet, we show the existence of a critical Strouhal number for which a symmetric instability mode leads to a pattern of lateral ejections of closed vortex loops. In addition, we show that at a higher Strouhal number a second symmetric mode develops which consists of a double number of lateral ejections. The aboved three-dimensional modes and their topological changes are analyzed in view of the three-dimensional inviscid induction of the two concentric array of vortex rings emanating from the jet's exit nozzle.

Introduction

The axisymmetric instability of a laminar shear layer of cylindrical shape is well understood, both from an analytical and experimental point of view, Michalke and Freymuth (1966), Wille (1963), Becker and Massaro (1968), Beavers and Wilson (1970), Crow and Chanpagne (1971), Cohen and Wygnanski (1987), among many others. This instability results in the roll up of the cylindrical vorticity sheet to form a periodic array of vortex rings. This evolution has also been corroborated as occuring in transitional and turbulent jets, Browand and Laufer (1975), Yule (1978), Dimotakis et al. (1983), Tso and Hussain (1989), Mungal and Hollingsworth (1989), and others. In essence, these high Reynolds number studies have shown that early in the evolution of a fully developed turbulent jet, the primary vorticity concentrates to form a coherent array of ring-like vortical structures.

Browand and Laufer (1975), Yule (1978), and others have shown that the transition to turbulent flow in jets involves a relatively orderly three-dimensional deformation of the initial vortex rings. Their results show that after the growth of the primary instability, the vortex rings further develop a high azimuthal structure whose breakdown may occur in a violent manner. During the breakdown process, they also observed that pieces of the core's vorticity can pinch off to form small vortex loops. Yule (1978) then argued the importance of this azimuthal instability mode of the vortex rings, suggesting that transition in round jets involves the entanglement of the street of coherent vortex rings which have developed along their circumference wave deformations.

The Global Geometry of Turbulence
Edited by J. Jiménez, Plenum Press, New York, 1991

The instability of the ring-like vortices observed in jets has further been associated with the short-wave instability found in isolated vortex rings.by Widnall and Sullivan (1973), Widnall, Bliss and Tsai (1974), Maxworthy (1972 and 1974), Didden (1977), and others. These analytical and experimental studies have shown that the instability manifests itself in the generation of waves over the circumference of the ring. The thinner the inner core of the vortex ring, the larger is the more unstable number of waves found to be amplified through this ring's instability.

In order to elucidate the mechanisms leading to the three-dimensional transition of the jet, and in particular to study the role that the ring's instability plays in the three-dimensional transition (and eventually in the turbulent transition), we conducted a set of experimental studies in which laminar, co-flowing jets were subjected to the combined effect of a pair of orthogonally oriented perturbations. The first one (hereafter referred to as the primary perturbation) consisted of a single wave, sinusoidal perturbation in the axial velocity component of the jet at the nozzle's exit section. The use of this primary perturbation was intended to lock to a given value (at least in the near region of the jet) the frequency and thus the wavelength of the jet's vortex rings. The second one (hereafter referred to as the secondary perturbation) consisted of a wavy, sinusoidal radial displacement of the cylindrical vorticity layer emanating from the nozzle. This was achieved by means of a corrugated nozzle exit whereby the cylindrical vorticity layer emanating from it is periodically displaced radially in and out along the jet circumference. Early experimental and numerical work on the effect of this second perturbation, Martin, Meiburg and Lasheras (1989, 1990) and Meiburg and Martin (1991), have shown that the growth and amplification of this secondary perturbation results in the formation of counter-rotating pairs of streamwise vortices which further interact with the vortex rings locking their azimuthal mode to a predetermined number of waves along its circumference. Thus, the intention of the secondary perturbation was to lock the ring's instability to a given azimuthal wavelength (i.e., number of waves along its circumference). The work is, in this respect, an extension of the axisymmetric configuration of similar studies conducted for free shear layers and wakes in the planar configuration, Lasheras and Choi (1988), Ashurst and Meiburg (1988), Meiburg and Lasheras (1988), Lasheras and Meiburg (1990), where we analyzed the combined growth of streamwise and spanwise perturbations.

Although our investigation has the general objective of obtaining a clearer understanding of the nature of the three-dimensional transitions which occur in the near region of co-flowing, round jets developing from thin, laminar boundary layers, this paper has a less broad scope. Rather, it is devoted to demonstrating that the structure of the near region of the jet can be altered considerably as a result of the growth and nonlinear interactions between the vortical structures formed through the jet's instabilities. Without intending to provide an exhaustive analysis, we will concentrate only on some preliminary results. Our aim is to show to the applied mathematics community interested in the general topic of vorticity dynamics the existence of several modes of three-dimensional vortical patterns resulting from the global induction of the jet's vorticity when subjected to the combined effect of axial and azimuthal forcing. For the particular case of co-flowing , low Reynolds number jets, we will show that depending on the jet's Reynolds number, the relative velocities between the two streams, and the wavelengths of the primary and secondary perturbations, the combined induction of the vortex rings and tubes forming in the jet results in a large variety of three-dimensional vortex patterns which lead to quite distinct.spreading and mixing characteristics of the jet.

The Experimental Set-up and Observation Techniques

The experiments were conducted in a highly versatile, atmospheric pressure, open-return wind tunnel where non-reacting (isothermal) as well as heat transfer or combustion experiments can be conducted. A layout of the flow facility is shown in Figure 1. The wind tunnel consists of three co-flowing, low-speed axisymmetric streams independently created of each other. The inner, or primary stream, is produced through a nozzle with an outer diameter of 24.5 mm. The nozzle is located downstream of a settling chamber where a combination of gases can be independently metered. To minimize the intensity and scale of the turbulent fluctuations, the settling chamber is equipped with an array of perforated plates, screens, and honeycomb. This primary jet is surrounded by a co-flowing, concentric, gas stream (secondary flow) which discharges into the test section through a round nozzle 160 mm. in diameter. Atmospheric air is drawn into the test section forming the tertiary co-flowing jet through the ejection-like effect produced by these two co-flowing streams discharging into the test section. The three co-flowing jets meet in a 440 x 440 mm. square cross-section that is 1,200 mm. high. Because of the cooling effect provided by the tertiary flow, the lateral walls of the test section (even for the combustion experiments) are built of transparent metacrilate. The test section exhausts to the atmosphere through a blower upstream of which a honeycomb flow straightener and a butterfly valve are located. Through the extensive use of honeycombs, screens, perforated plates and flexible joints we were able to produce steady and uniform conditions in the three streams, thus minimizing the effect of uncontrolled perturbations in the initial conditions. The velocity measurements obtained using LDV by Cuerno et al. (1989) confirmed that for any selected flow condition the turbulent levels are always well below 1%. The design is capable of producing flow conditions where the primary flow exit velocity can be varied from 0.1 to 10 m/sec, while both the secondary and tertiary streams can be operated with velocities ranging from 0.2 m/sec to 2 m/sec.

The wind tunnel co-flowing jets facility is mounted on a two axis computer-controlled stand. A micro-controller TL78 and IP 28 two axis stepping motor drive is used to position the test section relative to the instrumentation which is placed on a stationary optical bench. Furthermore, the design allows the test facility to be rotated around its axis.

The periodic streamwise forcing is produced through a vibrating membrane located in the settling chamber of the primary flow. At the base of a small chamber, a loudspeaker fed with a sinusoidal wave generates pressure pulses that result in the periodic displacement of the membrane in a piston-like type fashion. The membrane fluctuation adds a streamwise velocity perturbation of a given amplitude and frequency to the primary jet flow. The results reported here correspond to the case of a single sinusoidal wave with a frequency ranging from 0 to 20 Hz. The 6 inch bass loudspeaker used throughout these experiments generates a sinusoidal perturbation that, measured by LDV at the center of the primary cold jet, was found (in a range of 4 to 18 Hz forcing frequency) to have a very low level of harmonic distortion, up to perturbation amplitudes of the level of 100%. The instantaneous velocity signal was screened in the frequency domain by means of an Ono-Sokki CF 500 spectral analyzer. The harmonic content for the perturbation levels used was always less than -40db relative to the fundamental frequency.

The acoustic natural frequency of the Helmholtz resonator formed by the primary flow settling chamber and the exit tube is approximately 47 Hz. No attempt was made to determine the test section's natural frequencies as they are believed to be very low. The perturbation sensitivity measured as velocity amplitude over speaker input voltage, reaches a maximum at 14Hz. The smoothness of the sensitivity curve seems to indicate a loudspeaker-power amplifier electrical coupling effect more than any other fluid mechanics phenomena.

Using the above described axisymmetric flow forcing, we introduced azimuthal perturbations following the same techniques used in our early studies of plane, free shear flows, Lasheras and Choi (1988), Meiburg and Lasheras (1988) i.e., corrugated or indented nozzles. The effect of a gradual corrugation at the trailing edge of the nozzle is to periodically displace radially in and out the cylindrical vorticity layer emanating from the exit section. In the present case we will restrict the azimuthal perturbation to the case of the corrugated nozzle only. The perturbation amplitude was about +/-3mm in both cases, significantly smaller than the boundary layer thickness of the two streams. Owing to manufacturing limitations, nozzles with longer wavelengths have a larger perturbation amplitude. Although the number of waves per circumference was varied from four to nine, we will restrict the discussion here to the case of five waves per revolution only.

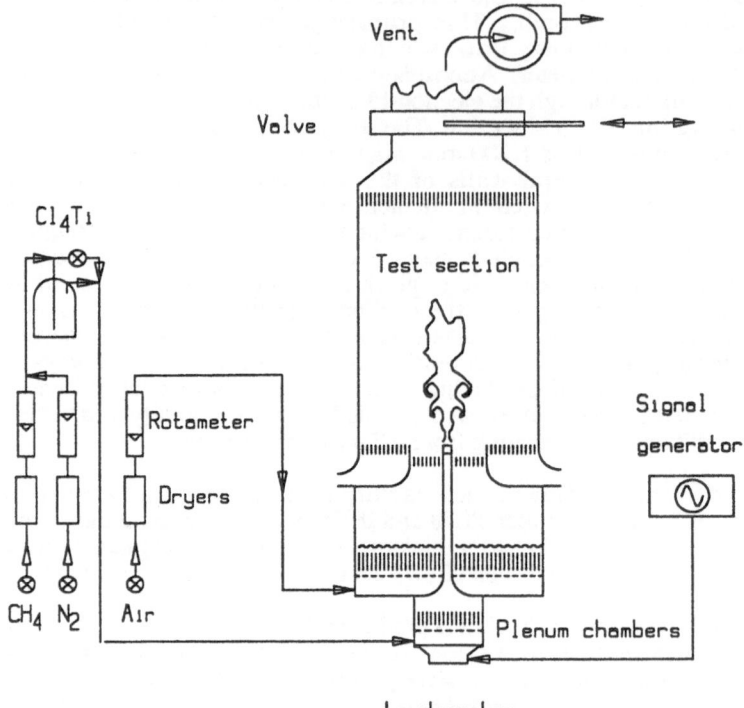

Figure 1. Schematics of the experimental set-up.

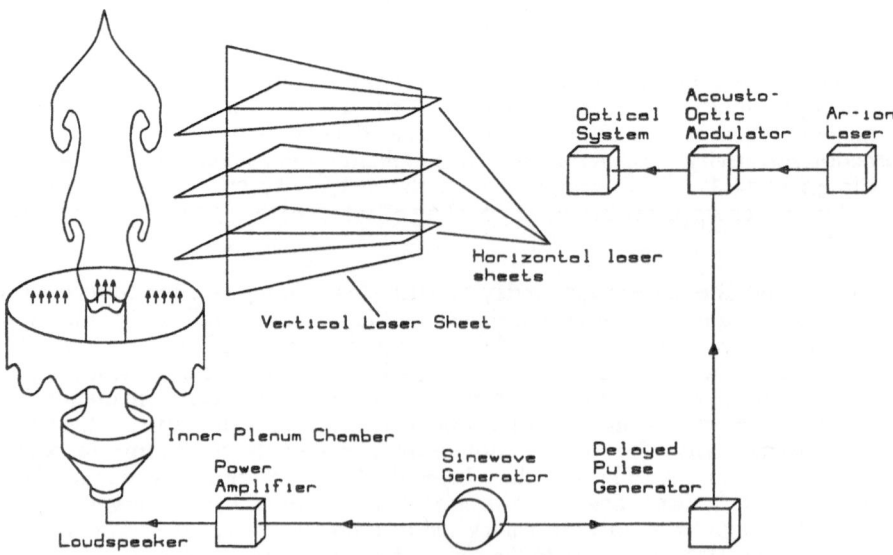

Figure 2. Flow visualization arrangement.

Flow Visualization Techniques

In order to reveal the three dimensional structure of the flow we used a combination of flow visualization techniques. Figure 2 shows a schematic of the visualization techniques employed. The inner, primary flow is first dried and subsequently saturated with TiCl4 vapors, meanwhile the secondary flow is supplied containing water vapor in a concentration equal to that existing in the ambient laboratory conditions. As the two streams meet at the edge of the nozzle the TiCl4 vapors react with the water vapor to form submicron size TiO_2 particles. This reaction continues to occur as the interface between the two streams is distorted under the combined effect of the instabilities triggered by the two perturbations. The reaction products (TiO_2) provide a continuous generation of low diffusivity scattering particles that will mark the instantaneous position of the interface. The visualization of the interface is then achieved by recording the light scattered by $TiO2$ particles by vertical and horizontal laser sheets which are created by means of cylindrical lenses. The combined use of horizontal cross-cuts at different downstream distances of the nozzle exit, and vertical cuts allows for the reconstruction of the three-dimensional structure of the jet.

At each downstream location we conducted both continuous as well as phase-locked visualization. In order to phase lock the particle image with the perturbation, the continuous laser beam was deflected using an A&A MTS 1200-16 & 80-B4605 acousto-optic modulator. This makes the green colored light pass through the sheet forming optics during a l ms TTL pulse, which is phase locked with the perturbation sine wave by means of a Feedback TW 6500 variable phase function generator. As video cameras and recorders usually have their own fixed frequency clock synchronization, phase locking is not possible with this equipment. Thus, in our case, non consecutive correct images were obtained by selecting an appropriate shutter speed. With the use of the photographic camera, a custom external delay circuit triggers the camera shutter opening for a time period slightly longer that the l ms illuminating flash. Phase average images can be obtained by opening asynchronously the camera shutter and free running the phase locked illuminating. Slides of Ektachrome P800/1600 film pushed up to 6400 ASA and Cibachrome prints were obtained with high reliability. In order to enhance images and obtain phase stacking, image processing was performed using a DT 2851 frame grabber working on a 80386 personal computer. The results are presented elsewhere.

Three-dimensional patterns in the near region of the forced co-flowing jets

The following,is a preliminary study intended to provide experimental evidence of the existence on a large variety of entrainment patterns that may result from the combined effect of the axial and azimuthal forcing. The work is exploratory in nature and no attempt will be made here to fully characterize the behavior of the jet for the whole range of parameters, i.e. the jet's Reynolds number, the frequency and amplitude of the axial forcing and the wavelength and amplitude of the azimuthal forcing. In particular, we will discuss only a set of experiments where a co-flowing jet has been strongly forced in the axial and in the azimuthal directions to show the development of several different vortical patterns resulting from changes in the initial conditions. A more detailed and complete parametric study including an exhaustive analysis of all the different three-dimensional vorticity modes and their analysis involving inviscid vortex dynamics simulations will be presented elsewhere, (Lasheras, Lecuona, Martin, Meiburg, and Rodriguez, 1990). There, we present an exhaustive comparison between Lasheras', Lecuona's and Rodriguez's experimentally found patterns and Martin's and Meiburg's inviscid dynamics simulations.

Early experimental and numerical studies on plane, free shear layers (Lasheras and Choi, 1988 and Ashurst and Meiburg, 1988) plane wakes (Meiburg and Lasheras, 1988; Lasheras and Meiburg, 1990) and jets (Meiburg, Martin, and Lasheras, 1989; Martin, Meiburg, and Lasheras, 1990; and Meiburg and Martin, (in this volume) subjected to the combined effect of streamwise and spanwise (azimuthal) forcing have shown that the smooth layer (or layers) of vorticity are unstable to this type of perturbations. Experimental and numerical results show that the nonlinear evolution of both perturbations leads to the formation of concentrated regions of vorticity (vortex tubes and vortex rings) which are oriented both in the streamwise as well as in the spanwise direction. For both planar and axisymmetric cases, the stretching of the perturbation vorticity, along the direction of the principal plane of positive strain (the braids between consecutive spanwise vortices or vortex rings) results in the formation of counter-rotating pairs of streamwise vortex tubes which further interact with the spanwise structure. The resulting topology of the vorticity field composed of a dual array of spanwise and streamwise vortices was shown to depend on both the geometry as well as on the initial conditions. Furthermore, these studies have shown beyond any doubt that many of the observed topological changes can be explained via inviscid vortex dynamics. For a detailed description of the vorticity dynamics observed in these free shear flows, the reader is referred to the above mentioned studies.

In a related study, Martin, Meiburg and and Lasheras (1990), using three-dimensional inviscid vortex dynamics simulations, have shown that a smooth, cylindrical vorticity layer when perturbed both with a single sinusoidal wave in the axial direction and an azimuthal, sinusoidal perturbation of a given wavelength results not only in the roll-up of the vorticity layer to form vortex rings, but also in an additional array of streamwise vortices located in the bridges connecting two consecutive rings (Figure 3). The observed evolution can be summarized in the following sequence of events sketched in Figure 4. A: the growth of the primary, axial instability leads to the periodic concentration of vorticity in vortex rings located in planes perpendicular to the axis of the cylinder. B: Under the effect of the evolving positive strain rate which is created between two consecutive rings, the perturbation vorticity is tilted, reoriented, and stretched to form counter-rotating pairs of vortex tubes that eventually wrap around the rings. Furthermore, as a result of the non-linear interaction between the evolving streamwise vortices and the vortex rings, the latter develop a wavy dislocation over their circumference with a number of waves equal to that of the initial perturbation, (Figure 3).

Since in the present study we will be looking at the evolution of strongly forced, co-flowing jets, the global vorticity dynamics may be considerably more complicated than the one described in Figure 4. However, an extrapolation of Meiburg's and Martin's calculations based on our early studies of three-dimensional plane wakes (Meiburg and Lasheras, 1988 and Lasheras and Meiburg, 1990) suggest that co-flowing forced jets should be modelled using two concentric vorticity cylinders of opposite sign with the appropriate relative strength between them given by the properties of the corresponding boundary layers. For large amplitudes of forcing, the effect of the primary perturbations can lead to alternate roll ups of the two vorticity layers so that a staggered array of vortex rings of opposite sign is formed, as shown in Figure 5. In addition, the growth and amplification of the secondary (azimuthal) perturbation now results in an array of counter-rotating vortex tubes located in the braids connecting two consecutive vortex rings of opposite sign (much as was the case with plane wakes, Lasheras and Meiburg 1990, their figure 10). As was the case in the plane wake, the topology of the resulting set of vortex tubes and rings would depend not only on the amplitude of the axial forcing, but also on the orientation of the azimuthal perturbation (indented or corrugated nozzle).

In general, depending on the amplitude and frequency of the axial forcing and on the boundary layer characteristics, the above described evolution will lead to two opposite sign, coaxial rings per wavelength located in a staggered configuration with a certain relative magnitude and core radius. As a consequence of the interaction with the counter-rotating pairs of streamwise vortices, each of these vortex rings will have a wavy dislocations in their cores of the same wavelength as the imposed azimuthal perturbation. However, the wavy dislocation in their cores will be either in phase or out of phase, depending on the orientation of the initial azimuthal perturbation, i.e. a co-flowing jet issuing from a corrugated nozzle when strongly forced, will develop a Karman-like vortex ring pattern with the vortices of each sign containing a circumferential wavy dislocation 180° out-of-phase between two rings of opposite sign. However, a jet issuing from an indented nozzle would produce a similar pattern, but the staggered array of opposite sign vortex rings would contain circumferential wavy dislocations in-phase with each other. The global induction of this highly complicated vorticity organization can subsequently lead to a large variety of topological changes. These topological changes and their effect in the entrainment and spreading of the jet is precisely the point intended to be addressed in this paper.

In the following, we elected to discuss four different modes of three-dimensional patterns observed in laminar, co-flowing jets with Reynolds numbers based on the inner jet's diameter and relative velocity of the order of 1000. The frequency of the axial forcing was varied between 4 and 20 Hz, and the number of lobes in the corrugated nozzle was 5.

Mode 1: In the case of small amplitude forcing and axial forcing frequencies resulting in wavelengths of the order of the jet diameter, the near field structure of the vorticity was found to evolve along the mechanism described in Figure 3. i.e., the dynamics can be captured with the inclusion of only one vorticity cylinder as shown by Martin, Meiburg and Lasheras (1990), and the wake-like effect of the existance of the outer boundary layer can be neglected. Figure 6 corresponds to a vertical cut of the jet along its axis showing that for a low amplitude of forcing, the cylindrical vorticity layer rolls up in a periodic array of vortex rings. In this visualization the dark product marks the position of the interface between the two co-flowing streams. A horizontal cross cut of the jet at a downstream distance of three jets' diameters (Figure 7) reveals the presence of counter-rotating pairs of streamwise vortices as well as the circumferential, wavy undulation in the cores of the vortex rings.This cross-cut has been made through the core of the vortex ring. The two concentric interfaces reveal the roll up of the vorticity sheet. A perspective view of the jet's interface corresponding to a jet length spanning between the first and fourth roll up is shown in Figure 8 where the lambda shape of the streamwise vortices can be clearly seen. As was the case with the visualization shown in Figure 7, and for the purpose of better clarity, we have selected the positive image whereby the bright (white in the figure) portion corresponds to the positions occupied by the TiO_2 particles (interface). At this downstream location, the instability of the vortex ring (Didden, 1977; Maxworthy, 1972 and 1977; Widnall, Bliss, and Tsai, 1974) has not grown appreciably.

Mode 2: For axial forcing frequencies higher than the most unstable one found for the natural case, and for low values of its amplitude, was found to lead to a new mode of three-dimensional vortex patterns consisting of a symmetric mode of lateral ejections. In Figure 10, we show a horizontal cross-cuts of the jet at a downstream distance of three jet diameters. Note the appearance of closed vortex loops forming in the outer portion of the vortex ring which occur at radial locations aligned with the valleys of the corrugated nozzle. For these high levels of the amplitude of the axial forcing, the ring structure of the jet is composed of a staggered array of pairs of vortex rings of opposite sign as shown in the vertical cut of Figure 9. Resulting from the amplification of the ring's instability, the outer-most weaker vortex ring is observed to deform to a stage by which vortex loops reconnect and pinch off as schematically shown in Figure 11. Under their own induction, the vortex loops pinching off from the outer ring, are propelled outwards; thus ejecting chunks of fluid from the jet into the outer stream. As a result of this, the spreading rate of the jet is locally enhanced at the location of the side ejection (compare Figures 6 and 9). In the horizontal cross-cut of Figure 9, observe that because of the 180º phase shift between wave perturbation of the inner and outer ring, the azimuthal location of the pinch-off corresponds to the valleys of the corrugated nozzle. It is important then to remark that although the streamwise vortex tubes, through their non-linear interaction with the rings, have locked the ring perturbation to the number of wavelengths imposed by the corrugation, the ejection-like mechanism observed does not involve streamwise vortex tubes, but only ring vorticity.

Mode 3: For amplitudes of forcing up to 50% of the initial streamwise velocity, we observe the development of a third, and more dramatic, three-dimensional symmetric mode. A horizontal cross-cut through the ring at a downstream distance of three jet diameters (Figure 13) now shows the appearance of closed vortex loops being ejected at circumferential locations coinciding with the crests of the corrugations. A possible mechanism consistent with all our visual data which could be responsible for this new three-dimensional vortex pattern is schematically shown in Figure 14. Note that in this case, the amplification of the perturbation of the instability of the primary, inner ring leads to the pinch-off of stronger vortex loops which are located at azimuthal positions aligned with the crests of the corrugated nozzle. Observe in the horizontal cross-cut of the jet (Figure 13) the reconnected inner vortex ring as well as the five strong vortex loops ejecting sideways. Observe also five weaker loops which had resulted from the reconnection of the outer ring also visible in between two of the strong loops. The

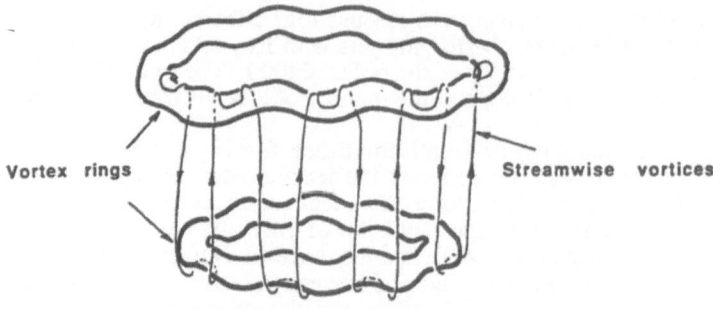

Figure 3. Vortex patterns resulting from the combined effect of axial and azimuthal perturbations. Note the presence of counter-rotating pairs of streamwise vortices which wrap around two consecutive vortex rings.

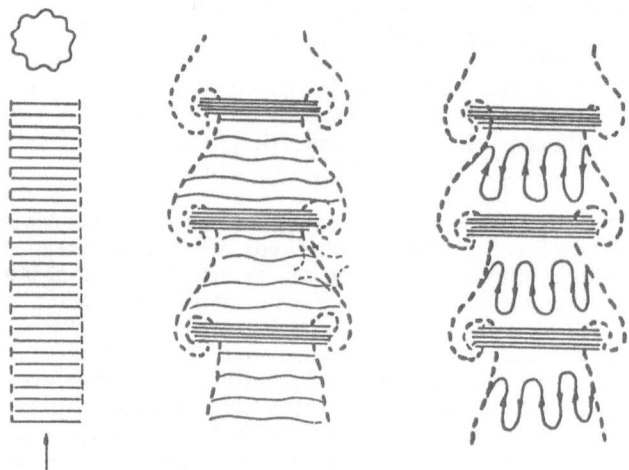

Figure 4. Schematic representation of the evolution of the forced jet showing the stretching of the perturbation's vorticity in the braids connecting two consecutive vortex rings.

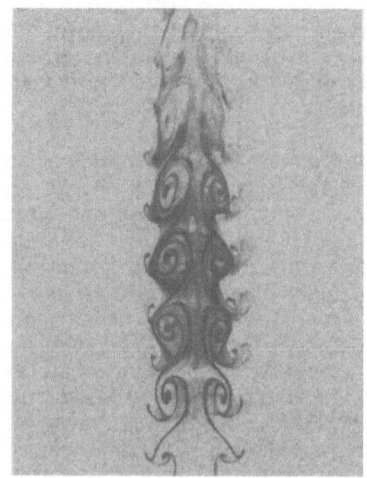

Figure 5. Vertical cross-cut of a strongly forced co-flowing jet. Note the formation of a staggered array of co-axial vortex rings of opposite sign. In this case, the strength of the outer ring is considerably smaller than that of the inner one.

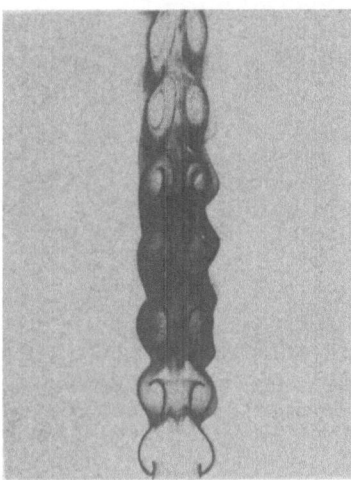

Figure 6. Mode 1. Vertical cross-cut of the jet. For these forcing conditions, the cylindrical vorticity layer rolls up in a periodic array of vortex rings of only one sign.

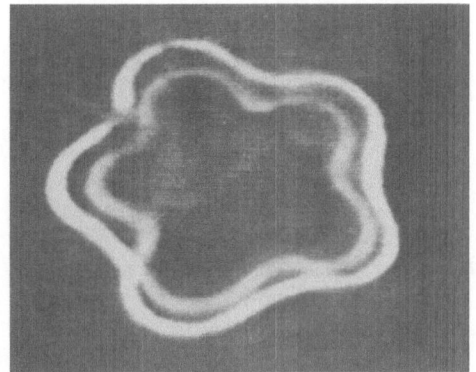

Figure 7. Mode 1. Horizontal cross-cut of the jet at a downstream distance of 3 jet diameters (Section A-A. in Figure 6.) The interface is marked by the white product (for the purpose of better visualization, we are showing the negative of the recorded image).

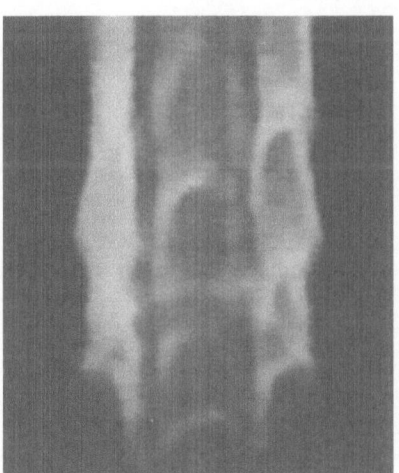

Figure 8. Mode 1. Perspective view of the jet from a position indicated by the arrow in Figure 6. Shown are 3 consecutive vortex rings and the array of streamwise vortices in the braids connecting them. As with Figure 7, the interface occupied by the scattering TiO_2 particles is shown in white.

Figure 9. Mode 2. Vertical cross-cut of the jet.

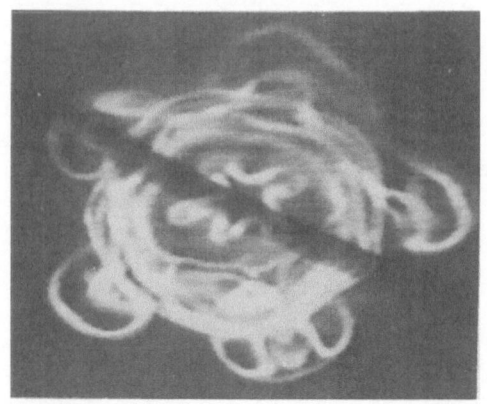

Figure 10. Mode 2. Horizontal cross-cut of the jet at a downstream distance of 3 jet diameters. Observe the presence of 5 closed vortex loops in the process of being pinched off from the jet. The vortex loops are located aligned with the valleys of the corrugated nozzle.

A

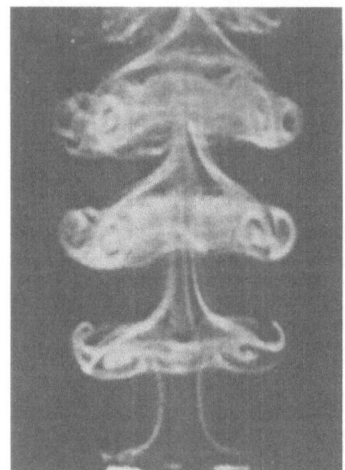

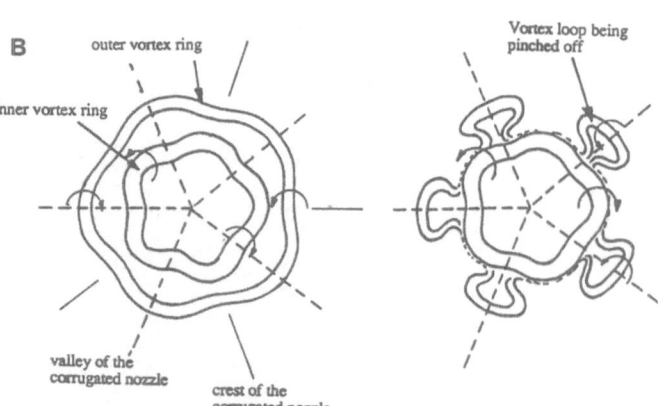

Figure 11. Mode 2. A. Side-view interface flow visualization. Notice the double roll-up of the vorticity layer to form a staggered array of co-axial vortex rings of opposite sign. Also noticeable are the streamwise vortices present in the stem (braids) connecting two consecutive vortex rings. B. Schematic representation of the combined evolution of the inner and outer ring leading to the pinch-off of closed vortex loops at azimuthal locations aligned with the valleys of the corrugated nozzles. The short wave instability of the outer, weaker ring, under the induction of the stronger inner one, leads to the pinch-off and reconnection of 5 closed vortex loops.

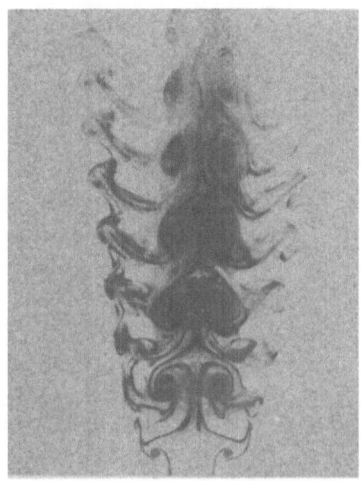

Figure 12. Mode 3. Vertical cross-cut of the jet.

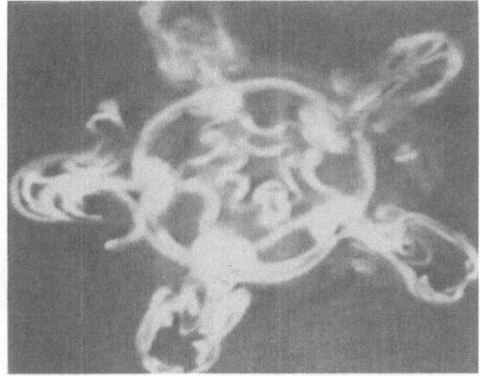

Figure 13. Mode 3. Hortizontal cross-cut of the jet at a downstream distance of 3 jet diameters. Note the existance of 5 closed vortex loops forming at azimuthal locations aligned with the crest of the corrugated nozzle. Observe also the reconnected inner ring after the pinch-off process.

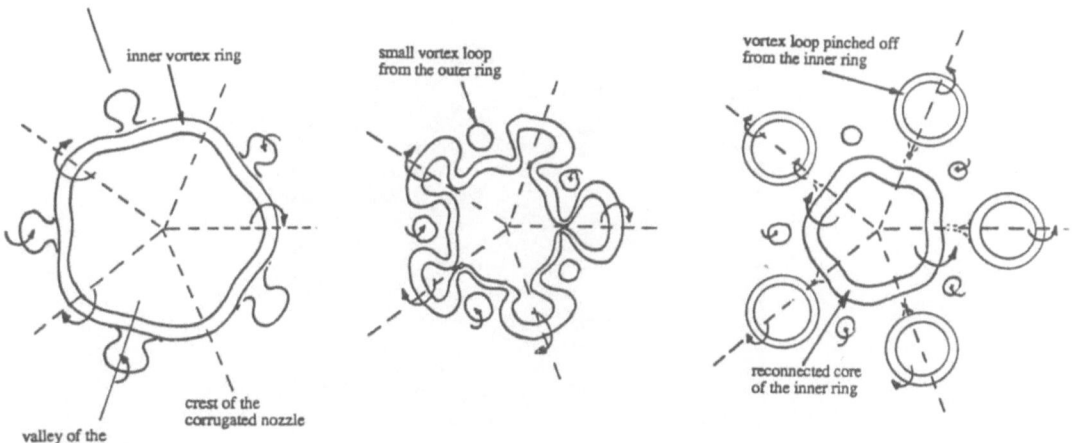

Figure 14. Mode 3. Schematic representation of the growth and amplification of the short wave instability of the inner vortex ring resulting now in a formation of closed vortex loops pinching off from azimuthal locations aligned with the crests of the corrugated nozzle, as shown in Figure 13.

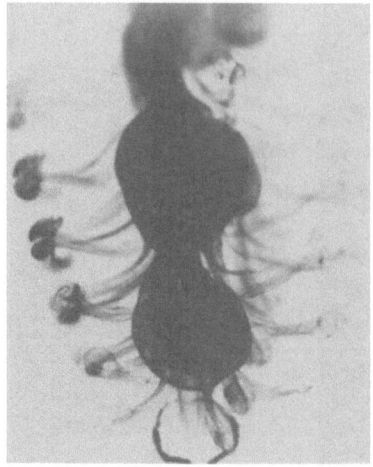

Figure 15. Mode 3. Prospective view of the interface from a location indicated by the arrow in Figure 12. For better clarity, we selected the positive image (dark product) as was the case with Figures 6, 9, and 12.

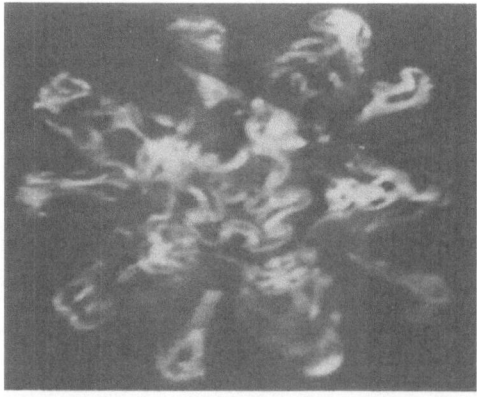

Figure 16. Mode 4. Horizontal cross-cut at a downstream distance of 3 jet diameters showing the existance of ten vortex loops being ejected (doubling the number of ejections).

spreading angle of the jet at the location aligned with the crests of the corrugation (location of the ejection of the strong loops) was found to be considerably enhanced with respect to the observed in the previous two modes (compare Figures 4, 10, and 12), indicating that the self-propelled motion of this new set of vortex loops is much stronger than the ones observed forming from the outer ring. A perspective view of the jet under these conditions is shown in Figure 15. Note that the self propelled motion of the loops is so strong that fluid is expelled outwards distances of several jet diameters.

Mode 4: As the axial forcing frequency is further increased above the most unstable one, we found the appearance of a new three-dimensional symmetric mode composed of a double number of lateral ejections (Figure 16). The doubling of the number of lateral ejections shown in the figure, is caused by the interaction between two consecutive ring structures. As each ring deforms, reconnects and closed vortex loops pinch off, the interaction with the upstream and downstream one results in their lateral displacement leading to the spatial organization shown in Figure 16.

In Lasheras et al, 1990, we present a detailed analysis of the parameter range leading to each one of the above four modes. In addition, an estimate of the mass and momentum ejected from the jet by the lateral vortex loops is also given.

For the case of the azimuthal perturbation consisting of the indented (serrated) nozzle, we found the appearance of additional symmetric modes topologically different from the above described ones.

Although the experiments described here were only concerned with laminar, co-flowing jets (Reynolds number = 1000), our preliminary results conducted with transitional and turbulent co-flowing jets confirm the existence of similar mechanisms by which the jets develop three-dimensional patterns of lateral ejections. Thus, it appears that our observed mechanisms described above for laminar flows could also be responsible for the lateral ejections reported in transitional and fully developed turbulent jets by Monkewitz, Lehmann, Barsikow,and Bechert, 1989; Monkewitz, Bechert, Barsikow, and Lehmann, 1990; Sreenivasan, Raghu, and Kyle, 1989; Subarao, 1987; and Liepmann, 1990.

Conclusions

We have presented experimental evidence of the existence of a large variety of entrainment patterns resulting from different modes of induction of the jet's vorticity. Through a set of well controlled experiments, we have shown that resulting from the combined effect of streamwise and azimuthal forcing, the instability of the vortex rings formed in the jet results in the formation of closed vortex loops that under their own induction are propelled sideways away from the jet's axis.

For the case of corrugated nozzles, we have shown the existence of several three-dimensional modes all of which involve the formation of a symmetric pattern of close vortex loops. The observed three-dimensional vortex patterns were interpreted in terms of the short wavelength instability of two arrays of co-axial vortex rings of opposite sign. Depending on the relative strength between these co-axial vortex rings and on the size of their cores, we have shown that through a process of reconnection, closed vortex loops may form from either or both arrays of co-axial vortex rings, thus resulting in chunks of jet fluid ejecting sideways.

Acknowledgements

This work was supported by a Gift from United Technologies Corporation and by the Spanish C.I.C.Y.T. under project PB86-0497. The help received through numerous discussions with Professor Amable Linan is gratefully acknowledged.

References

Ashurst, W. T. & Meiburg, E. 1988 Three-dimensional shear layer via vortex dynamics. *J. Fluid Mech.* **189**, 87-98.

Beavers, G.S. & Wilson, T.A. 1970 Vortex growth in jets. *J. Fluid Mech.* **44**, 97-112.

Becker, H.A. & Massaro, T.A. 1968 Vortex evolution in a round jet. *J. Fluid Mech.* **31**, 435-448.

Bradbury, L.J.S. & Khadem, A.H. 1975 The distortion of a jet by tabs. *J. Fluid Mech.* **70**, 801-813.

Browand, F.K. & Laufer, J. 1975 The role of large scale structure in the initial development of circular jets. *Proc. 4th Biennial Symp. Turbulence in Liquids*, University of Missouri-Rolla, pp 333-344. Princeton , New Jersey: Science Press.

Cohen, J. & Wygnanski, I. 1987 The evolution of instabilities in the axisymmetric jet. Part 1. The linear growth of disturbances near the nozzle. *J. Fluid Mech.* **176**, 191-202

Crow, S. & Champagne, F.M. 1971 Orderly structure in jet turbulence. *J. Fluid Mech.* **48**, 547-591.

Dimotakis, P.E., Miake-Lye, R.C. & Papantoniou, D.A. 1983 Structure and dynamics of round turbulent jets. *Phys. Fluids* **26**. 3185.

Eickhoff, H., Winandy, A. & Natarajan, R. 1989 Azimuthal instability in jets diffusion flames. *Experiments in Fluids*. **7**, 420-421.

Lasheras, J. C. & Choi H. 1988 Three-dimensional instability of a plane shear layer: an experimental study of the formation and evolution of streamwise vortices. *J. Fluid Mech.* **189**, 53-86.

Lasheras, J.C., Lecouna, A., Martin, J.E., Meiburg, E. & Rodriguez, P. 1990 Three-dimensional vorticity patterns in forced, co-flowing jets. *J. Fluid Mech.* (in preparation).

Lasheras, J. C. & Meiburg, E. 1990 Three-dimensional vorticity modes in the wake of a flat plate. *Phys. of Fluids A*. **2,3**, 371-380.

Lecuona, A., Rodriguez, P. & Lasheras, J.C. 1990 Three-dimensional structure of strongly forced jet diffusion flames: Flow visualization studies.

Liepmann, D. 1990 Ph.D. Thesis. University of California, San Diego.

Martin, J.E., Meiburg, E. & Lasheras, J.C. 1990 Three-dimensional evolution of axisymmetric jets: a comparison between computations and experiments. *IUTAM Symposium on Separated Flows and Jets*. July 9-13, Novosibirsk, USSR.

Maxworthy, T. 1972 The structure and stability of vortex rings. *J. Fluid Mech.* **51**, 15-32.

Maxworthy, T. 1977 Some experimental studies of vortex rings. *J. Fluid Mech.* **81**, 465-495.

Meiburg, E. & Lasheras, J. C. Experimental and numerical investigation of the three-dimensional transition in plane wakes. *J. Fluid Mech.* **190**, 1-30.

Meiburg, E., Lasheras, J.C. and Martin, J. E. 1989 Experimental and numerical analysis of the three-dimensional evolution of an axisymmetric jet. *Turbulent Shear Flows 7*. Springer-Verlag. F. Durst Ed.

Meiburg, E. & Martin, J. E. Three-dimensional evolution of an axisymetric jet: a numerical investigation. J. Fluid Mech., in press, (1991)

Michalke, A. 1964 On the inviscid instability of the hyperbolic-tangent velocity profile. *J. Fluid Mech.* **19**, 543-556.

Michalke, A. & Freymunth, P. 1966 The instability and the formation of vortices in a free shear layer. *AGARD, Conf. Proc. no. 4, paper 2*.

Michalke , A. & Hermann, G. 1982 On the inviscid instability of a circular jet with external flow. *J. Fluid Mech.* **114**. 343.

Michalke, A. 1988 Survey on jet instability theory. *Prog. Aerospace Sci.* **21**, 159.

Monkewitz, P.A., Lehmann B., Barsikow, B. & Bechert , D. 1989 The spreading of self-excited hot jets by sie-jets. *Phys. Fluids A*. **1**, 446-447.

Monkewitz, P.A., Bechert , D., Barsikow, B. & Lehmann B. 1990 Self-excited oscillations and mixing in heated round jets. *J. Fluid Mech.* **213**, 611-639.

Mungal, M.G. & Hollingsworth, D.K. 1989 Organized motions in a high Reynolds number jet. *Phys. Fluids A*. **1**, 1615-1616.

Sreenivasan, K.R., Raghu, S. & Kyle, D.1989 Absolute instability in variable density round jets. *Experiments in Fluids*, **7**, 309-317.

Subarao, E.R. 1987 Ph.D. Thesis. Stanford University.

Tso, J. & Hussain, F. 1989 Organized motions in a fully developed turbulent axisymmetric jet. *J. Fluid Mech.* **203**. 425.

Widnall, S.D., Bliss, D.B. & Tsai, C-Y. 1974 The instability of short waves on a vortex ring. *J. Fluid Mech.* **66**, 35-47.

Widnall, S.D. & Sullivan, J.P. 1973 On the stability of vortex rings. *Proc. Roy. Soc. A*, **332**, 335-353.

Wille, R. 1963 Growth of velocity fluctuations leading to turbulence in a free shear layer. *Heramn Fottinger Inst., Berlin, AFOSR Tech. Rep.* Contract AF 61(052)-412.

Yule, A. J. 1978 Large-scale structure in the mixing layer of a round jet. *J. Fluid Mech.* **89**, 413-432.

MODELING AND CONTROL OF KARMAN VORTEX STREETS

E. Roesch[+], F. Ohle[+*], H. Eckelmann[+], A. Hübler[*]

+ Institut für Angewandte Mechanik und Strömungsphysik
der Universität Göttingen,
D-3400 Göttingen, F.R.G.
 * Department of Physics, University of Illinois and Beckmann Institute,
Center of Complex Systems Research, Urbana-Champagne, IL 61801, USA

1. INTRODUCTION

One of the most important problems in hydrodynamics is the behaviour of low-dimensional dynamics in closed and open flows. Lorenz[1] and Ruelle & Takens[2] proposed that the dynamics of a hydrodynamic system at the onset of turbulence could be described by strange attractors and low-dimensional differential equations. This was in contrast to Landau's[3] hypothesis that turbulence is a quasiperiodic phenomenon with an infinite number of frequencies.

Brandstäter et al.[4] could experimentally verify low-dimensional dynamics with Lyaponov exponents and generalized dimensions in a closed hydrodynamic flow, namely in Taylor-Couette flow, and could thus disprove the hypothesis of Landau. The first work dealing with chaotic dynamics in an open hydrodynamic system stems from Olinger & Sreenivasan[5]. From the partly chaotic behaviour of a sinusoidally driven Kármán vortex street in this work it can be inferred that the original dynamics of a vortex street might be described by a van der Pol oscillator. This was postulated by Benaroya & Lepore[6]. Investigating experimentally mode-locking states of Kármán vortex streets Detemple[7] concluded from the response that a van der Pol oscillator might be an appropiate model for the dynamics of the vortex street. Nevertheless, there are essential differences between the response of a van der Pol oscillator and of the vortex street.

In this investigation differential equations are extracted from experimental time series obtained in Kármán vortex streets. It will be shown that the velocity signal measured in the regular range of a vortex street ($50 \leq Re \leq 150$) can be modeled by a second order differential equation with ten parameters. The parameters are nearly independent of the probe position and of the Reynolds number. With the knowledge of the constructed differential equation the response of the vortex street on pertubations can be predicted.

2. CONSTRUCTION OF DIFFERENTIAL EQUATIONS

The complete construction of a differential equation is based on a fit of the flow vector field by series of orthogonal polynomials. For most of the non-linear oscillators investigated so far, the flow vector fields can be described by Taylor expansions with a small number of non-vanishing coefficients. The parameters obtained by the fit are an approximation of the coefficients of the underlying differential equation. These parameters can be determined from an experimental set of data given as a time series. In a first step, these time series are numerically differentiated, thus providing the state space representation necessary to determine the complete flow vector field F as a function of the state space vector (x_1, x_2). The number of components of the flow vector field, and thus also the order of the differential equation, increase in proportion to the dimension of the state space representation. In the next step each component of the flow vector field is approximated seperately by a series of Legendre polynomials up to the order p:

$$F_1(x_1, x_2) = \sum_{i,j=0}^{i+j \leq p} a_{ij} L_i(x_1) L_j(x_2);$$

$$F_2(x_1, x_2) = \sum_{i,j=0}^{i+j \leq p} b_{ij} L_i(x_1) L_j(x_2).$$

Because of their orthogonality Legendre polynomials are appropriate for the approximation used here. The magnitude of the fit cofficients is nearly independent of the order p of the approximation which is based here on a general linear least square fit. Since the orthogonality of the Legendre polynomials is only valid for $|x| \leq 1$, all values have to be normalized to this range. The quality of a fit is estimated by the quantity χ which is defined by the equation:

$$\chi^2 = \sum_i \frac{|\underline{f}_j^{\,i} - \underline{f}_j(\underline{r}_i)|^2}{|\Delta \underline{f}_j^{\,i}|^2}.$$

Here $\underline{f}_j^{\,i}$ is the experimental value of the component j at the position $\underline{r}_i$, $\underline{f}_j(\underline{r}_i)$ is the adapted value and $\Delta \underline{f}_j^{\,i}$ is the error of the respective experimental data. The theoretical value of χ is the number of the data minus the number of parameters of the fit. The order of the fit is successively raised until χ has approximately reached the theoretical value.

It is only possible to approximate a flow vector field in the surrounding of the attractor if the flow vectors have been derived from this region of the state space. A fit is best, when the positions of the calculated flow vectors are homogeniously distributed in the region of interest. Therefore, it is also necessary to investigate the transient states if a system has a fixed point or a limit cycle as an attractor. In the case of a chaotic dynamic the fit is especially good since chaotic trajectories fill the state space quite regularly.

The quality of a fit is checked by a numerical integration of the constructed differential equation. In a state space representation the global geometry of the trajectories of the numerical and the experimental data are directly compared.

3. EXPERIMENTAL SETUP

The measurements were carried out in the open-return wind tunnel described by Detemple-Laake and Eckelmann[10]. The nozzle diameter of this tunnel is 180 mm. A circular cylinder of diameter d=1.5 mm was mounted horizontally at the nozzle exit. A hot-wire probe was placed in the wake of the cylinder at various $\tilde{x} = x/d$ locations downstream, $y/d = 1$ to one side and halfway between the suspension points of the cylinder. The Reynolds number was varied in the regular region ($50 \leq Re \leq 150$) of the vortex street. The velocity signal was amplified, sampled at 10 kHz and digitalized by a 12 bit A/D converter. Two loudspeakers working $180°$ out of phase were placed directly above and below the circular cylinder outside the flow to superpose sound. It should be noted that this loudspeaker arrangement is different from that used by Detemple-Laake and Eckelmann[9].

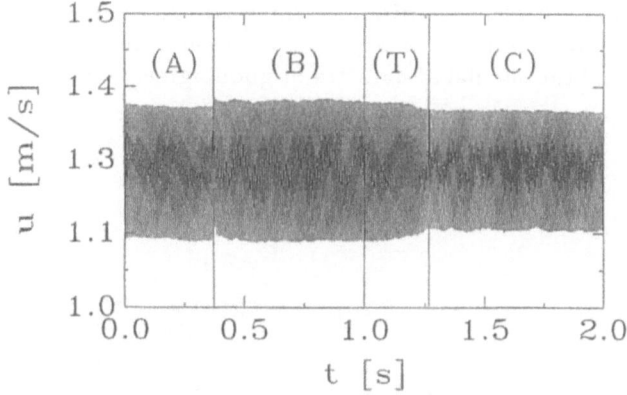

Fig.1. Time series from a Kármán vortex street at $Re = 114$. (A): Vortex street superposed with sound; (B):intermediate region after cutting off the sound; (T): transient state of the vortex street; (C): natural vortex street.

4. A MODEL FOR A KARMAN VORTEX STREET

A typical experimental time series showing the transient behaviour from the stimulated to the natural state of the vortex street is depicted in Fig.1. Four different parts, which can be attributed to different physical processes, are marked. The power spectra for three of these parts are shown in Fig.2. Part (A) in Fig.1 represents the vortex street with parallel vortex shedding stimulated by sound at the sound frequency of $f_s = 119.6 Hz$.

In order to increase the velocity fluctuation amplitude the sound frequency was chosen slightly higher than the natural Strouhal frequency of $f_{st} = 119.0 Hz$. This method of increasing the velocity fluctuation amplitude by stimulating the vortex street with a slightly different frequency to obtain the transient behavior was developed by the authors. Part (B) in Fig.1 defines the intermediate region between switching off the sound and the beginning of the transient state. Note that in this region, right after switching off the sound, the shedding frequency jumps from f to $f_{max} = 124.5 Hz$ (Fig.2 dashed line). Without sound excitation, the unforced parallel shedding of part (B) has a slightly higher frequency than the slantwise shedding of part (C), which is consistant with recent experiments (Ohle et al.[11]). The last part (C) in Fig. 1 represents the state of the natural vortex street with Strouhal frequency f_{st} and oblique vortex shedding. For the construction of differential equations, which are of the form

$$\frac{du_1}{dt} = u_2;$$

$$\frac{du_2}{dt} = \sum_{i,j=0}^{i+j \leq 3} a_{ij}(\tilde{x}) u_1{}^i u_2{}^j \qquad (1)$$

only the parts (T) and (C) of the time series were used. In this equation $u_1(t)$ defines the velocity fluctuation amplitude of the Kármán vortex street. Roesch, Eckelmann & Hübler[8] could verify that in the phase space representation the limit cycles of both the experimental and the simulated data are in good agreement.

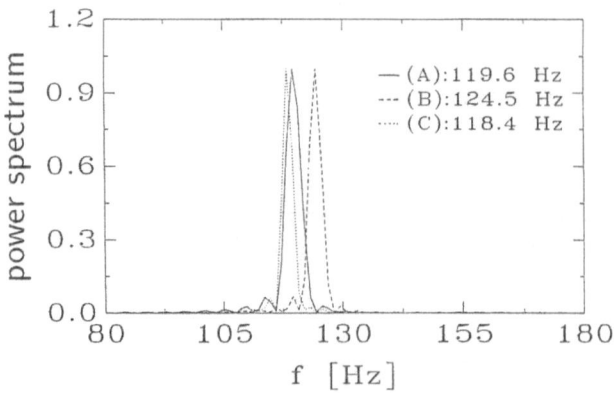

Fig.2. Power spectrum of the parts (A), (B) and (C) of the time series shown in Fig.1.

This guarantees that the dynamics of the vortex street is well represented by the constructed differential equation. The dependence of the normalized coefficients:

$$\alpha_{ij} = \frac{[a_{ij}(\tilde{x}) - a_{ij}(4.7)]}{a_{ij}(4.7)},$$

on the distance $\tilde{x}$ from the cylinder is shown in Fig.3. Since the values of the various

coefficients a_{ij} approach zero for $\tilde{x} \geq 2.5$, it can be concluded that the dynamics of the Kármán vortex street is developed beyond these locations[10].
The normalized coefficients

$$\beta_{ij} = \frac{[a_{ij}(Re) - a_{ij}(Re = 53)]}{a_{ij}(Re = 53)}.$$

in Fig.4a/b are nearly independant of the Reynolds number for the whole region $50 \leq Re \leq 150$. Special peculiarities like the Tritton[11] discontinuity at $Re = 87$, as well as the changes in the shedding modes, described by König, Eisenlohr & Eckelmann[12], are also reflected in the coefficients[13].

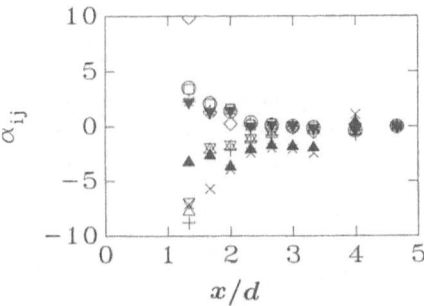

Fig.3. Normalized coefficients α_{ij} as a function of the nondimensional probe distance x/d from the cylinder. Here $\square : \alpha_{00}$; $+ : \alpha_{01}$; $\times : \alpha_{02}$; $\diamond : \alpha_{03}$; $\bigcirc : \alpha_{10}$; $\triangle : \alpha_{11}$; $\blacktriangle : \alpha_{12}$; $* : \alpha_{20}$; $\triangledown : \alpha_{21}$; $\blacktriangledown : \alpha_{30}$.

5. CONTROL OF A KARMAN VORTEX STREET

If it is possible to model the dynamics of an experimental system by differential equations, then it is also possible to predict its dynamics and its response. The basis for a resonant stimulation and specific control of a Kármán vortex street is the extensive knowledge of the differential equations at different Reynolds numbers in the complete wake of the cylinder. The response of such nonlinear oscillators to aperiodic driving forces can be essentially larger than the response to sinusoidal ones. Wagner et al.[14] succeeded experimentally in a resonant stimulation of nonlinear mechanical oscillators with aperiodic driving forces.

With the help of the constructed low-dimensional differential equation control of the Kármán vortex street is also possible. A problem, however, is that a complex system like the vortex street generally has an infinite number of degrees of freedom, whereas

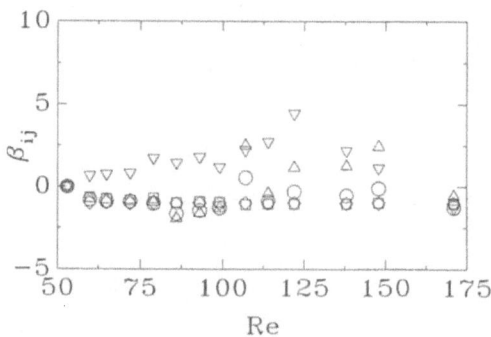

Fig.4a. Normalized coefficients β_{ij} as a function of the Reynolds number. Here $\triangle$: β_{00}; $\triangledown$: β_{01}; $\square$: β_{02}; $\diamondsuit$: β_{03}; $\bigcirc$: β_{10}.

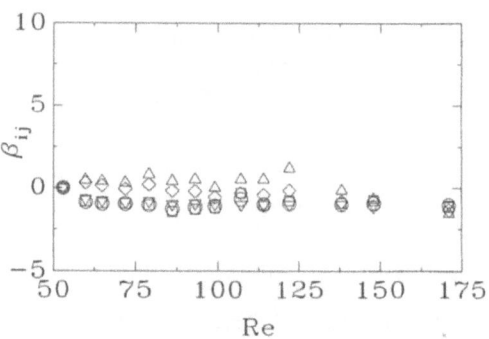

Fig.4b. Same as in Fig.4a. Here $\triangle$: β_{11}; $\triangledown$: β_{12}; $\square$: β_{20}; $\diamondsuit$: β_{21}; $\bigcirc$: β_{30}.

the constructed differential equation has only two. In spite of this difficulty it is possible to control a complex system such as a Kármán vortex street. It will be now explained by means of the Haken-Wunderlin-Zwanzig equation

$$\dot{u} = \epsilon u - su; \epsilon \ll 1$$
$$\dot{s} = s + u^2$$

what kind of assumptions a driving force has to satisfy in order to obtain a large and predictable response of the complex system. In this equation the unslaved and slaved modes are denoted as u and s respectively. Since $\epsilon \ll 1$ the unslaved modes u represent a very slowly and nearly constant motion. Therefore the slaved modes s relax very quickly to u^2, and the Haken-Wunderlin-Zwanzig equation can be reduced to a first order differential equation of the following form:

$$\dot{u} = \epsilon u - u^3.$$

This equation corresponds to our constructed differential equation. By adding a driving force $F(t)$ the equation

$$\dot{u} = \epsilon u - u^3 + F(t) \qquad (2)$$

is obtained.

If in contrast the driving force $F(t)$ is added to the original Haken-Wunderlin-Zwanzig equation

$$\dot{u} = \epsilon u - su + F(t); \epsilon \ll 1$$
$$\dot{s} = s + u^2 + F(t)$$

the unslaved modes u and the driving force $F(t)$ are very slow and nearly constant. Now the slaved modes s relax very quickly to $u^2 + F(t)$ and the equation can be reduced to

$$\dot{u} = (\epsilon - F(t))u - u^3 + F(t).$$

This equation is identical to Eq.(2) only when $F(t) \ll \epsilon$. Thus the constructed differential equation only predicts the correct response of the complex system for resonant, small and slowly driving forces.

Therefore a Kármán vortex street can only be controlled by small driving forces. If the driving force is to large, the slaved modes will also be stimulated, and hence the system can no longer be described by a low-dimensional differential equation. A typical example for the stimulation of slaved modes by a too large driving force was given by Detemple-Laake and Eckelmann[9]. They could stimulate a netting-pattern structure in the wake of a circular cylinder by superposing sound at $Re = 143$. Such a structure normally exists in the transition range ($150 \leq Re \leq 300$).

The experimental problem in the present investigation is that the constructed differential equation describes the situation of a vortex street with oblique shedding. If sound is superposed on such a vortex street parallel shedding will be obtained. Nevertheless, it is possible to control the vortex street to some extend with the constructed differential equation.

By using the constructed differential equation given by Eq.(1) it was found in a computer experiment, that a decrease in the amplitude of the vortex street can be achieved by a sinusoidal driving force, which is raised in frequency of about 10% per second.

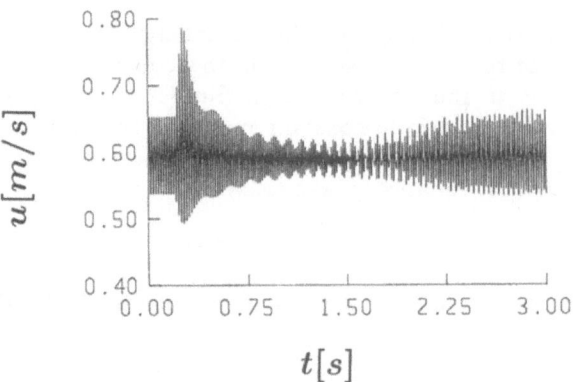

Fig.5: Simulated time series of the constructed differential equation superposed by a sinusoidal driving force with a frequency shift of about *10%* per second.

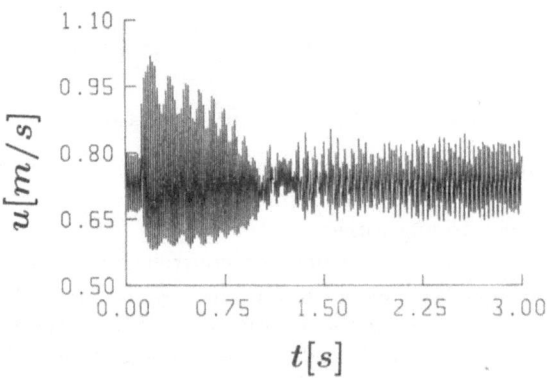

Fig.6: Experimental time series of the Kármán vortex street superposed by the same driving force as in Fig.5.

Fig.5 shows the relevant computer simulation. The typical features of the computer simulation, such as the beat, the decrease in amplitude down to a minimum, and finally the increase back to the natural amplitude can also be found in the experiment (Fig.6). Both, for the experiment, as well as for the computer simulation, the same kind of driving force was used. It should be noted, that the constructed differential equation describing oblique shedding leads to the same behaviour as for the experimental case, where under control conditions parallel shedding exists.

The results presented here can be considered as a first step of an attempt to control a Kármán vortex street without any feedback. For a resonant stimulation the case of parallel shedding should also be incorporated in the construction of the differential equation. Further work along these lines is in progress.

REFERENCES

1   Lorenz, E.N., J. Atmospheric Sci. 20, (1963), 130 +448

2   Ruelle, D. & Takens, F., Commun. Math. Phys., 20, (1971), 167

3   Landau, L.D., C.R. Acad. Sci. URSS, 44, (1944), 311

4   Brandstäter, A., Swift, J., Swinney, H.L., Wolf, A., Farmer, J.D., Jen, E. & Crutchfield, P.J., Phys. Rev. Letters 51, (1983), 1442

5   Olinger, D.J. and Sreenivasan, K.R., Phys. Rew. Let. 60, (1988), 797

6   Benaroya, H. and Lepore, J.A., J.S.V. 86(2), (1983), 159

7   Detemple, E., Diplomarbeit Göttingen (1983)

8   Roesch, E., Eckelmann, H. & Hübler, A., Max Planck Institut für Strömungs-forschung, Göttingen, Report No. 11/1988, (1988)

9   Detemple-Laake, E. & Eckelmann, H., Exp. Fluids 7,(1989), 217

10   Ohle, F., Lehmann, P., Roesch, E., Eckelmann, H. & Hübler, A., Phys. Fluids A, 4, (1990), 479

11   Tritton, D.J., J. Fluid Mech., 6, (1959), 547

12   König, M., Eisenlohr, H. & Eckelmann, H., to appear at Phys. Fluids A, 9, (1990)

13   Ohle, F., Max-Planck-Institut für Strömungsforschung, Göttingen, Report No. 110/1990, (1990)

14   Wagner, C., Stelzel, W., Hübler, A., Lüscher, E. & Altmann, W., Helv. Phys. Acta 61, (1987), 224

SHEAR FLOWS:

NUMERICAL EXPERIMENTS

ADVANCES AND SOME NOVEL EXPERIMENTS USING DIRECT NUMERICAL SIMULATION OF TURBULENCE

Parviz Moin

Stanford University & NASA-Ames Research Center

Stanford, CA 94305

Recent advances in direct numerical simulation methodology are reviewed. These include numerical methods for complex geometries, boundary conditions at the open boundaries, and applications to compressible flows. Some instructive numerical experiments that would be difficult to perform in the laboratory are described. The primary example is the isolation of the minimal flow unit near the wall and its application to the study of the flow over riblets, three-dimensional boundary layers, and active control of turbulent boundary layers.

1. INTRODUCTION

The utility of the direct numerical simulation technique as a powerful tool in turbulence research has been established in the last decade. Significant progress has been made in understanding the structure of turbulent shear flows using databases generated by direct numerical simulations. The availability of the flow fields in three space dimensions and time has been exploited to elucidate the nature of organized structures (e.g., Moin & Kim, 1985; Lesieur et al., 1988; Metcalfe et al., 1987, and the Proceedings of the Summer Programs of the Center for Turbulence Research), to compute the various terms that appear in the budgets of the Reynolds stresses which are needed in phenomenological turbulence modeling (Mansour, Kim & Moin, 1988), and to test turbulence theories (Moin & Moser, 1989; Domaradzki & Rogallo, 1990).

Despite its successes in revealing detailed turbulence physics, turbulence simulations have been largely limited to low Reynolds number, incompressible flows in simple geometries such as channels and temporally evolving mixing layers. For example, the highest Reynolds number, turbulent boundary layer that has been computed was by Spalart (1988) at the momentum thickness Reynolds number of 1400, and for temporally evolving mixing layer by Rogers & Moser (1990) at Reynolds number based on the vorticity thickness of 5000. The latter simulations were the first computations of the mixing layer that included the so-called mixing transition which is marked by significant onset of small scales. Recently, direct numerical simulations are being extended to compressible flows and to more complex flow domains. These extensions are particularly useful for phenomenological turbulence modeling. The standard models have been tuned against simple shear flows using the experimental data. The simulation databases in complex geometries could provide the critical data in flows that these models currently have difficulties predicting.

The Global Geometry of Turbulence
Edited by J. Jiménez, Plenum Press, New York, 1991

In attempting these computations, some issues with the numerical methods and boundary conditions had to be addressed. In the next section, these problems are discussed, and the progress made to date are outlined. This is followed by a description of some recent numerical experiments that have been very valuable in exploration of turbulence control strategies and the study of the mechanics of three- dimensional turbulent boundary layers.

2. NUMERICAL METHODS

2.1 Spatial Discretization

Most direct numerical simulations of turbulent and transitional flows have been performed using spectral methods. Spectral methods are highly accurate and efficient in applications involving simple geometries (see for example, Canuto et al., 1988). Their implementation in complex geometries with implicit time advancement algorithms leads to large full matrices that must be solved iteratively. In contrast, implicit formulations with finite-difference approximations for spatial derivatives result in banded matrices. Spectral element method (Patera, 1984) and spectral methods with domain decomposition are currently under extensive development for complex flow applications.

The main reason for the popularity of the spectral methods has been their capability to accurately resolve the small scale (short wave length) variation of the numerical solution with the fewest number of grid points. For sufficiently smooth functions, spectral approximations converge exponentially fast with an increasing number of terms in the expansion. However, in unsteady non-linear problems, this advantage in function representation may not be accompanied with accurate representation of the dynamics. The case in point is in turbulent flows. Although at a given instant the small scale motions can be represented accurately, their non-linear dynamics would involve scales beyond the numerical resolution and, therefore, would be computed inaccurately. In addition, the small scales are most susceptible to aliasing errors. Removal of the aliasing errors has been shown to be critical to successful direct simulation of turbulence (Kim, Moin & Moser, 1987; Spalart, 1988). Except in the simplest cases involving Fourier expansion, aliasing error removal involves a factor of 9/4 increase in the computation cost for the non-linear terms. Thus, the efficiency of spectral methods may be questioned for computation of turbulent flows with non-negligible energy at the smallest scales.

Recently, high-order finite-difference methods have been successfully employed in turbulent flow simulations (Rai & Moin, 1989; Browning & Kreiss, 1989; Lee, Lele & Moin 1991). Lee et al., simulated the decay of compressible isotropic turbulence using both spatially periodic boundary conditions and spatially developing simulation. Using the latter scheme, they also computed the passage of isotropic turbulence through a normal shock wave. A sixth-order Padé approximation was used for the spatially periodic computations and a fourth-order Padé scheme for the spatially evolving flows. No attempt was made to remove aliasing errors, but a skew symmetric form of the non-linear terms (Zang, 1990) had to be used to avoid nonlinear numerical instabilities. The normal shock wave was actually resolved in the computations with eight to ten grid points across the shock.

Rai & Moin simulated fully developed turbulent channel flow at the Reynolds number 3300 based on the centerline velocity and the channel half-width. The flow was identical to the spectral simulations of Kim et al. (1987) and was intended to provide a comprehensive comparison of finite-difference and spectral simulations. They used a fifth-order upwind biased scheme for the non-linear terms and a sixth-order central difference scheme for the viscous terms. The impetus for the use of the upwind biased scheme was to partially eliminate aliasing errors and to control non-linear instabilities. The dissipative nature of the upwind scheme results in the damping of the smallest scale motions. The non-linear terms were expressed in the convection form, $u_j \partial u_i / \partial x_j$, which is the most economical formulation of these terms. It is well known that the convection formulation leads to non-linear numerical instabilities if central differences are used without aliasing error control. Indeed, Rai & Moin's computation of channel flow suffered from numerical instabilities when the fourth-order central difference approximation was used in the convective formulation. When the aliasing errors were removed by the "3/2" rule, numerical instabilities disappeared. The popular second-order staggered mesh formulation (Harlow & Welsh, 1965) does conserve kinetic energy and, therefore, is stable. For applications in complex geometries which may involve body-fitted coordinates and imbedded grids, it may be difficult to construct kinetic energy conserving formulations or schemes with alias error control. For these applications, the use of *high-order* upwind-*biased* schemes appears to be an attractive candidate.

The results from channel flow calculations of Rai & Moin were most promising. Excellent agreement was obtained between the finite-differenced simulation results and the spectral calculations *with the same grid resolution*. Turbulent velocity fluctuation statistics up to third-order and second-order vorticity statistics were considered. Turbulent intensities and the root-mean square vorticity fluctuation profiles are shown in Figures 1 and 2, respectively. The agreement between the finite-difference and spectral computations is very good. This is particularly encouraging since small scale turbulent motions have an appreciable contribution to the vorticity fluctuations.

Based on these numerical experiments, it appears that most turbulence quantities of interest can be obtained from high-order finite-difference simulations with the same grid resolution as the comparable spectral computations. Simulations with second-order finite-difference schemes probably would require twice as many grid points in each spatial direction as required for spectral computations (Herring *et al.*, 1974; Browning & Kreiss, 1989).

2.2 Inflow and Outflow Boundary Conditions

One of the main difficulties with simulation of complex turbulent flows is the imposition of inflow "turbulent" boundary conditions. The inflow velocity field must be unsteady and stochastic with broad band spectra in the inflow plane and in time. Lee *et al.*, (1991b) gave a methodology for the generation of inflow conditions and successfully applied it to the problem of the decay of isotropic turbulence behind a grid. The Fourier amplitudes of the velocity field were obtained from a prescribed three-dimensional energy spectrum. Frequency spectra were obtained from the streamwise wave number spectra using Taylor's hypothesis. The total time of integration was divided into equal segments of duration T. For each wave number (frequency), a random phase angle was prescribed which changed only once in each period, T, at a random location and by a random amount. The downstream decay of turbulent kinetic energy and the evolution of the streamwise velocity derivative skewness factor are shown in Figures 3 and 4, respectively. Also shown are the results from a comparable temporally evolving simulation. The downstream distance for the temporal simulation was calculated using Taylor's hypothesis. The agreement between the two computations is very good. The agreement of the other turbulent statistics is similarly good, verifying the adequacy of the procedure at least to the same degree as the

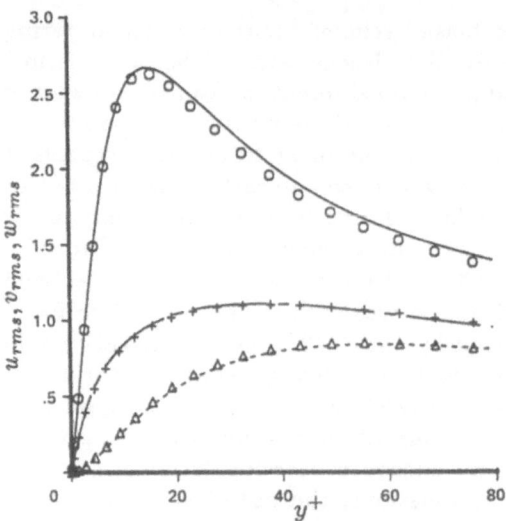

FIGURE 1. Turbulent intensities in fully developed turbulent channel flow. Symbols are from spectral simulations of Kim *et al.* (1987). ∘ , streamwise, u; △ , wall-normal, v; and +, spanwise, w, components. Lines are from the finite difference simulations of Rai & Moin (1989).

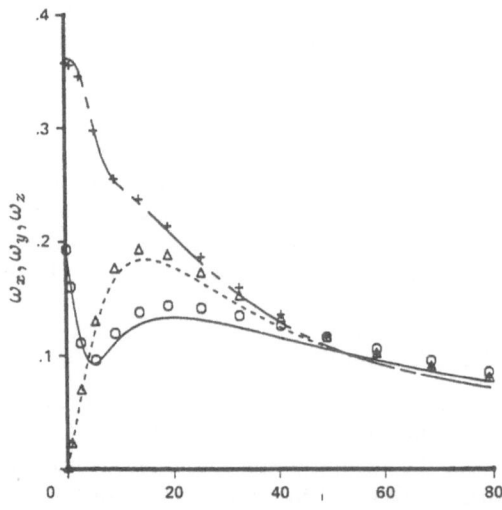

FIGURE 2. Root-mean square vorticity fluctuation normalized by the mean wall-shear. Symbols are from spectral simulations of Kim *et al.* (1987). ∘ , streamwise, ω_x; △ , wall-normal, ω_y; and +, spanwise, ω_z, components. Lines are from the finite difference simulations of Rai & Moin (1989).

prescription of random initial conditions for temporally developing computations.

In isotropic turbulence the skewness of the velocity derivatives is measured to be about -0.4 to -0.5. From Figure 4, it can be deduced that turbulence reaches a realistic state in a distance corresponding to travel with the mean flow for a period of about $0.3l/u'$, where l and u' are the integral and the rms velocity fluctuations at the inflow.

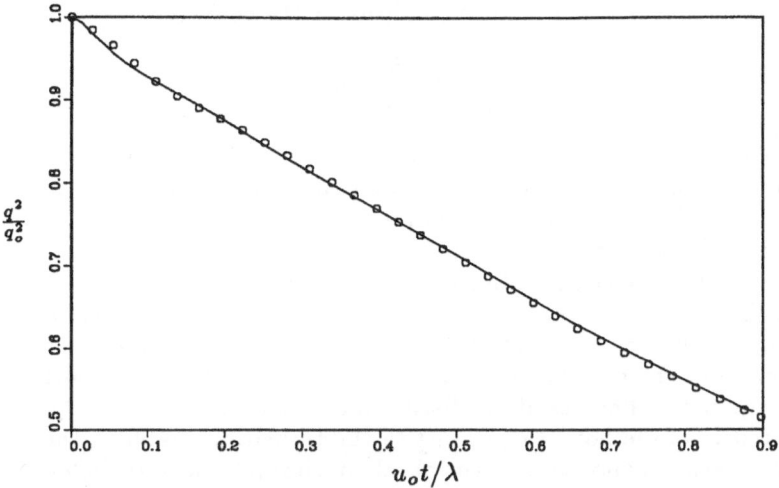

FIGURE 3.   Decay of kinetic energy in isotropic turbulence. Symbols are from temporal simulation, and the line is from spatially developing simulation (from Lee *et al.*, 1991b).

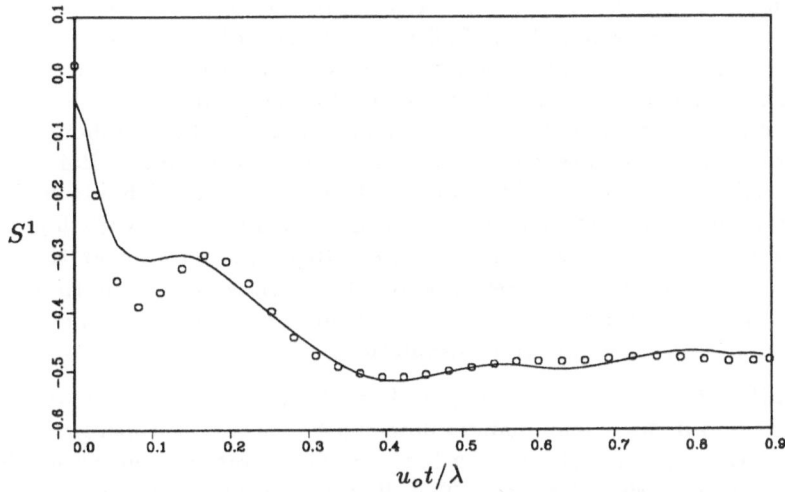

FIGURE 4.   Development of the skewness of the derivative of the streamwise velocity in isotropic turbulence. Symbols are from temporal simulation, and the line is from spatially developing simulation (from Lee *et al.*, 1991b).

Le, Moin & Kim, (1990) applied a similar procedure to the entry section of turbulent channel flow. Although at the entrance the prescribed velocity field had the same mean velocity, turbulent intensities, and Reynolds shear stress profiles as in the fully developed channel flow, about 20 channel widths were required for the flow to recover its initial friction coefficient. Thus, longer distances appear to be necessary to achieve structural equilibrium in sheared turbulence. It should be noted that experimentally, at moderate Reynolds numbers, the mean flow near the wall reaches equilibrium in about the same downstream distance from the wind tunnel entrance.

127

The most commonly used outflow condition is the convective boundary condition

$$\frac{\partial u_i}{\partial t} + c\frac{\partial u_i}{\partial x} = 0,$$

where c is a constant convection velocity. Pauley, Moin & Reynolds (1990) found this condition to be adequate in computations of unsteady laminar separated flows. Vortical structures travelling downstream were found to exit the computational domain without noticeable distortion, and the results appeared to be insensitive to the streamwise extent of the computational domain. A similar conclusion was reached by Le *et al.* in simulations of unsteady laminar flow over a backward facing step. However, in their simulations of a spatially evolving mixing layer, Buell & Huerre (1988) demonstrated that the convective boundary condition produces disturbances at the upstream boundary. The resulting disturbances altered the hydrodynamic stability characteristics of the flow. In contrast to the transition problem, the downstream boundary feedback may not play a crucial role in simulations of turbulent flows where the entrance of the computational domain is dominated by large amplitude random disturbances.

3. EXPERIMENTS WITH THE MINIMAL FLOW UNIT

Despite the progress in the numerical methodology for simulation of complex flows, direct numerical simulation of all but a handful of laboratory flows remains beyond the reach of the current computers. The computer time required for simulation of even relatively simple perturbations to the canonical flows is prohibitive. The loss of spatial statistical sample due to additional directions of flow inhomogeneity is one factor leading to the increased computational cost. For example, Rai & Moin (1991) used over seventeen million grid points and several hundred hours of CRAY-YMP to compute the laminar/turbulent transition on a flat plate with high free-stream turbulence levels. The momentum thickness Reynolds number at the exit of their computational domain was merely about 2000. They point out that their results probably suffer from inadequate grid resolution and about four times as many grid points are required for an accurate simulation.

Recently Jimenez & Moin (1990) showed that the essential dynamics of the near wall-region in turbulent channel flow can be reproduced using a computational box extending only 100 wall-units in the spanwise direction and about 200-300 wall units in the streamwise direction. This flow volume was named the minimal flow unit. The computed mean velocity profile and turbulent intensities in the wall region were in good agreement with the experimental data. Due to the small size of the computational domain, simulations using the minimal flow unit are significantly less computer-intensive than the full simulations. In addition, the minimal flow units contain only a few (essential) flow structures which provide an ideal setting for the study of the dynamics of these structures. These advantages have recently been exploited in several investigations of the structure and control of turbulent boundary layers. It is believed that owing to the simplicity of the minimal flow, in some cases a better understanding of the mechanics of the flows involved will be gained, and in other cases, tentative solutions can be obtained where the full simulation would have been prohibitive.

Sendstad & Moin (1990) used the minimal channel flow to investigate the mechanics of three-dimensional turbulent boundary layers. It has been shown experimentally that when a two-dimensional turbulent boundary layer is subjected to a

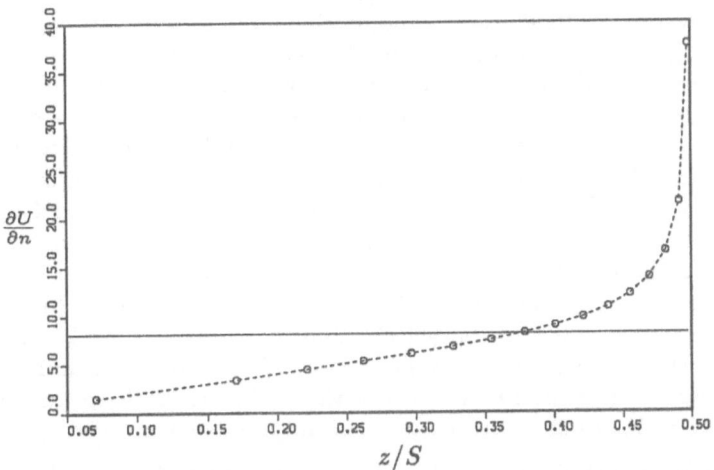

FIGURE 5. Spanwise variation of the wall-normal derivative of the mean velocity in turbulent channel flow with riblets. The outmost left point is at the riblet valley and the tip is at the right of the figure. The solid line is the mean velocity derivative from a channel flow without riblets and the same mass flux.

FIGURE 6. Contours of the streamwise vorticity in a cross-stream plane in turbulent channel flow with riblets. The lower wall has riblets (not shown) with spacing, $S^+ = 20$. Dashed lines are used for negative contours.

transverse pressure gradient, the Reynolds shear stress and turbulent kinetic energy drop. Currently available turbulence models cannot predict these effects. Sendstad & Moin imposed a sudden spanwise pressure gradient in a channel flow and reproduced the experimentally observed reductions of the Reynolds stresses. Although all turbulence statistics were computed using the full channel simulations, the detailed dynamics of the turbulence structures responsible for the statistical changes could only be isolated in the minimal channel flow. It was found (with the help of J. Jimenez) that the generated spanwise mean flow shifts the streamwise vortices with respect to the wall-layer streaks. In the two-dimensional channel, these vortices are generally located in between the high- and low-speed streaks, pumping low-speed fluid away from the wall and high-speed fluid towards the wall. With the sudden imposition of the spanwise pressure gradient and the spanwise shift of the vortices, low-speed fluid is often pumped towards the wall and high-speed fluid is pumped away from the wall, which is contrary to the Reynolds shear stress producing events.

Simulation of turbulent boundary layer over riblets is an example of a calculation that is practically impossible to perform on the current supercomputers, but it is feasible in the minimal flow unit. Choi, Moin & Kim (1990) simulated a turbulent channel flow with longitudinal riblets on one wall and a flat other wall. They used a finite-difference spatial differencing with body fitted coordinates. Careful grid refinement studies showed that at least 32 grid points per riblet are required. They considered different riblet spacings and ridge angles. With riblet spacing of $S^+ = 40$ wall units, the total drag was increased about 10%, whereas with $S^+ = 20$, apparently a slight drag decrease was obtained. The precise account of the computed drag in the latter simulations is awaiting further computations for improvement of the statistical sample. The variation of the skin friction with the spanwise direction is shown in Figure 5. The drag is minimum in the valleys of the riblets and is maximum at the tips. Contours of instantaneous streamwise vorticity in the channel cross section are shown in Figure 6. It is remarkable that the skin friction has slightly decreased despite the apparent increase in the turbulence activity near the wall with riblets and about 40% increase in the wetted area. Choi *et al.* have also successfully used the minimal flow unit to explore active control concepts for drag reduction. They showed that significant drag reduction can be achieved if sensors are placed near the wall but within the flow. Control algorithms based on placing sensors at the surface, which would be more amenable to practical implementation, led to lower drag reductions. Work is currently in progress for development of more practical strategies.

4. SUMMARY

High-order finite-difference schemes are attractive candidates for direct or large eddy simulation of complex turbulent flows. Random inflow boundary conditions appear to reach equilibrium in less than one eddy turnover time in isotropic turbulence. Longer periods appear to be required for structural equilibrium in sheared turbulence. The minimal flow unit has been shown to be a very useful tool in simulation and studies of wall turbulence.

The results reported herein were obtained with the support of the U.S. Air Force Office of Scientific Research.

References

Browning, G. L. and Kreiss, H. -O., 1989, Comparison of numerical methods for the calculation of two-dimensional turbulence, *Mathematics of Computation*, **52**, 369-388.

Buell, J. C., and Huerre, P., 1988, Inflow/outflow boundary conditions and global dynamics of spatial mixing layer, in:"Proceedings of the 1988 Summer Program of the Center for Turbulence Research," Stanford University and NASA Ames Research Center.

Canuto, C., Hussini, M. Y., Quarteroni, and Zang, T. A., 1988, Spectral methods in fluid dynamics, *Springer-Verlag*.

Choi, H., Moin, P. and Kim, J., 1990, Turbulence control for drag reduction in wall-bounded flows, Bull. of the American Physical Society, **35**, 10, 2232.

Domaradzki, J. A. and Rogallo, R. S., 1990, Local energy transfer and non-local interactions in homogeneous, isotropic turbulence, *Phys. Fluids A*, 2:413-426.

Harlow, F. H. and Welch, J. E., 1965, Numerical calculation of time-dependent viscous incompressible flow of fluid with free surface, *Phys. Fluids*, 8:2182.

Herring, J. R., Orszag, S. A., Kraichnan, R. H. and Fox, D. G., 1974, Decay of two-dimensional homogeneous turbulence, *J. Fluid Mech.*, 66:417-444.

Jimenez, J. J. and Moin, P., 1990, The minimal flow unit in near wall turbulence, *CTR Manuscript 105*, Center for Turbulence Research, Stanford University and NASA-Ames Research Center. Also to appear in *J. Fluid Mech.*

Kim, J., Moin, P. and Moser, R. D., 1987, Turbulence statistics in fully developed channel flow at low Reynolds number, *J. Fluid Mech.*, 177:133-166.

Le, H., Moin, P. and Kim, J., 1990, Direct numerical simulation of turbulence over a backward facing step, Bull. of the American Physical Society, **35**, 10, 2228.

Lee, S. S., Moin, P. and Lele, S. K., 1991, *AIAA Paper 91-0523*.

Lee, S. S., Moin, P. and Lele, S. K., 1991b, to be published.

Lesieur, M., Staquet, C., Le Roy, P., and Comte, P., 1988, The mixing layer and its coherence examined from the point of view of two-dimensional turbulence, *J. Fluid Mech.*, 192:511-534.

Metcalfe, R. W., Orszag, S. A., Brachet, M. E., Menon, S. and Riley, J. J., 1987, Secondary instability of a temporally growing mixing layer, *J. Fluid Mech.*, 184:207-243.

Mansour, N. N., Kim, J. and Moin, P., 1988, Reynolds stress and dissipation rate budgets in turbulent channel flow, *J. Fluid Mech.*, 194:15-44.

Moin, P. and Kim, J., 1985, The structure of the vorticity field in turbulent channel flow. Part 1. Analysis of the instantaneous fields and statistical correlations, *J. Fluid Mech.*, 155:441-464.

Moin, P. and Moser, R. D., 1989, Characteristic eddy decomposition of turbulence in a channel, *J. Fluid Mech.*, 200:471-509.

Moser, R. D. and Rogers, M. M., 1990, Mixing transition and cascade to small scales in a plane mixing layer, *IUTAM Symposium on Stirring and Mixing*, La Jolla, CA, Aug. 20-24, 1990.

Patera, A. T., 1984, A spectral element method for fluid dynamics: laminar flow in a channel expansion, *J. Comput. Phys.*, 54:468-488.

Pauley, L. L., Moin, P. and Reynolds, W. C., 1990, The structure of two-dimensional separation, *J. Fluid Mech.*, 220:397-411.

Rai, M. M. and Moin, P., 1989, Direct simulation of turbulent flow using finite-difference schemes, *AIAA Paper 89-0369.*

Rai, M. M. and Moin, P., 1991, AIAA 10th Computational Fluid Dynamics Conference, Honolulu, HI, June 24-26, 1991.

Sendstad, O. and Moin, P., 1990, On the mechanics of 3D turbulent boundary layers, Bull. of the American Physical Society, **35**, 10, 2266.

Spalart, P. R., 1988, Direct simulation of a turbulent boundary layer up to $Re_\theta = 1410$, *J. Fluid Mech.*, 187:61-98.

Zang, T. A., 1990, On the rotation and skew-symmetric forms for incompressible flow simulations, to appear in *Applied Numerical Mathematics.*

BUBBLE FORMATION IN DENSE FLUIDISED BEDS

Juan A. Hernández and Javier Jiménez

School of Aeronautics
Universidad Politécnica of Madrid
28040-Madrid, Spain

A linear analysis of the stability of the two fluid model equations for a dense gas fluidized bed is carried out for a generalized closure relation. The classical slug forming instability is confirmed, but a new instability, leading to the formation of streamwise channels, is also found when the particle pressure is allowed to depend on the slip velocity between phases. The outcome of these instabilities is then followed using a fully nonlinear numerical code. Slugs are unstable to a secondary corrugation, which is shown to result in the formation of gas bubbles in the bed. Whether or not these bubbles coalesce into larger elongated units is shown to depend on the presence or absence of the channelling instability.

INTRODUCTION

Fluidization is a process in which a bed of solid particles is suspended in an upward moving stream of fluid, with the weight of the particles balanced by their drag. Dense fluidized beds are known to be subject to a wide range of instabilities. As the fluid velocity is increased, beds undergo a series of transitions [1,2], which change substantially their mixing and transport properties. Typically there is an initial stable regime in which fluidization is steady, followed by the breaking of the bed into parallel horizontal slugs, and finally by the formation of large gas bubbles which rise through the bed. At still higher velocities, these bubbles coalesce into larger units, or partial channels, and a new regime develops which is sometimes referred to as turbulent [3]. Depending of the type of suspended particles and of the fluidizing medium, some of these stages may be missing and others might appear, but this sequence is typical of fine powder beds in air. The origin of the slugs and of the bubbles has been studied extensively. Most of the studies have used some type of continuum model in which the two phases are represented as interpenetrating fluids, each of which is endowed with its own stress tensor, and which are coupled by an interphase friction force.

The modelling of these forces and, in particular, of the stress tensor associated to the "fluid" which represents the dispersed phase, is particular troublesome, and we are aware of few investigators that attempt a rigorous derivation (see however [4,5] and references therein). The usual approach is phenomenological, with generalized models being fitted to represent particular experiments. Still, this approach has had some success, and it was shown in [6,7,8] that the two fluid model with appropriate closure assumptions can be used to predict instabilities of a uniform bed, leading eventually to the formation of slugs. The situation is less clear with bubbles, but [9,10] showed that the same equations are subject to a secondary instability of the horizontal concentration interfaces created in the trailing edge of the rising slugs, whose initial character and length scale are consistent with those observed in bubble formation. The same was shown in [11] for the slightly more general case of sharp concentration gradients.

In this paper we carry further the study of the formation of bubbles in a uniform gas fluidized bed, with special emphasis on their nonlinear development and coalescence. We use a generalization of the closure model used by most previous investigators, in that the parameters of the particle cloud stress tensor are assumed to depend on the local slip velocity between phases, in addition to their usual dependence on the volume fraction. We show that this generalization leads to a new instability, resulting in the formation of *vertical* channels. We then proceed to show numerically that this instability, which is usually weaker than the one leading to the formation of the slugs, is necessary for the process of bubble coalescence. The nonlinear flow regime is explored by means of a full numerical simulation of the continuum equations, which allows us to follow the process of the formation of the slugs, of their deformation into bubbles by the action of the secondary instability, and of their eventual coalescence.

EQUATIONS FOR PARTICLES FLUIDISED BY GAS

We consider a periodic domain $(0, L_x) \times (0, L_y)$, with gravity acting on the $-x$ direction. The gas density will be assumed to be much smaller than the particle density, but the volumetric loading will be taken as $O(1)$. Under those circumstances, the momentum carried by the gas is negligible, and the interaction between neighboring particles becomes dominant. We will model that interaction by an effective stress tensor associated to the dispersed phase, and neglect the viscous forces in the gas itself. The equations of motion of the two phases become,

$$\frac{\partial \alpha}{\partial t} + \nabla.(\alpha \mathbf{v}_p) = 0,$$

$$\frac{\partial (\alpha \mathbf{v}_p)}{\partial t} + \nabla.(\alpha \mathbf{v}_p \mathbf{v}_p) = \nabla.\sigma + \alpha B \left(\mathbf{v}_g - \mathbf{v}_p\right) - \frac{\alpha}{F} \mathbf{i}_1 \tag{1}$$

$$\frac{\partial (1 - \alpha)}{\partial t} + \nabla.\{(1 - \alpha)\mathbf{v}_g\} = 0,$$

$$0 = -\nabla p + \alpha B \left(\mathbf{v}_g - \mathbf{v}_p\right)$$

where $\mathbf{v}_p$ and $\mathbf{v}_g$ are the macroscopic velocity fields of particles and gas, p is the macroscopic pressure associated to the gas, α is the volume fraction of the dispersed phase, $\mathbf{i}_1$ is the unit vector in the x direction, B is a friction coefficient between phases, and σ is the stress tensor of the dispersed phase, which we will assume to be of the form

$$\sigma = -\Pi \delta_{ij} + \frac{1}{R} \left(\frac{\partial v_{pi}}{\partial x_j} + \frac{\partial v_{pj}}{\partial x_i}\right) \tag{2}$$

These equations are made nondimensional with the equilibrium velocity difference between phases, V_0, the particle density, ρ_p, and some length, $L \sim O(L_x, L_y)$. The Froude number, $F = V_0^2/gL$ is a constant, but the Reynolds number, $R = \rho_p V_0 L/\mu_p$, is part of the closure relations for the particle cloud, and is a function of the local flow conditions. The detailed closure will be left open at this stage. For now, we will only assume that, in equation (2), Π and R are non-decreasing functions of α and of $V = |\mathbf{v}_p - \mathbf{v}_g|$, the local slip velocity. For the friction coefficient, B, we will just assume a non-decreasing function of α,

$$\frac{\partial R(\alpha, V)}{\partial \alpha}, \frac{\partial \Pi(\alpha, V)}{\partial \alpha}, \frac{\partial B(\alpha)}{\partial \alpha} > 0, \tag{3}$$

$$\frac{\partial \Pi(\alpha, V)}{\partial V}, \frac{\partial R(\alpha, V)}{\partial V} \geq 0.$$

Under these conditions, there is an equilibrium solution of uniform fluidization that satisfies

$$\mathbf{v}_g - \mathbf{v}_p = \mathbf{i}_1, \qquad \alpha = \alpha_0,$$

$$F B(\alpha_0) = 1. \tag{4}$$

Throughout the paper, $()_0$ subscripts will refer to quantities evaluated at the conditions of this solution. In the next section we address its stability to infinitesimal perturbations.

INSTABILITIES OF THE UNIFORM STATE

Assume that a infinitesimal perturbation of the form

$$\alpha = \alpha_0 + \tilde{\alpha} \exp[i(\kappa_x x + \kappa_y y) + \mu t] \qquad (5)$$

is applied to the equilibrium solution, with similar perturbations for the rest of the flow variables. We will be concerned with the temporal stability problem, and κ_x, κ_y will be assumed real, while the eigenfrequency μ will in general be complex. The bed will be unstable for those cases in which the real part of μ is positive.

After linearizing the equations of motion, and setting the characteristic determinant to zero, three eigenvalues are obtained. One of them

$$\mu_3 = \frac{-\kappa^2}{R_0}, \qquad \kappa^2 = \kappa_x^2 + \kappa_y^2, \qquad (6)$$

is always stable, while the other two satisfy the equation

$$\mu^2 + m\,\mu + \kappa^2 h - i\kappa_x q = 0, \qquad (7)$$

where

$$m = \frac{2\kappa^2}{R_0} + \frac{1}{\alpha_0(1-\alpha_0)}(\alpha_0 B_0 - i\kappa_x \Pi'_{0V}),$$

$$q = B_0 - \left(\frac{1}{\alpha_0} + \frac{1}{1-\alpha_0} + \frac{B'_{0\alpha}}{B_0}\right)(\alpha_0 B_0 - i\kappa_x \Pi'_{0V}), \qquad (8)$$

$$h = \Pi'_{0\alpha} - \left(\frac{1}{\alpha_0} + \frac{B'_{0\alpha}}{B_0}\right)\Pi'_{0V},$$

and the operators $()'_\alpha$ and $()'_V$ stand for differentiation with respect to α and V. In general, for models satisfying the monotonicity conditions (3), only one of these eigenvalues may become unstable.

It can easily be shown, from analytic function theory considerations, that any extremum of the real part of these eigenvalues lies either on the $\kappa_y = 0$ axis, or on $\kappa \to \infty$, so that the most unstable wavenumbers have to be searched at those locations. There is usually a most unstable point at a finite wavenumber ($\kappa_x \neq 0$, $\kappa_y = 0$), which corresponds to the slug forming instability discussed in [6,7,8]. The growth rate for this instability vanishes at $\kappa = 0$ and is always negative for $\kappa_x \to \infty$. When this is the only instability present, a schematic representation of the (κ_x, κ_y) plane is shown in figure 1.b. There is no qualitative change to this instability by the introduction of the generalized model used here.

Under some conditions, however, the κ_y axis becomes unstable, and the marginal boundary takes the form in figure 1.a. This represents a new instability which tends to break the bed into alternating vertical channels and dust streamers. Because of the argument cited above, its maximum growth rate is always at $\kappa_y \to \infty$, and it can be shown that the eigenvalue behaves in that limit like

$$\mu_1 \to -R_0 h/2. \qquad (9)$$

The condition, $h < 0$ is therefore the criterion for instability and it is easy to see from the definition of h in equation (8), and from the conditions in (3), that it can only be satisfied if $\Pi'_{0V} \neq 0$, larger than some finite positive value.

The reason for this requirement is clear from the physical mechanism of this new instability. Assume that an incipient longitudinal channel is formed in the bed, along which the volume fraction of particles is lower than in the surrounding. The gas velocity increases in the channel as the interphase friction decreases, and the resulting increase in Π creates a gradient that pushes more particles away from the channel, reinforcing the instability. It is clear that the increase in particle pressure with V is an intrinsic part of this loop. It is also clear that the same mechanism will act on nonuniform beds, and will tend to form channels in any situation in which such formation leads to an increase of the gas velocity.

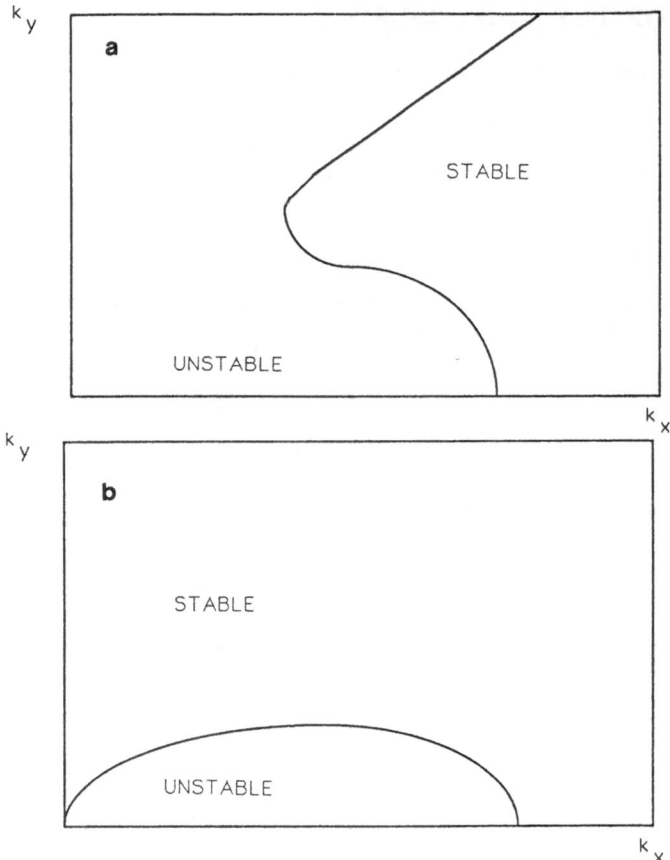

Figure 1. Schematic representation of instability region in wavenumber space when channeling instability is *a)* active or *b)* absent.

Note that this mechanism is different from those of the two "convective" instabilities described in [13,14]. Those instabilities also give rise to longitudinal channeling, but are due to coupling of the particle volume fraction with the gas supply mechanism, either through the distributor plate, or by internally generated gas. The new instability described here does not rely on such coupling, and should be active in externally driven beds, independently of the driving mechanism.

The inclusion of a velocity dependence in the particle stress tensor is unconventional, and has generally not been used by previous investigators. The model used by most of them can be traced to an empirical relation derived in [12] on the basis of stability considerations for experimental beds. However, the basic assumption that Π is only a function of α is taken as given in that paper without discussion. An attempt to derive the form of the particle pressure term from first principles was made by Buyevich in [5] and a series of related papers. Although these are difficult papers, and many of their assumptions are probably not applicable to dense beds, the conclusion that the stress tensor should be related to V is based on the general physical argument that the stresses originate by the interaction among neighboring particles through the perturbations that they introduce in the surrounding gas, and that these perturbations increase with V. From rough dimensional arguments, his conclusion that $\Pi \sim V^2$ seems plausible, and will be used here in the later numerical analysis.

Note that the growth rate for the new longitudinal instability does not decay to zero at infinity. This is not inconsistent with the well posedness of the equations, since the eigenvalue behaves at infinity only like a constant. It does mean, however, that there is no mechanism to limit the density gradients in the edges of the channels, and that the volume fraction profile will tend to become very sharp. In real beds, this tendency will presumably be stopped at scales of the order of the distance between particles by the breakdown of the continuum assumption.

NONLINEAR EVOLUTION OF THE DISTURBANCES

We study the nonlinear evolution of the disturbances by means of a fully spectral numerical simulation of the time evolving initial value problem. The spatial discretization uses fully dealiased Fourier expansions, while time integration is accomplished by an explicit fourth order Runge-Kutta scheme. Two extra flux conditions are needed to supplement the periodicity. They are chosen so that the total mass flux in both the x and y directions remain constant in time. Further details of the numerical code can be found in [11].

The computational box is always chosen to be of length $L_x = 2\pi$, but the spanwise dimension, L_y, is varied to explore the effect of the aspect ratio of the perturbation. Initially the particles are considered to be at rest, and the particle volume fraction is perturbed slightly away from its equilibrium value. A typical form for the initial perturbation is,

$$\alpha(x,y) = H(x)\{1 + \epsilon_y \cos(2\pi y/L_y)\}, \tag{10}$$

$$H(x) = \begin{cases} \alpha_0(1 + \epsilon_x), & 0 < x < b \\ \\ \alpha_0, & b < x < 2\pi \end{cases}$$

To avoid numerical problems with the spectral expansion, the discontinuities in the function H are smoothed initially over a few grid points using a gaussian window.

A more restricting choice is the form of the closure relations used for the model, which had been left fairly generic up to this moment. The relations used for the numerical experiments are,

$$B(\alpha) = \delta(1 - \alpha)^{-3}, \tag{11}$$

$$\Pi(\alpha, V) = c_0^2 \alpha^3 V^2 e^{m\alpha/(\alpha_p - \alpha)}, \tag{12}$$

$$R(\alpha, V) = \frac{R_0}{\alpha} V^{-1/4}, \tag{13}$$

where δ, c_0, R_0 and m are adjustable constants that were used to explore the parameter space, and $\alpha_p = 0.6$ is the maximum packing density of particles, chosen to represent a bed of identical spheres.

Equation (11) was taken from [12] and is the equation with the most experimental support of the three. The functional dependence of the particle pressure Π on V was discussed on the previous section and, from the linear stability results, is probably the most critical aspect of the model. The cubic dependence on α is taken from theoretical considerations in [5], which should hold for dilute beds. The exponential factor is just a convenient way of modelling the expected rise to infinity of the particle pressure as it approaches the maximum particle packing density. Considerable numerical experimentation was spent in finding a suitable form for this factor. In summary, a rapidly increasing pressure was found to be necessary to prevent the nonlinear bed behavior from resulting into overshoots above the maximum density, but the particular form chosen was not found to be too important in determining the qualitative behavior of the bed. Typical values for m were 0.25 to 1.

Finally, the bed viscosity, R, was derived from a reevaluation of experimental values from several sources, whose details can be found in [11]. The linear stability analysis suggest that its detailed form is not very important to the qualitative bed behavior, and several numerical experiments carried with different closures confirm that view.

In view of the uncertainties associated with the closure model, it should be emphasized that the purpose of the numerical experiments reported here is to explore the effect of different closure parameters on the nonlinear behavior of the bed perturbations. The use of this code to

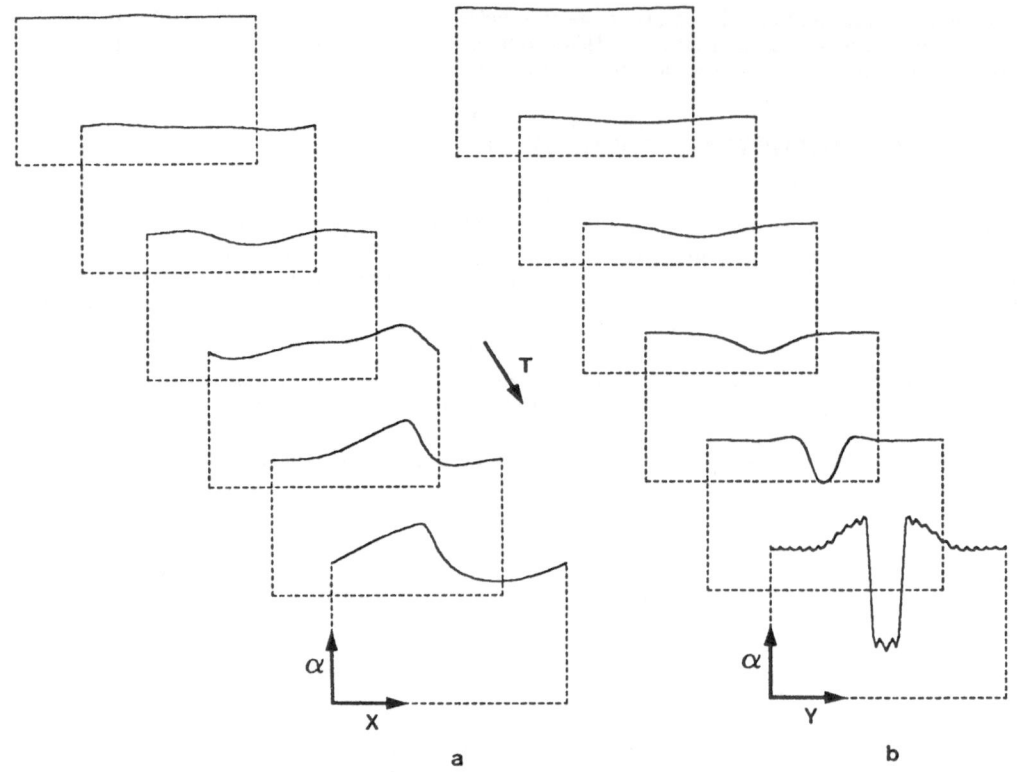

Figure 2. Distribution of particle volume fraction, α. Times: 0, (3.7), 18.5. Model parameters: $\alpha_0 = 0.4$, $c_0 = 1$, $\delta = 0.5$, $m = 0.4$, $R_0 = 5$. Resolution: 50×50. *a)* One dimensional slug formation. Flow from left to right. Initial conditions: $\epsilon_x, \epsilon_y = (0.1, 0.)$; *b)* One dimensional channel formation. Flow into the page. Initial conditions: $\epsilon_x, \epsilon_y = (0., 0.01)$.

simulate realistic beds must wait until more experimental or theoretical information becomes available on the right closure relations to use in particular situations.

ONE DIMENSIONAL SOLUTIONS

A first experiment was to study the nonlinear behavior of each of the two one dimensional instabilities taken in isolation. This was done by forcing only the horizontal or vertical modes of the initial condition (10). The numerical code is clean enough that no subsequent development of transversal modes is possible.

Figure 2.a shows the results for the evolution of slugs. The initial perturbation grows to form a moving region of high concentration, whose upper edge steepens into a relatively sharp horizontal front. Eventually, the waveform becomes steady and the bed is broken into a uniform wavetrain of identical horizontal bands. The formation of concentration shocks had already been reported in [6,7], and the computation of the steady nonlinear wavetrains, independently of the initial value problem, was accomplished in [8]. We have carried out a fairly extensive scan of the behavior of these solutions as a function of the model parameters and of the initial conditions. It appears that the uniform wavetrain is a strongly attracting solution, and that fairly arbitrary initial conditions will eventually evolve into uniform trains. We mentioned in the previous section that this instability has a most unstable finite wavelength. Solutions started from very weak initial perturbations grow into wavetrains whose wavenumber is close to this most unstable one. It is curious, however, to note that it is possible to initiate solutions with the "wrong" wavelength which grow eventually into stable wavetrains with wavelengths very far from that most favored one.

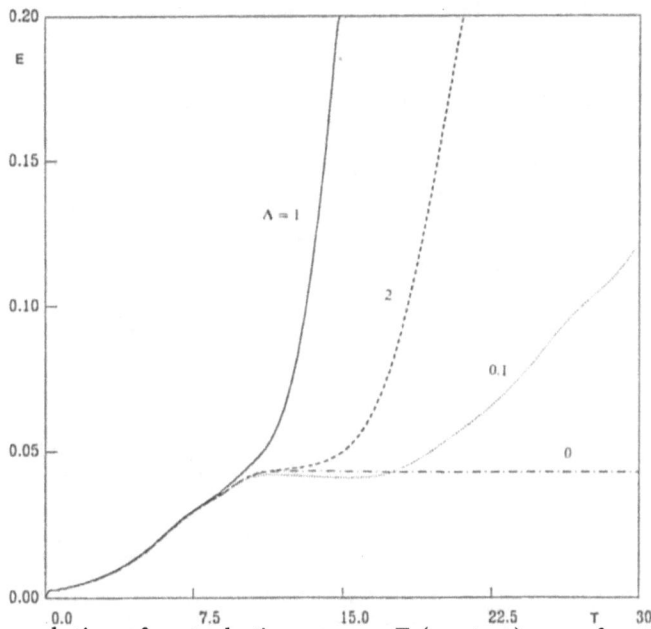

Figure 3. Time evolution of perturbation energy, E (see text), as a function of aspect ratio, $\Lambda = L_x/L_y$. Model parameters: $\alpha_0 = 0.4$, $c_0 = 0.3$, $\delta = 1$, $m = 1$, $R_0 = 5$. Resolution: 50×4. $\epsilon_x, \epsilon_y = (0.1, 0.001)$.

Figure 2.b shows the equivalent evolution of a vertical channel. The horizontal axis is now the transverse coordinate y. The initial cosine perturbation deepens quickly into a well defined channel whose sharp edges are a consequence of the lack of a high frequency cut-off for the linear stability. The velocity of the gas, and of the particles, along the channel axis is much larger those on the denser lateral regions which, depending on the frame of reference, will appear to fall back to preserve continuity. Contrary to the slug instability, there is no wavelength selection mechanism for channels, and the shape of the final channelling structure is very sensitive to initial conditions. There is also no amplitude limiting mechanisms for the depth of the channel, which will deepen until the particle concentration at their center drops to zero. At this point the bed would be broken into separated vertical bands of gas and of particles, rising at very different velocities.

TWO DIMENSIONAL EVOLUTION AND BUBBLES

When both longitudinal and transversal perturbations are imposed on the initial conditions for the bed, it becomes apparent that the horizontal slugs are subject to a secondary instability. This is clear from figure 3, which presents the evolution in time of the perturbation "energy",

$$E = \left(\int_0^{L_x} \int_0^{L_y} (\alpha u_p)^2 \, dx \, dy \right)^{1/2},$$

as a function of the aspect ratio, $\Lambda = L_x/L_y$, of the computational box. For $\Lambda = 0$ (horizontal slugs), the perturbation grows to a given amplitude, and settles into a stable one dimensional wavetrain. For fairly high or low values of Λ, however, a secondary instability appears after the primary one has saturated, and the growth of the perturbation resumes at a much faster pace. The evolution of the volume fraction fields during this early part of the instability is similar to that shown in the first two frames of figure 4, although the latter is run for different model parameters. In essence, the lower edges of the slugs corrugate and break into falling "tears" leaving empty gas bubbles in between. The growth rate of this instability depends on the aspect

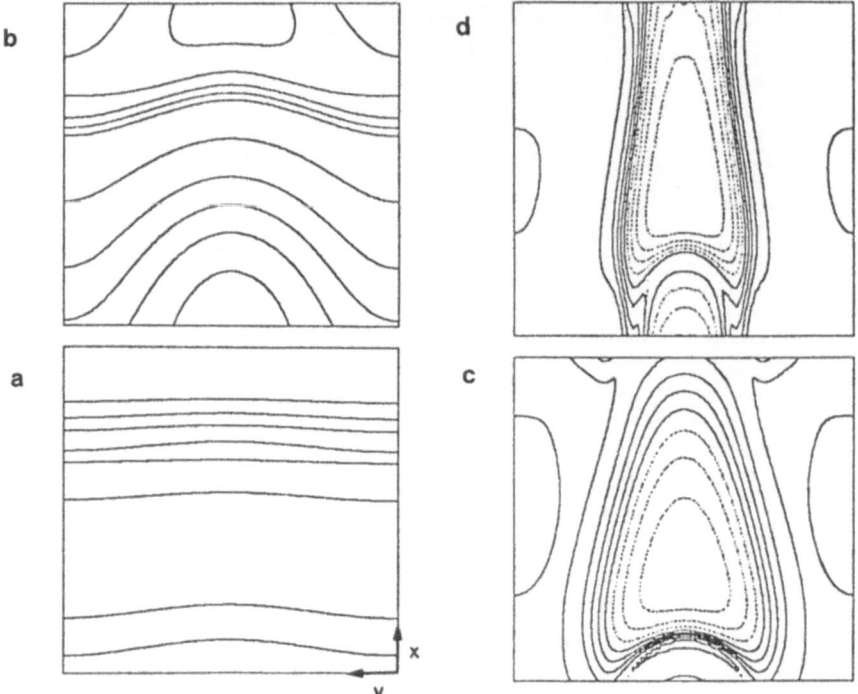

Figure 4. Volume fraction evolution in time, showing successive formation of slugs, bubbles and channels. Times, a: 13.9, b: 19.4, c: 24., d: 25.9. Model parameters: $\alpha_0 = 0.4$, $c_0 = 1$, $\delta = 0.5$, $m = 0.4$, $R_0 = 5$. Resolution: 50×50. $\epsilon_x, \epsilon_y = (10^{-1}, 10^{-3})$. Gas flow bottom to top.

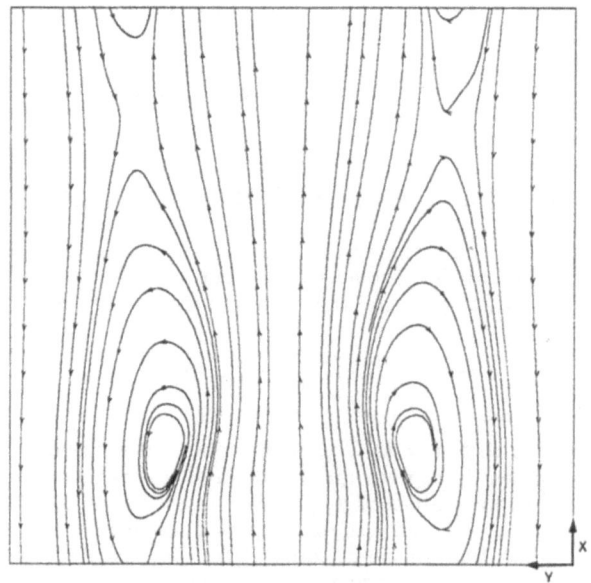

Figure 5. Streamlines for $\mathbf{v}_p$ during late stages of bubble formation. Flow field is the same as $t = 24$ in figure 4.

ratio, and is highest when $\Lambda \sim 1$. Its outcome is always the breaking of the low density regions left between the slugs into empty bubbles surrounded by denser areas. A similar instability was reported by [9,10], who suggested that it was responsible for bubble formation, although they did not followed its nonlinear behavior.

The later evolution of the bubbles depends on the linear stability properties of the bed. The simulations in figure 3 were run in a bed for which the channeling instability was inactive. Under these conditions the bubbles grow to a stage similar to that of the third frame in figure 4, and stop growing. Note that, at this stage, a high density band remaining from the original slug separates consecutive bubbles which are therefore prevented from merging. The shape of the bubbles also stays constant and the gas remains trapped in them, rising with a relatively slow velocity with respect to the denser parts of the bed.

Figure 4 is computed in a bed for which both the channeling and slug instabilities are active. The first three frames show bubble formation, and have already been described. They seem to be relatively independent of channeling. The final evolution of the bubbles, however, is different. The horizontal walls that separate consecutive bubbles now disappear, and the bubbles fuse into a continuous channel. Both the gas and the few particles left in the channel core move at a much higher velocity than those outside it, and all that remains of the original slug instability is a series of mixing vortices in the high shear interface separating both regions (figure 5).

CONCLUSIONS

We have presented a linear stability analysis of a two fluid model representing a dense fluidized bed in gas. We have shown that the transverse instability leading to the formation of slugs, that had already been described in the literature, remains a property of a fairly generalized family of closure models, including cases in which the particle pressure is made to depend on interphase slip velocity. We have also shown that the addition of this new dependence on V leads to the appearance of a new instability, resulting in the formation of longitudinal channels in the bed.

The nonlinear behavior of these instabilities was followed numerically, using a full simulation code for the model equations. Under the conditions of our experiments, the transverse instability results in the formation of horizontal slugs, in accordance with previous investigators. We have confirmed the presence of a secondary instability of these slugs, also previously reported, and we have shown that its nonlinear outcome is the formation of rising gas bubbles. The subsequent evolution of these bubbles depends on the presence or absence of the channeling instability. When this new instability is present, the bubbles coalesce into longer units or vertical channels, through which the gas passes at high speed. When the channelling instability is absent, this coalescence does not occur, and the bubbles rise stably, with a lower overall gas flow.

We are not aware of many experiments in which channel formation is clearly documented. Since their origin is the coalescence of bubbles, channels should only be expected to form in the upper part of the bed, and this mechanism has recently been suggested as an explanation of anomalously high gas flows in bubbling beds [15]. Spouting of beds, characterized by gas jets breaking to the bed surface, has been known for a long time, but is generally attributed to instabilities associated with the distributor bottom plates [13]. Bubble coalescence has also been observed or suggested for a long time as a necessary step for the transition from bubbling to turbulent flows [3]. We suggest that the channeling instability reported here might be connected with some of those phenomena, but more experimental work is needed to confirm that. In particular, much more experimental and theoretical work is needed to clarify the adequacy of the modelling assumptions about the form of the particle cloud stress tensor. The linear stability analysis presented here shows that some of the qualitative behavior of the instabilities is independent of the modelling assumptions, but that other features important depend strongly on the model.

This work was supported in part by a grant from the Spain-U.S.A. Committee for Technological and Scientific Cooperation. One the authors (J.A.H.) was supported in part by a

Fellowship from the Spanish Ministry of Education and Science. Computer time for the simulations was provided by the IBM Madrid Scientific Center.

REFERENCES

[1] J.P.Couderc, in *Fluidization*, edited by J.F.Davidson, R.Clift and D.Harrison (Academic Press, London, 1985) p.39.

[2] A.K.Didwania and G.M.Homsy, Int. J. Multiphase Flow, **7**, 563 (1981).

[3] J.Yerushalmi and A.Avidan, in Ref 1, p.233.

[4] D.A.Drew, Ann. Rev. Fluid Mech. **15**, 261 (1983).

[5] Y.A.Buyevich, J. Fluid Mech. **56**, 313 (1972).

[6] D.J.Needham and J. H. Merkin, J. Fluid Mech. **131**, 427 (1983).

[7] J.B.Fanucci, N. Ness and R. -H. Yen, J. Fluid Mech. **94**, 353 (1979).

[8] D.J.Needham and J.H.Merkin, ZAMP **37**, 322 (1986).

[9] A.K.Didwania, Ph.D. thesis, Stanford U. (1981).

[10] G.M.Homsy, M.M.El-Kaissy and A.K.Didwania, Int. J. Multiphase Flow. **6**, 305 (1980).

[11] J.A.Hernández, Ph.D. thesis, School of Aeronautics, Madrid (1990).

[12] S.K.Garg and J.W.Pritchett, J. Appl. Phys. **46**, 4493 (1975).

[13] R.Jackson, in Ref 1, p.47.

[14] D.Green and G.M.Homsy, Int. J. Multiphase Flow. **13**, 459 (1987).

[15] T.W.Yule and L.R.Glicksman, AIChE Symposium Series, **262**, 1 (1988).

THREE-DIMENSIONAL NUMERICAL SIMULATIONS
OF COHERENT STRUCTURES IN FREE-SHEAR FLOWS

M. LESIEUR, P. COMTE, Y. FOUILLET and A. SILVEIRA

Institut de Mécanique de Grenoble*

Institut National Polytechnique de Grenoble and

Université Joseph Fourier, Grenoble
BP 53 X - 38041 Grenoble-Cedex, France

Abstract

We investigate, with the aid of three-dimensional direct-numerical simulations, the origin and dynamics of coherent structures in the following mixing-layers: temporal case, flow behind a backwards-facing step, and spatially-developing case.

In the periodic case, the calculation is done using pseudo-spectral methods. The basic velocity field is a hyperbolic-tangent profile $U \tanh 2y/\delta_i$, with an initial Reynolds number $U\delta_i/\nu = 100$. The initial velocity field is the basic velocity, above which have been superposed two small isotropic random perturbations of wide spectrum peaking at the fundamental mode: a three-dimensional one, of kinetic energy $\epsilon_{3D} U^2$ and a two-dimensional one, of kinetic energy $\epsilon_{2D} U^2$. According to the value of the initial three-dimensionality rate $r = \epsilon_{3D}/\epsilon_{2D}$, two regimes are found: for $r = 0.1$, quasi two-dimensional large coherent Kelvin-Helmholtz billows are formed. They slightly oscillate in phase, as in the translative instability proposed by Pierrehumbert and Widnall (1982). The vortex lines plots show also in this case the presence of longitudinal vortices stretched between the big rollers. They might originate from vortex lines oscillating about the stagnation line, as proposed by Lasheras and Choi (1988). For $r = \infty$, the fundamental billows which appear have strong spanwise oscillations which are not in phase. Pairings between the primary vortices lead to reconnections of the billows, giving rise to a vortex-lattice structure of the Kelvin-Helmholtz billows. This corresponds to the helical-pairing instability discovered by Pierrehumbert and Widnall (1982).

Afterwards, we have looked at two spatially-growing mixing layers: an unforced flow behind a backwards-facing step, and a slightly-compressible layer developing downstream of a hyperbolic-tangent velocity profile, and forced by a small three-dimensional random perturbation: these two flows offer examples of the two regimes discussed above.

---

* Unité associée CNRS

The Global Geometry of Turbulence
Edited by J. Jiménez, Plenum Press, New York, 1991

1 Introduction

An important issue in the study of turbulent plane free-shear layers concerns the two-dimensionality of the large coherent Kelvin-Helmholtz vortices. Can these vortices be studied on the basis of two-dimensional Navier-Stokes equations? This is much easier and cheaper from a computational point of view: it has been shown for instance by Lesieur et al. (1988), that an enstrophy-cascading like spectrum develops in the small scales of a two-dimensional temporal mixing layer after the pairing of fundamental vortices. Figure 1, taken from Normand (1990), shows a two-dimensional direct-numerical simulation of a spatially-developing mixing layer, originating from a hyperbolic-tangent velocity profile forced upstream by a small white-noise perturbation. The calculation is done in the same conditions as in the incompressible uniform density laboratory experiments reported in Brown and Roshko (1974): Figure 1-a shown a passive scalar plot, which is very similar to the pictures of the laboratory experiment. Figure 1-b shows the corresponding vorticity field, displaying how coherent structures do correspond to localized concentrations of vorticity (with respect to a passive contaminant). This is of course a great advantage of the numerical experiment upon its laboratory counterpart to be able to determine extremely precisely the vorticity field. If one forgets about the numerical errors, these two-dimensional calculations represent a sort of ideal physical reference state for laboratory experiments trying to investigate the dynamics of two-dimensional flows.

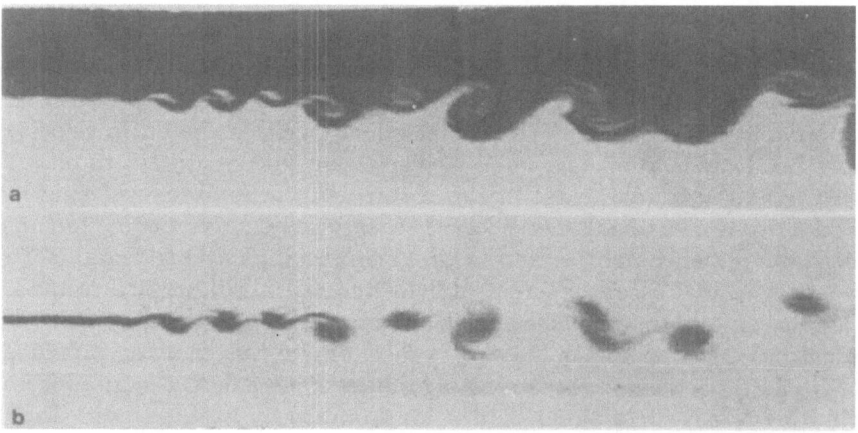

Figure 1. two-dimensional numerical simulation of a spatially growing mixing layer: a) passive scalar; b) vorticity.

However, losing the third dimension is extremely questionable: all the natural plane shear layers are known to develop three-dimensionality, both in the small scales (when turbulence is developed[1]) and in the large scales. In the plane mixing la-

---

[1] We recall that one of the major differences between three-dimensional and two-

yers for instance, the laboratory experiments of Breidenthal (1981) show longitudinal streaks strained downstream. Bernal and Roshko (1986) have reconstructed from their measurements a hairpin vortex filament winding around the two-dimensional billows: indeed, laser cross-section of the flow across the braids show mushroom-type vortices, in good agreement with the hairpin model. These vortices are shown also in the direct-numerical simulations carried out by Metcalfe et al. (1987).

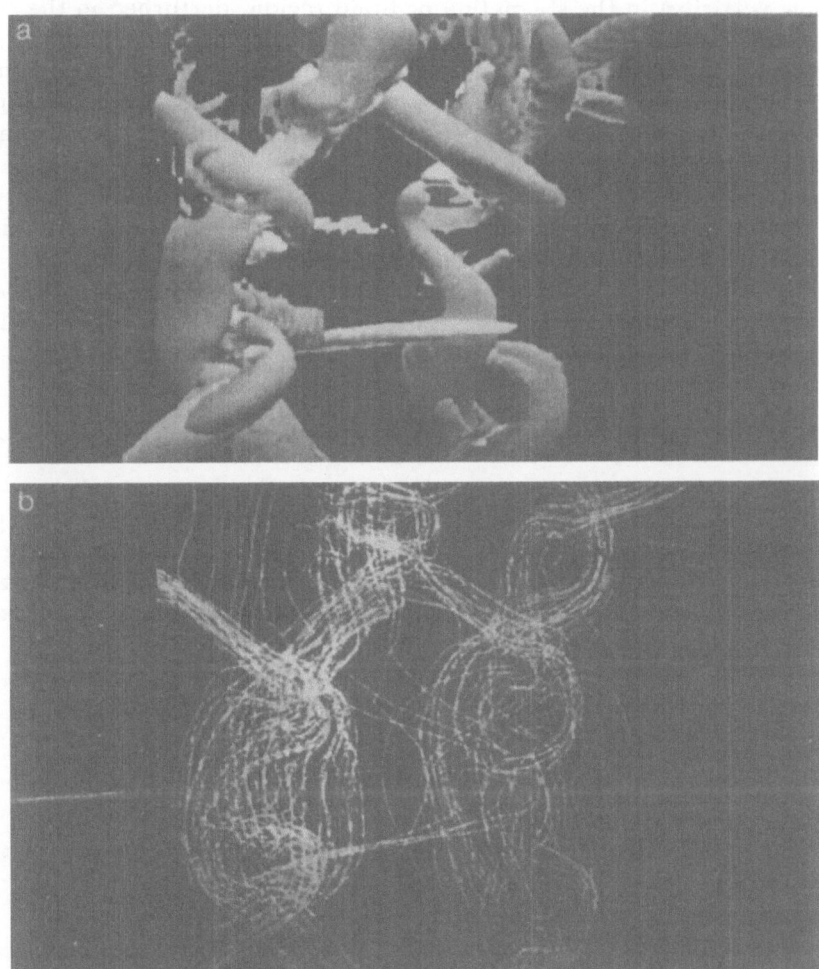

Figure 2. temporal mixing layer forced by a three-dimensional perturbation at $t = 30\ \delta_i/U$; top view of, a) vorticity topology, b) vortex lines.

There are several explanations for the existence of these vortex filaments. Pierrehumbert and Widnall (1982) associate them with an instability they discovered, called

dimensional turbulence is that, when viscosity goes to zero, the former dissipates kinetic energy at a finite rate, while the latter conserves it (see Lesieur, 1987).

the translative instability: the latter results from a secondary instability analysis performed on Stuart vortices[2]. Such an instability is characterized by a global in-phase spanwise oscillation of the primary billows. The laboratory experiments of Bernal and Roshko (1986) show a spanwise wavelength of the hairpin vortex filament of the order of $2\ \lambda/3$, where λ is the longitudinal wavelength of the Kelvin-Helmholtz vortices, in good agreement with the most-amplified spanwise wavelength of the translative instability. However, this does not explain how thin longitudinal vortices are formed from the global oscillation of the big billows. Therefore, another mechanism has been proposed by Corcos and Lin (1984) and Lasheras and Choi (1988), where vortex filaments (carrying low vorticity) in the stagnation or braid region, perturbed in the spanwise direction, would be strained longitudinally by the basic shear, yielding the hairpin vortex structure. This was confirmed by a three-dimensional numerical simulation using vortex method carried out by Ashurst and Meiburg (1988): here, the spanwise wavelength of the translative instability of the big rollers imposes the hairpin vortex spanwise wavelength.

Since all the above-quoted direct-numerical simulations involved both a small number of fundamental Kelvin-Helmholtz vortices (generally 2) and deterministic initial perturbations, we decided to carry out a high-resolution calculation (128^3 Fourier modes) with 4 fundamental billows[3], forcing initially the mixing layer with a three-dimensional random perturbation, more apt to model the residual turbulence in natural mixing layers.

2 Numerical simulations of the temporal mixing layer

We solve Navier-Stokes equations (with uniform density) in a box (x, y and z being respectively the longitudinal, transverse and spanwise directions. Periodicity is assumed in the x and z directions, with free-slip boundary conditions on the lateral boundaries. Using classical pseudo-spectral methods developed by Orszag, the equations are written

$$\frac{\partial}{dt}\underline{\hat{u}}(\vec{k},t) = \Pi(\vec{k}) \circ F[F^{-1}(\underline{\hat{u}}) \times F^{-1}(\underline{\hat{\omega}})] - \nu\ k^2 \underline{\hat{u}}(\vec{k},t)$$

$$\underline{\hat{\omega}} = i\ \vec{k} \times \underline{\hat{u}} \quad ; \quad \vec{k}.\underline{\hat{u}}(\vec{k},t) = 0\ , \qquad (2-1)$$

where $\underline{\hat{u}}$ and $\underline{\hat{\omega}}$ stand respectively for the velocity and vorticity vectors in Fourier space. F is the Fast-Fourier Transform operator. The transport equation of a passive scalar θ is solved as well:

$$\frac{\partial}{dt}\hat{\theta}(\vec{k},t) = -i\vec{k}.F[F^{-1}(\hat{\theta})F^{-1}(\underline{\hat{u}})] - \kappa\ k^2 \hat{\theta}(\vec{k},t) \quad , \qquad (2-2)$$

with $\kappa = \nu$, corresponding to a Schmidt number equal to 1. This passive scalar will be used as a numerical dye. Other fields allowing to investigate the geometry and the vortex topology of the flow will be the vorticity components. These various fields will be visualized with the aid of the FLOSIAN (FLOw SImulation ANalysis) software developed in the Turbulence Modelling Group at the Institut de Mécanique de Grenoble.

---

[2] The same study has been carried out by Corcos and Lin (1984) on Kelvin-Helmholtz vortices.

[3] That is, in a domain of streamwise length equal to $4\ \lambda_a$, where λ_a is the most-amplified fundamental longitudinal wavelength.

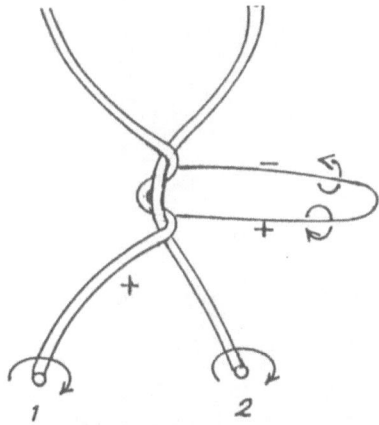

Figure 3. schematic representation of the vortex structure shown on Figure 2a.

The initial velocity consists in a basic hyperbolic-tangent velocity profile $U \tanh 2y/\delta_i$, to which are superposed two random isotropic perturbations of wide spectrum, peaking at the fundamental mode $k_a = 2\pi/7\delta_i$: the first perturbation is two-dimensional (in the x, y plane), of kinetic energy $\epsilon_{2D}U^2$. The second perturbation is three-dimensional, of kinetic energy $\epsilon_{3D}U^2$. The initial Reynolds number is $R_e = U\delta_i/\nu = 100$, and the initial passive scalar distribution is identical to the basic velocity.

If $\epsilon_{3D} = 0$, the problem is two dimensional, and corresponds to the study done in Lesieur et al (1988). Now, we show calculations with $\epsilon_{2D} = 0$ and $\epsilon_{3D} = 10^{-4}$, that is, with a purely three-dimensional perturbation. The following figures correspond to a time of 30 δ_i/U which is characteristic of the pairing in the two-dimensional analogous simulations (see Lesieur et al., 1988): Figure 2a, corresponding to a top view of the various components of the vorticity field, show that the Kelvin-Helmholtz billows are severely distorted in the spanwise direction, and pair only in the central region of the span. From the vortex lines shown on Figure 2b, and the determination of the longitudinal vorticity signs on the vortex lines, it is also clear that the thin longitudinal vortices appearing on Figure 2a come from the straining by the flow of a vortex filament initially located in the periphery of the billow numbered one on Figure 3: the strong torsion of this billow in the region of pairing is certainly responsible for the detachment of this vortex filament.

Figure 4 shows the interface of the mixing layer, visualized by the isosurface $\theta = 0$ of the passive scalar, and corresponding to Figure 2. The top view wears the signature of the vortex structure shown on Figure 2. However, the side view presented in Figure 2c does not indicate at all the violent three-dimensional distorsion of the Kelvin-Helmholtz billows.

This highly three-dimensional Λ-shaped structure of the Kelvin-Helmholtz billows[4] may be explained with the aid of the helical-pairing instability, proposed by Pierrehumbert and Widnall (1982): assume that the Kelvin-Helmholtz billows oscil-

---

[4] Notice that a structure presenting analogies with the present results (but without pairings) has been found by Sandham and Reynolds (1990) in three-dimensional direct-numerical simulations of a temporal compressible mixing layer for a convective Mach number larger than 0.6.

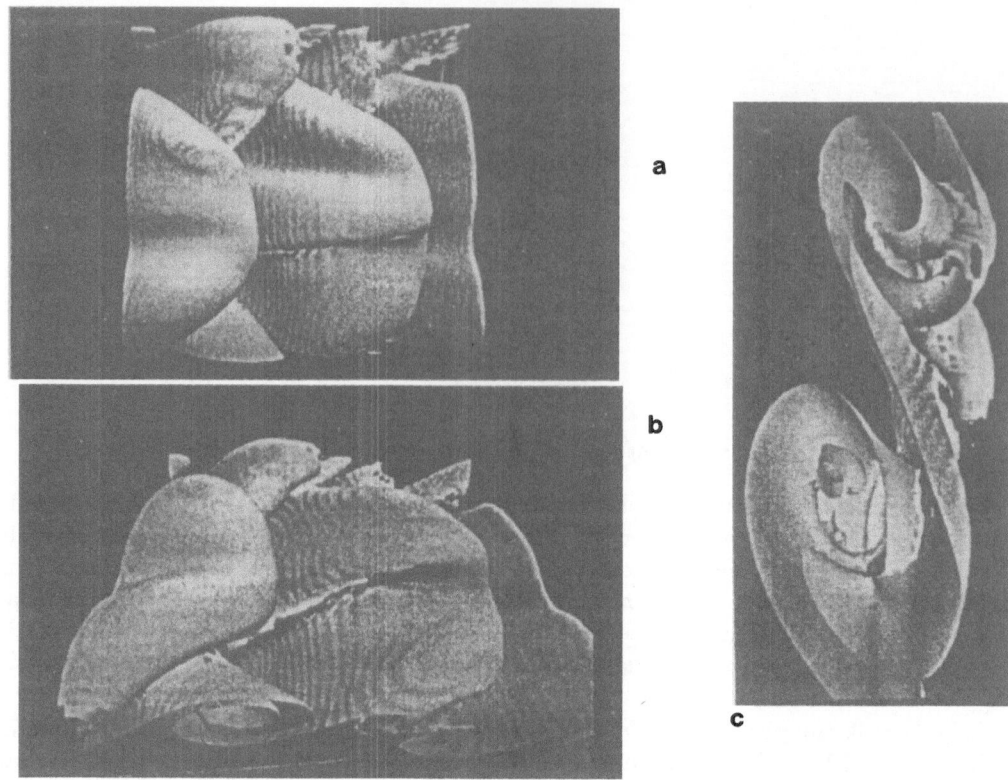

Figure 4. interface of the mixing layer for the same calculation as in Figure 2; a) top view, b) perspective, c) side view in the x, y plane.

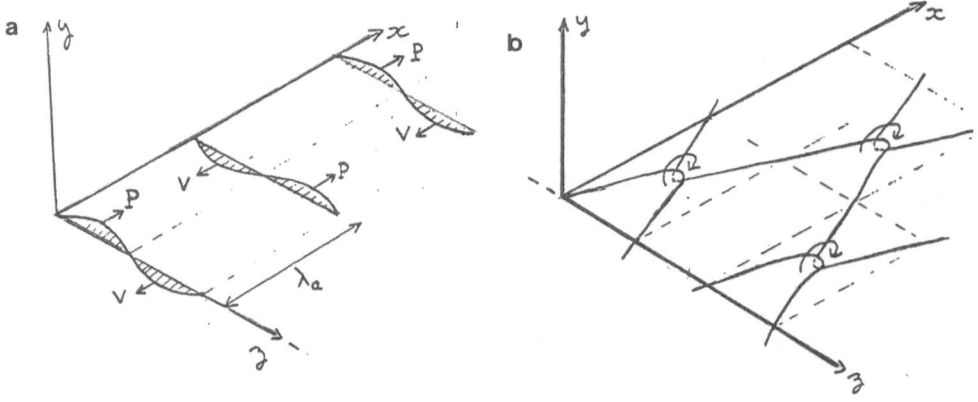

Figure 5. schematic representation of the helical-pairing instability, leading to a vortex-lattice structure.

late in vertical y, z planes and out of phase[5], as indicated on Figure 5a. Thus, the vorticity of the peaks P will be (to the lowest order) convected along the upper flow, while that of the valleys V will be convected by the lower flow. This will lead to pair-

---

[5] which corresponds to a spanwise sine modulation of the longitudinal subharmonic

ings as indicated on Figure 5b, resulting into a vortex-lattice structure which is highly three-dimensional. It seems that the same mechanism is at work in the calculations of Figure 2. To our knowledge, this is the first numerical evidence of the helical-pairing instability. The latter was named after Chandrsuda et al. (1978), who discovered this type of pairing in laboratory experiments with a high level of turbulence. Notice that a one-mode spanwise spectral expansion of the three-dimensional Navier-Stokes equations, of the form

$$\vec{u}(x, y, z, t) = \vec{u}_{2D}(x, y, t) + \vec{u}_{3D}(x, y, t) \sin(k_z z) \quad , \qquad (2-3)$$

where $\vec{u}_{2D}(x, y, t)$ and $\vec{u}_{3D}(x, y, t)$ are two non-divergent velocity fields parallel to the x, y plane, has been developed by Lesieur et al. (1988). It corresponds to a secondary-instability analysis, but without any linearization. When substituted into the equations of motion, the model yields two independant two-dimensional Navier-Stokes equations for $\vec{u}_1 = \vec{u}_{2D} + \vec{u}_{3D}$ and $\vec{u}_2 = \vec{u}_{2D} - \vec{u}_{3D}$. If $\vec{u}_{2D}$ is for instance a row of Kelvin-Helmholtz vortices aligned on the x-axis, and $\vec{u}_{3D}$ a small subharmonic perturbation[6], the two fields $\vec{u}_1$ and $\vec{u}_2$ will be out of phase, as far as their subharmonic perturbation is concerned. Hence, they will decorrelate very quickly during the pairing, in such a way that $\vec{u}_{3D} = (\vec{u}_1 - \vec{u}_2)/2$ will grow, corresponding to three-dimensionality growth. As far as one accepts the spanwise spectral truncation, our analysis generalizes the helical-pairing instability. There is no spanwise preferred mode, but in Pierrehumbert and Widnall's (1982) analysis, the maximum amplification rates depend very weakly upon k_z.

Afterwards, we have redone the calculation with a quasi two-dimensional perturbation, taking $\epsilon_{2D} = 10^{-4}$ and $\epsilon_{3D} = 10^{-5}$. Figure 6a shows, at the same time as in Figure 2, a perspective of the vorticity field: it is clear that the 4 initial Kelvin-Helmholtz billows have undergone a quasi two-dimensional pairing: the two remaining billows oscillate slightly in phase, as shown on Figure 6b, while hairpin vortex filaments are stretched between the primary vortices. This is reminiscent of the translative instability discussed above, and the associated vortex stretching mechanisms.

It seems thus that the two types of instabilities (translative or helical-pairing) may arise in a temporal mixing layer with a large number of initial fundamental vortices[7], if the initial perturbation is random (with a wide spectrum): for a quasi two-dimensional perturbation, two-dimensional rollers will form and pair, while being submitted to the translative instability. For a three-dimensional isotropic calculation, the helical-pairing instability is excited. It is feasible that, for a perturbation having the same amount of energy in two and three-dimensions, the flow would remain quasi two-dimensional, since it has been shown by Pierrehumbert and Widnall (1982) that the two-dimensional pairing (resulting from a two-dimensional subharmonic perturbation) is more amplified than the helical pairing.

3 Spatially-growing mixing layers

3.1 Flow behind a backwards-facing step

We have used a finite-volume numerical simulation code (TRIO, developed at the

---

perturbation

[6] This is the initial configuration leading to the helical-pairing instability considered above.

[7] Calculations with two initial vortices done by Comte (1989) have never displayed the helical-pairing instability, whatever the nature of the initial perturbation.

Commissariat à l'Energie Atomique, Grenoble), to study the flow in a channel behind a backwards-facing step. The flow is not forced upstream, and it is possible that it develops from pressure perturbations caused by the outflow boundary condition (where the pressure is taken uniform). Figure 7 shows the vorticity field in a two-dimensional calculation, indicating clearly the formation of coherent structures in the mixing layer behind the step.

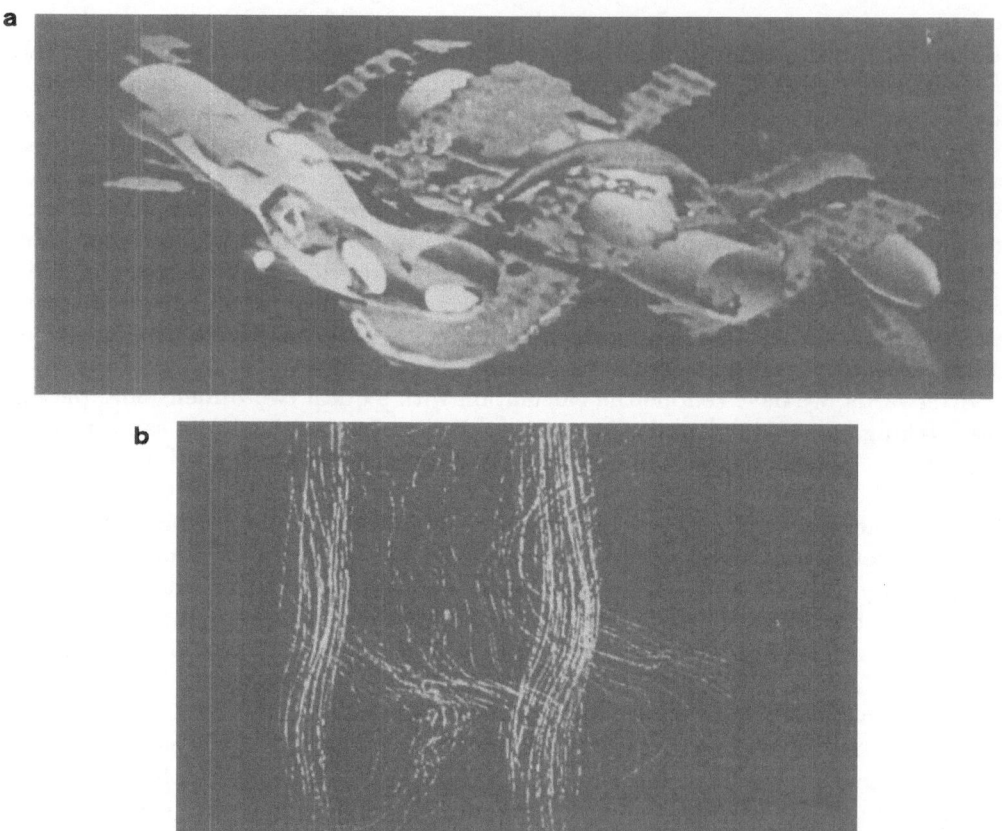

Figure 6. temporal mixing layer forced by a quasi two-dimensional perturbation at $t = 30\ \delta_i/U$; a) perspective of the vorticity field, b) top view of the vortex lines.

Figure 8 shows the vorticity field in a three-dimensional simulation: quasi two-dimensional billows are shed behind the step: as in the above quasi two-dimensional periodic mixing layer, weak longitudinal vortices of alternate sign are stretched, and seem to merge downstream, where the big rollers pair. This is in agreement with the experimental measurements of Huang and Ho (1990) in a mixing layer behind a splitter plate.

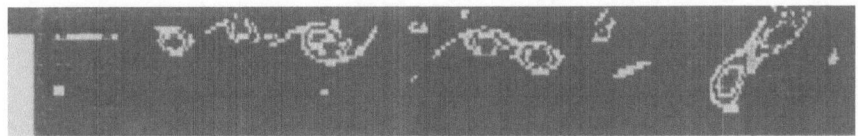

Figure 7. vorticity field in a mixing layer behind a backwards-facing step; two-dimensional simulation.

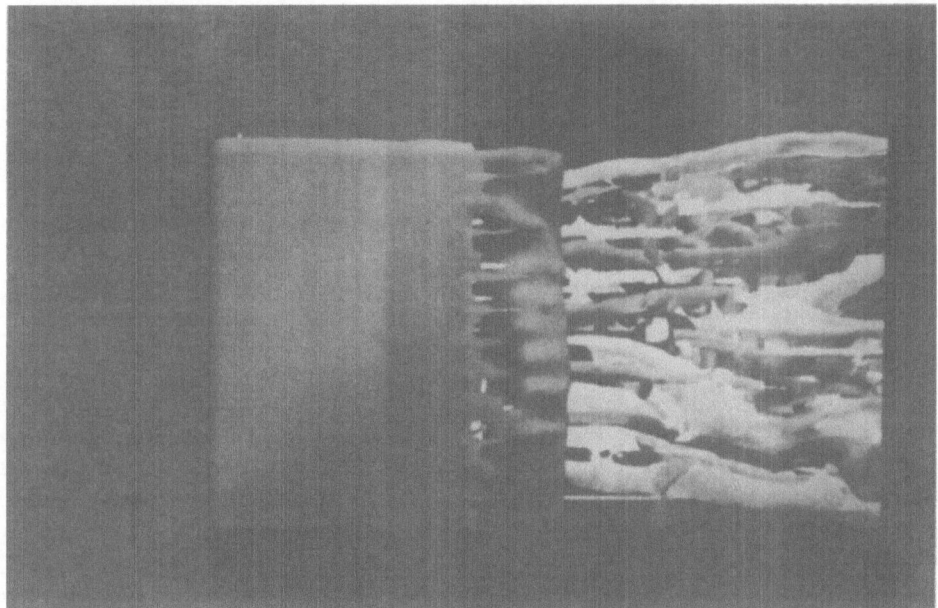

Figure 8. vorticity field in the three-dimensional numerical simulation of the flow behind a backwards-facing step.

3.2 Spatially-growing mixing layer

In the case of a slightly-compressible spatially-growing mixing layer, developing downstream of a hyperbolic-tangent velocity profile, and forced by a small three-dimensional random perturbation, the pressure plots show branchings of the Kelvin-Helmholtz billows, which might be due to the helical-pairing instability found in the temporal case. This is shown in Figure 9, representing the low-pressure contours. This is reminiscent of the vortex-method simulations presented by Fiedler (1990) in the same volume.

4 Conclusion and discussion

We have shown, using three-dimensional direct-numerical simulations, that a periodic mixing layer, developing from a hyperbolic-tangent velocity profile, could be subject to a violent three-dimensional instability leading to a vortex-lattice structure

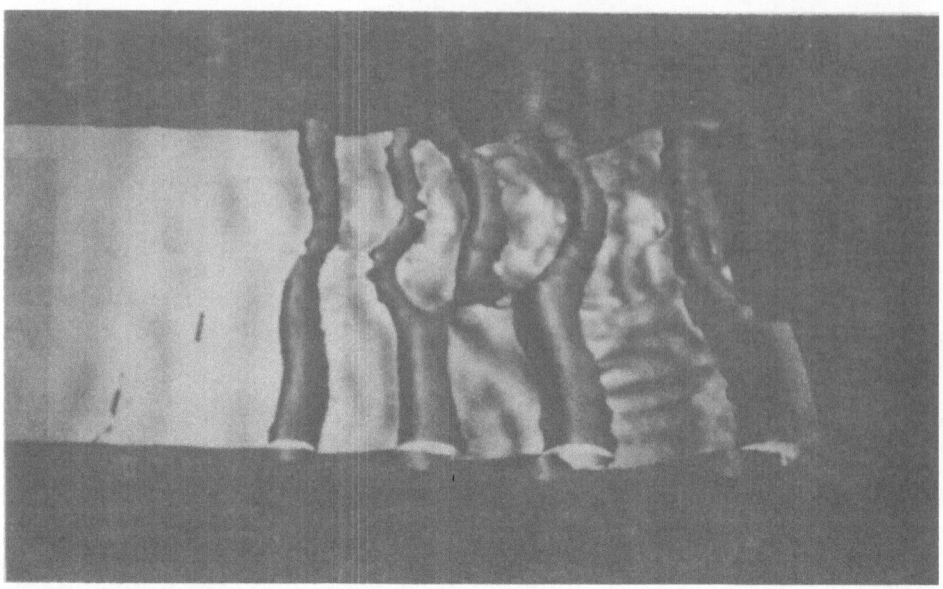

Figure 9. low pressure contours in the direct-numerical simulation of a slightly-compressible spatially growing mixing layer.

if the initial perturbation was random and three-dimensionally isotropic. This behaviour may be explained in terms of the helical-pairing instability, discovered by Pierrehumbert and Widnall (1982), and is triggered by a spanwise periodic oscillation of the subharmonic perturbation. As noted in the above-quoted paper, this could explain the apparent dislocations of the large billows found by Browand and Troutt (1980).

Calculations with a quasi two-dimensional random perturbation show on the contrary the development of the translative instability of the large billows, while thinner longitudinal vortices are stretched between the billows. In the spatially-growing case, numerical simulations of a mixing layer respectively unforced (behind a backwards-facing step) or forced randomly in the three dimensions, seem to offer an illustration of these two extreme situations. It seems therefore that the geometry of turbulence in a mixing layer may depend heavily upon the nature of the residual incoming turbulence.

Aknowledgements

This work has been supported by D.R.E.T. under contract 88/150, by CNES/Avions Marcel Dassault, by Commissariat à l'Energie Atomique (Grenoble), and by GDR-CNRS *Mécanique des Fluides Numérique*. Part of the calculations were done on a grant of the Centre de Calcul Vectoriel pour la Recherche.

References

Ashurst, W.T. and Meiburg, E., 1988, *Three-dimensional shear layers via vortex dynamics*, J. Fluid Mech., **189**, pp 87-116.

Bernal, L.P. and Roshko, A., 1986, *Streamwise vortex structure in plane mixing layers*, J. Fluid Mech., **170**, pp 499-525.

Breidenthal, R., 1981, *Structure in turbulent mixing layers and wakes using a chemical reaction*, J. Fluid Mech., **109**, pp 1-24.

Brown, G.L. and Roshko, A., 1974, *On density effects and large structure in two-dimensional mixing layers*, J. Fluid Mech., **64**, pp 775-816.

Chandrsuda, C., Mehta, R.D., Weir, A.D. and Bradshaw, P., 1978, *Effect of free-stream turbulence on large structures in turbulent mixing layers*, J. Fluid Mech., **85**, pp 693-704.

Comte, P., 1989, *Etude par simulation numérique de la transition à la turbulence en écoulement cisaillé libre*, Thèse de l' Institut National Polytechnique de Grenoble.

Corcos, G.M. and Lin, S.J., 1984, *The mixing layer: deterministic models of a turbulent flow. Part 2. The origin of the three-dimensional motion*, J. Fluid Mech., **139**, pp 67-95.

Huang, L.S. and Ho, C.M., 1990, *Small-scale transition in a plane mixing layer*, J. Fluid Mech., **210**, pp 475-500.

Lasheras, J.C. and Choi, H., 1988, *Three-dimensional instability of a plane free shear layer: an experimental study of the formation and evolution of streamwise vortices*, J. Fluid Mech., **189**, pp 53-86.

Lesieur, M., 1987, *Turbulence in Fluids*, Martinus Nijhoff Publishers; revised edition 1990, Kluwer Publishers, Dordrecht.

Lesieur, M., Staquet, C., Le Roy, P. and Comte, P., 1988, *The mixing layer and its coherence examined from the point of view of two-dimensional turbulence*, J. Fluid Mech., **192**, pp 511-534.

Metcalfe, R.W., Orszag, S.A., Brachet, M.E., Menon, S. and Riley, J., 1987, *Secondary instability of a temporally growing mixing layer*, J. Fluid Mech., **184**, pp 207-243.

Normand, X., 1990, *Transition à la turbulence dans les écoulements cisaillés libres ou pariétaux*, Thèse de l'Institut National Polytechnique de Grenoble.

Pierrehumbert, R.T. et Widnall, S.E., 1982, *The two and three-dimensional instabilities of the spatially periodic shear layer*, J. Fluid Mech., **114**, pp 59-82.

Sandham, N. D. and Reynolds, W.C., 1990, *Three-dimensional simulations of the compressible mixing layer*, J. Fluid Mech., to appear.

LARGE-EDDY SIMULATION OF TURBULENT SCALAR: THE INFLUENCE OF INTERMITTENCY

O. MÉTAIS

Institut de Mécanique de Grenoble*

Institut National Polytechnique de Grenoble and
Université Joseph Fourier, Grenoble
BP 53 X - 38041 Grenoble-Cedex, France

Abstract

We perform a spectral large-eddy simulation of decaying isotropic turbulence convecting a passive temperature at a resolution of 128^3 collocation points. The temperature spectrum tends to follow at large scale a k^{-1} range. The temperature variance and kinetic energy are found to decay respectively like $t^{-1.37}$ and $t^{-1.85}$ (when self-similar spectra have developed). The spectral eddy-viscosity and diffusivity are recalculated explicitly from the large-eddy simulation: although the eddy-viscosity exhibits a constant value at low wavenumbers, the eddy-diffusivity is found to decay logarithmically with increasing wavenumbers. The probability density functions for the velocity field, the temperature field and their derivatives show that the temperature is more intermittent than the velocity.

We then consider stably-stratified turbulence, within the frame of the Boussinesq approximation. The dynamical coupling between temperature and velocity fields greatly modifies the temperature behaviour. The level of temperature intermittency is found to be much lower than in the passive case.

Finally, we propose a generalization of the spectral eddy-viscosity to highly-intermittent situations in physical space, based upon a local second-order velocity structure function.

1 Introduction

In the atmosphere, the turbulence generated, for instance, by strong convective motions or wave-breaking is often submitted to the stabilizing effects of vertical temperature stratification. This is the case in the stratosphere or closer to the ground in the presence of an inversion profile. In the ocean, wave-breaking occuring at the thermocline level generates small-scale turbulence which then strongly influences the exchanges atmosphere-ocean. For neutral conditions, the temperature is no longer dynamically coupled to the turbulent velocity field: the prediction of the characteristics of this passive scalar diffusion is of prime importance in an environmental context.

---

* Unité associée CNRS

The Global Geometry of Turbulence
Edited by J. Jiménez, Plenum Press, New York, 1991

In order to relate results of numerical simulations with geophysical or experimental observations, very high Reynolds number flows need to be simulated. Even with the fastest existing computers, high Reynolds number turbulence cannot be computed explicitly at all scales. In many cases, the attention is directed towards the large scales, since they contain most of the energy of the flow. Thus, the small scales have to be modelled through a proper subgrid-scale parameterization.

When working in physical space with finite-difference methods: the large scales are defined with the aid of an adequate spatial filtering of the equations of motion, and the subgrid-scale terms are generally expressed in terms of eddy-coefficients (see, e.g., Rogallo and Moin, 1984, for a review). Another approach consists in working in Fourier space, which is possible only if the boundary conditions are simple enough (periodic or free-slip for instance). In this case, the large scales correspond to low wavenumbers, and the explicitly resolved scales can be defined by retaining the Fourier wave vectors $\underline{k}$ such that $k = |\underline{k}| \le k_c$, where k_c is the cutoff wavenumber: this corresponds to a sharp filter in Fourier space. The time-evolution of the modes within the interior of the sphere of radius k_c is coupled with that of the truncated modes lying outside of the sphere. Hence, a subgrid-scale parameterization is needed.

We present here the results of large eddy simulations (L.E.S) of three-dimensional decaying turbulence, firstly isotropic and convecting a passive temperature and secondly stably-stratified. We focus on the behaviour of the turbulent scalar. The concepts of spectral eddy-viscosity and eddy-diffusivity are used to parameterize the subgrid-scale terms (Kraichnan, 1966, 1968). The calculations are performed at high resolution (128^3 collocation points) in order to have a precise description of the large scales. High resolution calculations of isotropic turbulence with passive scalar done by Lesieur and Rogallo (1989), Lesieur et al. (1989) and Métais and Lesieur (1989) showed several "anomalous" characteristics of the temperature, with respect to the predictions of statistical the ories of turbulence like the Eddy-Damped Quasi-Normal Markovian theory (E.D.Q.N.M. , see e.g. Lesieur, 1987, for details). The aim of the present paper is to:

a) present and discuss the anomalous behaviour of the passive temperature, on the basis of a 128^3 L.E.S. calculation with initial spectra peaking at $k_i = 20$. We will in particular examine the probability density functions of the velocity and the temperature.

b) perform a stratified decay calculations with the same code, in order to investigate whether the above passive scalar results persist when the temperature is dynamically coupled with the velocity.

2 Three-dimensional isotropic turbulence

2.1 Methodology

The velocity fluctuation $\hat{\underline{u}}(\underline{k}, t)$ and the passive temperature fluctuation $\hat{T}(\underline{k}, t)$, respective Fourier transforms of $\underline{u}(\underline{x}, t)$ and $T(\underline{x}, t)$, satisfy the L.E.S. modified spectral Navier-Stokes equations. Using classical pseudo-spectral methods developed by Orszag, the equations are written:

$$\left(\partial/\partial t + (\nu + \nu_t(k|k_c)) \ k^2 \right) \hat{\underline{u}} = \Pi[F[F^{-1}(\hat{\underline{u}}) \times F^{-1}(\hat{\underline{\omega}})]] \qquad (2-1)$$

$$\left(\partial/\partial t + (\kappa + \kappa_t(k|k_c)) \ k^2 \right) \hat{T} = -i\underline{k}.F[F^{-1}(\hat{T})F^{-1}(\hat{\underline{u}})] \qquad (2-2)$$

$$\underline{k}.\hat{\underline{u}}(\underline{k}, t) = 0 \qquad (2-3)$$

where F stands for the discrete Fourier transform operator, Π is the projector on the plane perpendicular to $\underline{k}$, and $\hat{\underline{\omega}} = i\underline{k} \times \hat{\underline{u}}$ is the vorticity in Fourier space. ν and κ are the molecular viscosity and conductivity.

$\nu_t(k|k_c)$ and $\kappa_t(k|k_c)$ are respectively the spectral eddy-viscosity and diffusivity coefficients. Their formulation has been derived by Kraichnan (1976), Chollet and Lesieur (1981) and Chollet (1985) from two-point closure techniques. For k_c located in infinite (or very wide) Kolmogorov inertial, Chollet and Lesieur normalize the eddy-viscosity by $[E(k_c)/k_c]^{1/2}$, where $E(k)$ is the kinetic energy spectrum at wavenumber k. This yields:

$$\nu_t(k|k_c, t) = \nu_t^+(k/k_c)[\frac{E(k_c, t)}{k_c}]^{1/2} \quad , \qquad (2-4)$$

where $\nu_t^+(k/k_c)$ can be approximately expressed as (Chollet, 1985):

$$\nu_t^+(k/k_c) = 0.267 + 9.21 e^{-3.03(k_c/k)} \quad . \qquad (2-5)$$

Furthermore, in very wide Corrsin-Oboukov inertial-convective ranges, Chollet found for $\kappa_t(k|k_c)$ expressions analogous to (2-4), that is

$$\kappa_t(k|k_c, t) = \kappa_t^+(k/k_c)[\frac{E(k_c, t)}{k_c}]^{1/2} \quad , \qquad (2-6)$$

with the non-dimensional eddy-diffusivity $\kappa_t^+(k/k_c)$ nearly proportional to the eddy-viscosity, with a turbulent Prandtl number $\nu_t(k|k_c)/\kappa_t(k|k_c)$ close to a constant of the order of 0.6. In the large-eddy simulation, the molecular coefficients are negligible in comparison with the eddy-coefficients: it is in this sense that we will refer to infinite Reynolds number large-eddy simulations.

The initial kinetic and temperature spectra are identical and equal to:

$$E(k, 0) = E_T(k, 0) = A k^8 e^{-4[k/k_i(0)]^2} \qquad (2-7)$$

and peak at $k = k_i(0) = 20$. Calculations are carried out with 128^3 wave vectors, and the cutoff wave number is $k_c = 60$. A is a dimensional constant chosen such that the total kinetic energy $(1/2)v_0^2 = \int E(k, 0) dk$ and the total scalar energy ("potential energy") $(1/2) < T^2(\underline{x}, 0) > = \int E_T(k, 0) d\underline{k}$ are equal to 1.5.

2.2 Numerical results

We take $k_i(0) = 20$. In an initial phase, the temperature cascades faster than the velocity field towards small scales, in qualitative agreement with closures results described in Lesieur et al. (1987). Afterwards, the kinetic energy and the scalar energy respectively decay like $t^{-1.37}$ and $t^{-1.85}$. The agreement with the $EDQNM$ predictions ($t^{-1.38}$ for an initial kinetic energy spectrum $E(k, 0) \propto k^s$ when $k \to 0$, with $s \geq 4$) is excellent for the kinetic energy. However, the statistical theory yields a very different law for the decay of the passive scalar energy ($t^{-1.48}$).

Figure 1 shows, at $t = 60/v_0 k_i(0)$, the kinetic energy and temperature spectra. It is obvious that the potential energy has decayed much faster than the kinetic energy. The temperature peak, initially equal to k_i, has migrated much faster towards low wavenumbers. The kinetic energy spectrum at the cutoff is close to $k^{-5/3}$, with a Kolmogorov constant, measured from the compensated spectrum $k^{5/3}E(k)$, of the order of 1.5. However, the kinetic energy seems somewhat constrained by k_c, resulting in a slope close to k^{-2} in the vicinity of k_c. The temperature spectrum agrees quite well with Corrsin-Oboukov's law $E_T(k) \sim \eta\, \epsilon^{-1/3}\, k^{-5/3}$ at the cutoff (where η is the temperature dissipation rate), with a constant equal to 0.9. But the most striking feature of this spectrum is the formation, for $k < 30$, of a range close to a k^{-1} power law. Lesieur and Rogallo (1989) propose that this spectral behaviour is due to a temperature diffusion controlled by the shearing due to velocity gradients at scales $\sim k_i^{-1}$. This yields for the scalar variance spectrum:

$$E_T(k,t) = C_T\, \eta\, \frac{<\underline{u}^2>}{\epsilon}\, k^{-1} \quad , \qquad (2-8)$$

where C_T is a constant. When renormalized by (3-7), the whole time-evolution of the temperature spectrum eventually nicely collapses on a plateau of nearly one decade, with a value of the constant $C_T \approx 0.1$.

Using a fictitious cutoff wave number $k'_c = k_c/2 = 30$, it is possible to evaluate directly from the numerical simulation eddy-viscosity and diffusivity coefficients (Domaradzki et $al.$, 1987; Lesieur and Rogallo, 1989; Lesieur et $al.$, 1989).

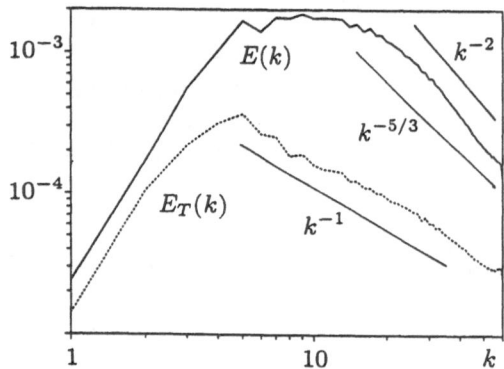

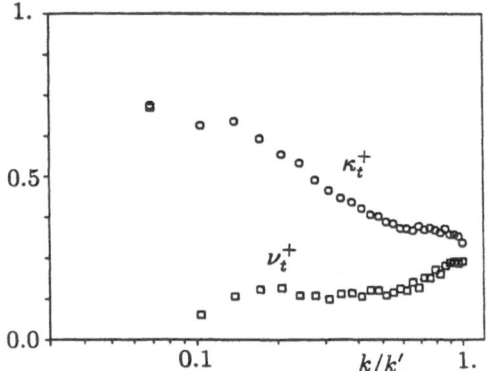

Figure 1. Kinetic energy spectrum $(E(k))$ and passive temperature spectrum $(E_T(k))$ at $t = 60/v_0 k_i(0)$, large-eddy simulation.

Figure 2. Spectral eddy-viscosity ν_t^+ and eddy-diffusivity κ_t^+ (see 2-4 and 2-6) corresponding to figure 1.

Figure 2 shows, at the same time as in Figure 1, the normalized coefficients ν_t^+ and κ_t^+ (see 2-4 and 2-6) as functions of k/k'_c. The eddy-viscosity does display a plateau of intensity 0.15 far from k'_c. It can be shown (Métais and Lesieur, 1990) that this plateau has to be multiplied by 1.66 in order to take into account the contribution from the wavenumbers larger than k_c: the corrected intensity 0.25 is close to the E.D.Q.N.M. value (0.267). The eddy-diffusivity, on the contrary, has no plateau, and decays logarithmically with k in the range where ν_t^+ is constant. In this range, the eddy-Prandtl number increases from 0.2 to 0.8.

2.3 Discussion

The *anomalous* characteristics displayed by the passive scalar have to be considered cautiously since they arise from numerical simulations where the subgrid-scales are modelled. However, direct numerical simulations, with $k_i(0) = 8$ (Prandtl number

= 1), exhibit the same trends (see; Métais and Lesieur, 1990, for details). Furthermore, moderate Reynolds number experiments by Yeh and Van Atta (1973) and by Warhaft and Lumley (1978) show a temperature spectral range of slope $-3/2$, although no inertial range is apparent for the corresponding kinetic energy spectrum. In high Reynolds number E.D.Q.N.M. calculations, the temperature spectrum also displays a well-defined $k^{-3/2}$ slope before establishing a $k^{-5/3}$ inertial-convective range and decaying self-similarly (Herring *et al.*, 1982; Lesieur *et al.*, 1987). Furthermore, the Test Field Model (T.F.M.) closure calculations of Herring (1990) indicate a fairly good agreement with the L.E.S. of Lesieur and Rogallo (1989) except for a steeper temperature slope given by the closure. This suggests that the *anomalous* temperature slopes observed in the experiments and in the closure calculations are the analogue of the k^{-1} temperature slope observed in the present simulations, and could be attributable to the shearing of the scalar by the large-scale velocity gradients. However, as indicated by the statistical theory results, this spectral regime in the large scales might only be transient and the temperature spectrum could eventually return to a self-similar decaying state. In this transient phase, the fast decay of the temperature variance of the explicit scales may just be the signature of a fast scalar transfer towards subgrid scales.

Atmospheric and oceanic measurements of small-scale turbulence seem to confirm the universal nature of the velocity spectrum in the Kolmogorov inertial range. However, they suggest that the Corrsin-Oboukov theory does not provide a universal description of temperature fluctuations in water nor in air (see, e.g., Williams and Paulson, 1977; Mestayer, 1982; Gargett, 1985). One hypothesis to explain this apparent contradiction with the theory could be the degree of intermittency in the velocity and the scalar fields. The study of intermittency effects on the scalar spectrum was first investigated by Kraichnan (1968) and pursued by Van Atta (1971, 1973). Information on the level of the field intermittency is given by the probability density functions (pdf). Therefore, we examine pdf $\mathcal{P}(X)$ for both velocity and scalar fields and their derivatives at the end of the direct simulation ($t = 17/v_0 k_i(0)$, $k_i(0) = 8$). We introduce the skewness (S) and flatness factors (F) of the distribution f: $S_f = \langle f^3 \rangle / \langle f^2 \rangle^{\frac{3}{2}}$, $F_f = \langle f^4 \rangle / \langle f^2 \rangle^2$; x, y, z are the components of $\underline{x}$ in a right-handed Cartesian coordinate system, u, v, w are the corresponding components of $\underline{u}$. All pdf are normalized to have a variance equal to one and are compared to the corresponding Gaussian distribution. As in grid turbulence experiments (Batchelor, 1953), figure 3a shows that $\mathcal{P}(X)$, $X = u$ is close to gaussian ($S_u \approx 0; F_u \approx 3$). However, T exhibits near exponential ranges in the wings of its distribution (figure 3b): the scalar-variance flatness is 4. This departure from Gaussianity for the scalar fluctuation pdf indicates a large-scale intermittency. Indeed, the temperature presents very strong fluctuations in small spatial regions, with quiescent regions being more likely the result of random sampling.

Small scale intermittency is illustrated by the distributions of the velocity and scalar derivatives ($X = \partial u / \partial z$ and $X = \partial T / \partial z$; figure 3c and 3d). The velocity derivative pdf seems to be close to an exponential in the tails. However, the dynamical model for the evolution of the pdf of the transverse velocity derivative developed by Kraichnan (1990) suggests that, in the tails, the function could have the following form: $|X|^{-1/2} \exp \left(-const |X| / < X^2 >^{1/2} \right)$. The departure from Gaussianity for the scalar derivative is even more pronounced: $F_{\partial T/\partial z} = 5.47$ when $F_{\partial u/\partial z} = 4.59$. Larger values for the scalar derivative flatness, as compared to the velocity derivative flatness, are in good agreement with experimental values (Antonia *et al.*, 1978; Sreenivasan *et al.*, 1980; Antonia and Chambers, 1980). The weaker velocity intermittency could be

159

attributable to the nonlocal effect of the pressure which tends to redistribute the velocity fluctuations. In figures 3a, 3b, 3c, 3d are also plotted the pdf of $u, \partial u/\partial z, T, \partial T/\partial z$ as given by the large eddy simulation ($k_i(0) = 20$). As compared to the ones given by the direct numerical simulation, the distribution is less intermittent since only the largest scales of the flow are simulated.

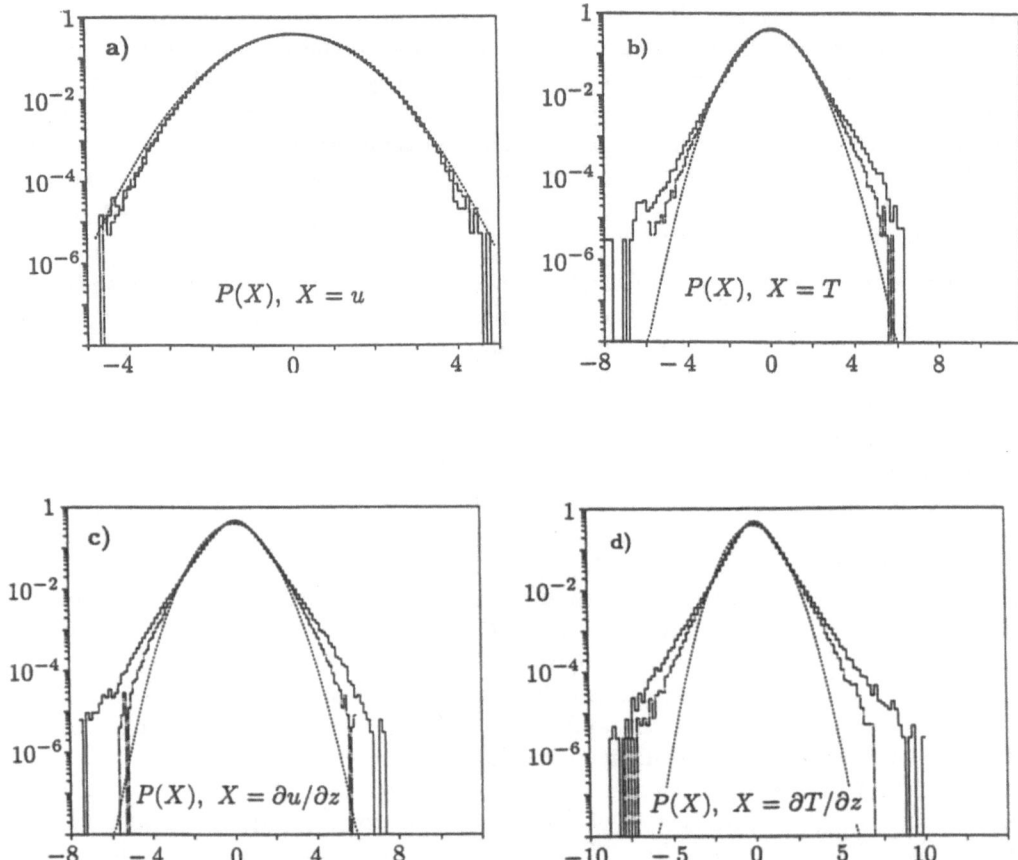

Figure 3. Probability density function $P(X)$: a) $X = u$; b) $X = T$; c) $X = \partial u/\partial z$ d); $X = \partial T/\partial z$. Large eddy-simulation: _ . _, direct simulation _____. The dashed line corresponds to a gaussian distribution.

3 Stably-Stratified Turbulence

The scalar, which is still called the temperature (but it may be the potential temperature for an ideal gas) is no longer passive but now coupled with the velocity field. The velocity and temperature fields satisfy the LES-modified Boussinesq set:

$$(\partial/\partial t + (\nu + \nu_t(k|k_c))\ k^2)\ \hat{\underline{u}} = \Pi\ [F[F^{-1}(\hat{\underline{u}}) \times F^{-1}(\hat{\underline{\omega}})] + \underline{z}\ \hat{T}] \qquad (3-1)$$

$$(\partial/\partial t + (\kappa + \kappa_t(k|k_c))\ k^2)\ \hat{T} = -N^2\hat{w} - i\underline{k}.F[F^{-1}(\hat{T})F^{-1}(\hat{\underline{u}})] \qquad (3-2)$$

$$\underline{k}.\hat{\underline{u}}(\underline{k}, t) = 0 \qquad (3-3)$$

$\underline{z}$ is directed towards the vertical, and w is the vertical velocity. T is proportional to the temperature deviation θ' from the mean temperature profile $\bar{\theta}(z)$; $T = g\theta'/\theta_0$, with θ_0 the volume-averaged value of $\bar{\theta}(z)$ and g the acceleration due to gravity (gravity vector: $\underline{g} = (0, 0, -g)$). N is the (constant) Brunt-Väisälä frequency:

$$N = \left(\frac{g}{\theta_0}\frac{d\bar{\theta}}{dz}\right)^{1/2} \qquad (3-4)$$

We have assumed in eqs. (3-1) and (3-2) that the isotropic subgrid-scale modelling of momentum and temperature is still valid in the stratified case, provided that turbulence should be isotropic for $k > k_c$.

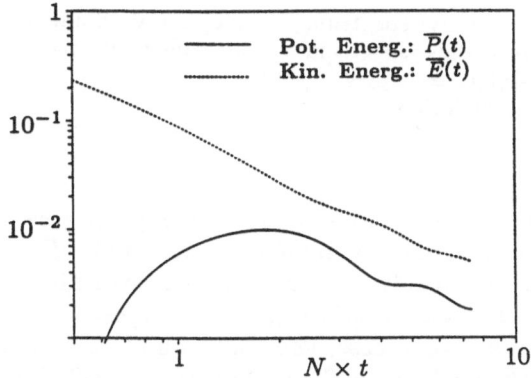

Figure 4. Time evolution of the kinetic energy ($----$), and potential energy (_____). Stratified turbulence; the time is normalized by N^{-1} (N Brünt-Väissäla frequency).

Starting with the same initial velocity spectrum as in the isotropic case, we let the turbulence evolve until it reaches a fully developed state. Then, at this time t_0 ($t_0 = 17/v_0 k_i(0)$) , the stratification is turned on. N is chosen such that the initial inertial effects are dominant over the stratification effects (large Froude number regime): $N = 1.047$, $F_r(t_0) = 3.7$. The initial temperature fluctuations are taken equal to zero and will build up due to the mean stratification. In the presence of stable stratification, the problem is complicated by the appearance of internal gravity waves. The velocity field has then to be decomposed along its turbulent and wave parts (Riley *et al.*, 1981; Métais and Herring, 1989). For brevity such decomposition is not considered here (see Métais and Lesieur (1990), for details). We focus on the influence on the temperature field of the dynamical coupling with the velocity.

Figure 4 shows the time evolution of the kinetic energy and potential energy (proportional to the square of the temperature fluctuations). The time is normalized by N^{-1}. Due to buoyancy flux, the potential energy grows from zero until it is of the same order as the kinetic energy. After this initial phase and contrary to the passive scalar case, it is then dissipated at approximately the same rate as the kinetic energy. Because of internal waves, kinetic and potential energy decays exhibit oscillations.

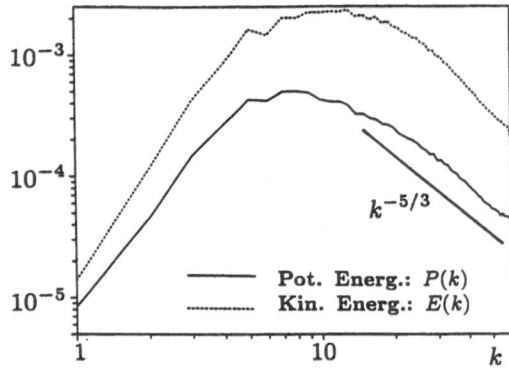

Figure 5. Kinetic energy spectrum ($E(k)$) and potential energy spectrum ($P(k)$) at $t = 43/v_0 k_i(0) = 1.23 N^{-1}$, large-eddy boussinesq simulation.

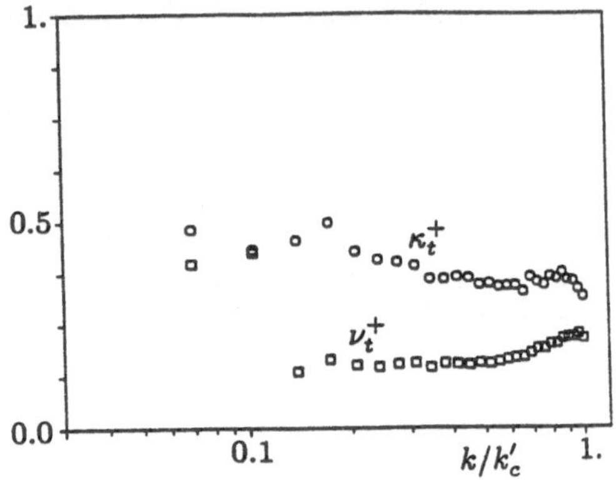

Figure 6. Same as figure 2 in the stably-stratified case (same time as figure 5).

We next consider the kinetic and potential energy spectra at $t = 1.23 N^{-1} = 43/v_0 k_i(0)$. At this stage, the Froude number is not small ($F_r = 1.3$) and the turbulence can be considered as weakly influenced by the stratification and its statistical properties unchanged. The temperature spectrum (figure 5) has lost the anomalous character it had in the isotropic case, and looks quite similar to the kinetic energy spectrum. Figure 6 is the analogue of figure 2 for the stratified case. The eddy viscosity does not differ from the isotropic eddy viscosity previously found. The normalized eddy diffusivity now displays a plateau in the small wavenumbers. The eddy-Prandtl number exhibits a near constant value of 0.45.

As opposed to the isotropic case, both velocity and temperature pdf ($\mathcal{P}(X), X = u$ and $X = T$) exhibit a Gaussian behaviour (see figure 7a). This indicates the disappearance of large scale temperature intermittency. Furthermore, the pdf for $X = \partial T/\partial z$ (given here for a direct Boussinesq simulation with $N = 1.047$, $Pr = 1$) exhibits a very asymmetrical shape with a skewness factor of 1.027 (figure 7b). One can propose the following explanation. Since we start with zero temperature fluctuations, the dominant term of the right hand side of (3-2) is initially the buoyancy term. Neglecting the diffusive effects, the equation of evolution of $T(\underline{x})$ in physical space simply writes:

$$\frac{\partial T(\underline{x}, t)}{\partial t} = -N^2 w(\underline{x}, t) \qquad (3-5)$$

Differentiating (3-5) with respect to z, we get

$$\frac{\partial}{\partial t}\left(\frac{\partial T}{\partial z}\right) = -N^2 \frac{\partial w}{\partial z} \qquad (3-6)$$

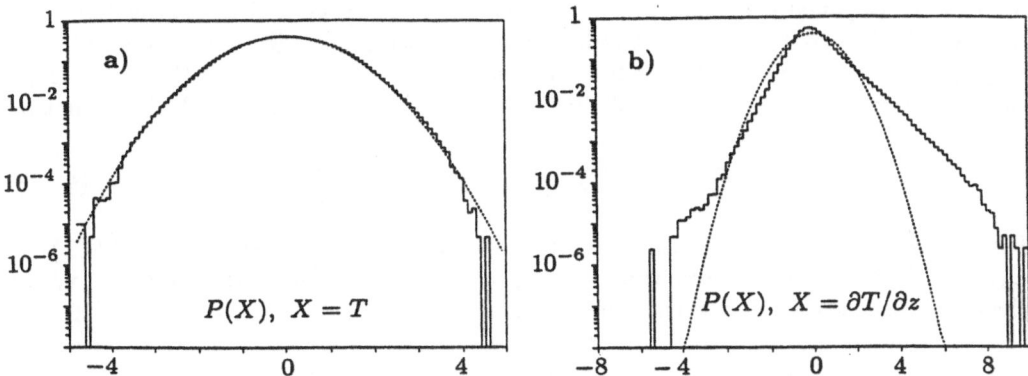

Figure 7. Probability density function $P(X)$: a) $X = T$; b) $X = \partial T/\partial z$. Stably-stratified direct simulation: $t = 1.23N^{-1}$. The dashed line corresponds to a gaussian distribution.

Furthermore, from (3-6) we get:

$$\frac{1}{3}\frac{\langle(\partial T/\partial z)^3\rangle}{\partial t} = -N^2\langle(\partial w/\partial z)(\partial T/\partial z)^2\rangle \qquad (3-7)$$

We have:

$$\langle(\partial w/\partial z)(\partial T/\partial z)^2\rangle = S_{Tw} \times \left[\langle(\partial w/\partial z)^2\rangle^{1/2}\langle(\partial T/\partial z)^2\rangle\right] \qquad (3-8)$$

where S_{Tw} is the mixed-derivative skewness. For isotropic turbulence, its negative is a measure of temperature-variance spectral transfer to large wavenumber (Herring and Kerr, 1982; Kerr, 1985): S_{Tw} is negative. Therefore, $\langle(\partial T/\partial z)^3\rangle$ will grow to positive finite value, and so will $S_{\partial T/\partial z}$. The skewness of the temperature fluctuation derivative in the direction of the mean temperature gradient was also found non zero in the laboratory experiment by Sreenivasan and Antonia (1977).

4 Eddy-viscosity in physical space

When considering complex boundary conditions, spectral methods are less adapted and the calculation has then to be performed in physical space using finite differences or finite volumes methods for instance. Let us now consider how the spectral eddy-viscosity given by (2-4) can be used in physical space. Neglecting ν_t^+ dependence on k, the eddy-viscosity (4-1) is replaced by:

$$\nu_t(t) = \nu_* \left[\frac{E(k_c,t)}{k_c}\right]^{1/2} \qquad (4-1)$$

ν_* is determined by balancing (in the inertial range) the subgrid-scale flux with the kinetic energy dissipation rate ϵ as proposed by Leslie and Quarini (1979): $\nu_* = 0.402$. An eddy-viscosity of this type can be used in physical-space large-eddy simulation equations (see e.g. Lilly (1967)) in so far as $E(k_c)$ can be determined. If it is estimated by averaging over the whole computational domain, the eddy-viscosity is then uniform in physical space. However, the eddy-viscosity should obviously take into account the intermittency and inhomogeneity of turbulence: in (4-1), the kinetic energy spectrum at the cut-off should be replaced by a local spectrum $E(k_c, \underline{x}, t)$. How can we determine such a spectrum? We introduce the second-order velocity structure function

$$F_2(\underline{r}, t) = \left\langle (\underline{u}(\underline{x} + \underline{r}, t) - \underline{u}(\underline{x}, t))^2 \right\rangle . \tag{4-2}$$

For isotropic turbulence with an infinite Kolmogorov inertial range ($E(k) = C_k \epsilon^{2/3} k^{-5/3}$), one can check that $F_2(r, t) = 4.82 \ C_k (\epsilon r)^{2/3}$. Taking $k_c = \pi/\Delta x$, the eddy viscosity (4-1) can then be written:

$$\nu_t(\Delta x, t) = 0.0396 \ \Delta x \ F_2(\Delta x, t)^{1/2} \tag{4-3}$$

This expression can now be estimated locally in physical space by averaging over the 6 points situated at a distance Δx from $\underline{x}$ (for a regular grid of mesh Δx). However, one can only determine the grid-scale structure function corresponding to the contribution of scales larger than Δx. It does not take into account the velocity fluctuations in the subgrid scales. Métais and Lesieur (1990) propose a correction allowing to derive the structure function of the whole signal from the filtered one ($\overline{F}_2$). With this correction, (4-3) becomes:

$$\nu_t(\underline{x}, t) = 0.063 \ \Delta x \ \overline{F}_2(\underline{x}, \Delta x, t)^{1/2} \tag{4-4}$$

Low resolution large-eddy simulations using this subgrid scales model have been performed by Métais and Lesieur (1990): as compared to the spectral edyy-viscosity previously mentioned, they yield a spectrum closer to the Kolmogorov spectrum. We are in the process of testing this local eddy-viscosity for non-homogeneous flows.

5 Conclusion

In the large-eddy simulations of decaying turbulence reported here, a convected passive scalar displays several characteristics, which are at variance with the predictions of classical phenomenological theories and of statistical theories of turbulence: the rapid decay of the passive scalar energy, the appearance of $\approx k^{-1}$ spectral range for the scalar at large scale, the strong rise of the eddy-conductivity with decreasing wavenumber. The coupling between temperature and velocity fields due to the presence of stratification greatly modifies the temperature behaviour. The lack of universality of the passive-scalar spectrum in the inertial-convective range and the strong differences with the stably-stratified case are confirmed by the Oceanic measurements of Gargett (1985). A possible explanation could come from the level of intermittency for the temperature field which is much higher when the latter is passive than when buoyancy forces are present. The differences between the pdf for the two cases could indicate régimes of *hard* and *soft* turbulence similar to those observed in Rayleigh-Bénard convection by Castaing *et al.* (1989).

Finally, we have proposed here a formulation of the eddy-viscosity in physical space directly derived from Chollet and Lesieur's (1981) spectral formulation and based upon the second order structure function. This space varying eddy-viscosity gives good results in the case of homogeneous isotropic turbulence. The first applications to the case of shear flows are very encouraging (see e.g. Normand; 1990).

Aknowledgements

This work has been supported by the D.R.E.T. under contract 87/238. The calculations were done on a grant of the Centre de Calcul Vectoriel pour la Recherche.

References

Antonia, R.A. & Chambers, A.J., 1980: On the correlation between turbulent velocity and temperature derivatives in the atmospheric surface layer. *Boundary-Layer Met.*, **18**, pp. 399-410.

Antonia, R.A., Chambers, A.J., Van Atta C.W., Friehe C.A. & Helland K.N., 1978: Skewness of temperature derivative in a heated grid flow. *Phys. Fluids*, **21** (3), pp. 509-510.

Batchelor, G.K. 1953 *The Theory of Homogeneous Turbulence*, Cambridge University Press., 197 pp.

Castaing, B., Gunaratne, G., Heslot, F., Kadanoff, L., Libchaber, A., Thomae, S., Wu, X.-Z.,Zaleski, S. & Zanetti, G., 1988: Scaling of hard thermal turbulence in Rayleigh-Bénard Convection. *J.Fluid Mech.*, **204**, pp. 1-30.

Chollet, J.-P. 1985 Two-point closure used for a sub-grid scale model in large eddy simulations. *Turbulent Shear Flows IV* (eds. L.J.S. Bardbury et al.), 62-72, Springer-Verlag.

Chollet, J.-P. & Lesieur M. 1981 Parameterization of small scales of three-dimensional isotropic turbulence utilizing spectral closures. *J. Atmos. Sci.* **38**, 2747-2757.

Domaradzki, J.A., Metcalfe, R.W., Rogallo R.S. & J.J. Riley 1987 Analysis of subgrid-scale eddy viscosity with the use of results from direct numerical simulations. *Phys. Rev. Lett.* **58**, pp 547-550.

Gargett, A.E. 1985: Evolution of scalar spectra with the decay of turbulence in a stratified fluid. *J. Fluid Mech.* **159**, pp. 379-407.

Herring, J.R. 1990 Comparison of closure to the Lesieur-Rogallo direct numerical simulations. *Phys. Fluids* **2**, pp. 979-983.

Herring, J.R. & Kerr, R.M., 1982: Comparison of direct numerical simulations with predictions of two-point closures for isotropic turbulence convecting a passive scalar. *J. Fluid Mech.*, **118**, pp. 205-219.

Kerr, R.M., 1985: Higher-order derivative correlations and the alignement of small-scale structures in isotropic numerical turbulence. *J. Fluid Mech.*, **153**, pp. 31-58

Kraichnan, R.H. 1966 Isotropic turbulence and inertial-range structure. *Phys. Fluids* **9** (9), pp. 1728-1752.

Kraichnan, R.H. 1968 Small-scale structure of a scalar field convected by turbulence. *Phys. Fluids* **11** (5), pp. 945-953.

Kraichnan, R.H. 1976 Eddy viscosity in two and three dimensions. *J. Atmos. Sci.* **33**, 1521-1536.

Kraichnan, R.H. 1990 Models of intermittency in hydrodynamic turbulence. *Submitted to Phys. Rev. Letters*.

Lesieur, M. 1987 *Turbulence in Fluids*, Nijhoff Publishers, Dordrecht, Boston, Lancaster, 286 pp; revised edition, 1990, Kluwer Publishers, Dordrecht.

Lesieur, M., Métais, O. & Rogallo, R.S., 1989 Etude de la diffusion turbulente par simulation des grandes échelles. *C.R. Acad. Sci. Paris*, **308**, Série II, pp. 1395-1400.

Lesieur, M., Montmory, C. & Chollet, J.-P. 1987 The decay of kinetic energy and temperature variance in three- dimensional isotropic turbulence. *Phys. Fluids* **30**, pp. 1278-1286.

Lesieur, M. & Rogallo, R.S. 1989 Large-eddy simulation of passive scalar diffusion in isotropic turbulence. *Physics of Fluids A* **1** (4), pp. 718-722.

Leslie, D.C. & Quarini, G.L., 1979: The application of turbulence theory to the formulation of subgrid modelling procedures. *J. Fluid Mech.*, **91**, pp. 65-91.

Lilly, D.K., 1967 The representation of small-scale turbulence in numericam experiments. *Proc. IBM Sci. Comput. Symp. Environ. Sci., IBM Data Process. Div., White Plains, NY*, pp. 195-210.

Métais, O. & Herring, J.R., 1989: Numerical simulations of freely evolving turbulence in stably stratified fluids. *J. Fluid Mech.*, **202**, pp. 117-148.

Métais, O. & Lesieur M. 1989 Large eddy simulations of isotropic and stably-stratified turbulence. *Advances in Turbulence 2* (eds. H.H. Fernholz and H.E. Fiedler), pp. 371-376, Springer-Verlag.

Métais, O. & Lesieur M. 1990 Spectral large-eddy simulation of isotropic and stably-stratified turbulence. *Submitted to J. Fluid Mech.*.

Mestayer, P. 1982 Local isotropy in a high-Reynolds-number turbulent boundary layer. *J. Fluid Mech.* **125**, pp. 475-503.

Normand, X. 1990 Transition à la turbulence dans les écoulements cisaillés compressibles libres ou pariétaux. *Thèse de l'Institut National Polytechnique de Grenoble.*

Riley, J. J., Metcalfe R.W. & Weissman, M.A., 1981: Direct numerical simulations of homogeneous turbulence in density stratified fluids. Proc. AIP Conf. *Nonlinear properties of internal waves.* Bruce J. West, Ed., 79-112.

Rogallo, R.S. & Moin, P. 1984 Numerical simulation of turbulent flows. *Ann. Rev. Fluid Mech.* **16**, 99-137.

Sreenivasan, K.R. & Antonia, R.A., 1977: Skewness of temperature derivatives in turbulent shear flows. *Phys. Fluids*, **20** (12), pp. 1986-1988.

Sreenivasan, K.R., Tavoularis, S., Henry, R. & Corrsin, S., 1980: Temperature fluctuations and scales in grid-generated turbulence. *J. Fluid Mech.*, **100**, pp. 597-621.

Van Atta, C.W. 1971 Influence of fluctuations in local dissipation rates on turbulent scalar characteristics in the inertial subrange. *Phys. Fluids* **14**, pp.1803-1804.

Van Atta, C.W. 1973 Erratum: Influence of fluctuations in local dissipation rates on turbulent scalar characteristics in the inertial subrange. *Phys. Fluids* **16**, p. 574.

Warhaft, Z. & Lumley, J.L. 1978 An experimental study of the decay of temperature fluctuations in grid-generated turbulence. *J. Fluid Mech.* **88**, pp. 659-684.

Williams, R.M. & Paulson, C.A. 1977 Microscale temperature and velocity spectra in the atmospheric boundary layer. *J. Fluid Mech.* **83**, pp. 547-567.

Yeh, T.T. & Van Atta, C.W., 1973, Spectral transfer of scalar and velocity fields in heated-grid turbulence. *J. Fluid Mech.*, **58**, pp. 233-261.

THERMAL CONVECTION AT HIGH RAYLEIGH NUMBERS IN TWO DIMENSIONAL SHEARED LAYERS

Stéphane Zaleski

Laboratoire de Physique Statistique,
C.N.R.S., E.N.S., 24 rue Lhomond 75231 Paris Cedex 05

1. Introduction

Rayleigh-Bénard convection is a system of great interest for the understanding of irregular fluid motion and turbulence. These systems are attractive in part because very high Rayleigh numbers may be reached in well-controlled laboratory conditions. There, the flow displays an interesting complexity together with some simple and yet unexplained features such as the scaling behavior of the heat flux (Castaing et al. 1989). New theories for this scaling have been recently proposed. They point to an interesting structure of the thermal boundary layers (Shraiman and Siggia, preprint). Thus, the role boundary layers play together with the large scale flow around the box demonstrates the importance of global geometry in thermal turbulence.

In order to help understand these features, numerical simulations of a model thermal boundary layer have been performed by the author. In section 2 a short review of the recent experimental and theoretical results on thermal turbulence is given with a special emphasis on the issue of the stability of the boundary layers. In section 3 results at moderate Rayleigh number with shear are presented for various wavelengths. In section 4 results at higher Rayleigh number are shown. At high enough Rayleigh number calculations without shear were also performed. These are made and interpreted from a different point of view: they display effects characteristic of turbulence in the entire box. In particular, the probability distributions of temperature are discussed.

2. Boundary layers in thermal turbulence experiments

Typically, we consider a cavity of height L heated from below with a temperature difference Δ accross it. The Rayleigh number is

$$Ra = \frac{g\alpha\Delta L^3}{\kappa\nu} \tag{1}$$

where g is the gravity acceleration, α the expansion coefficient, κ is the thermal diffusivity and ν the kinematic viscosity. Of particular interest is the heat transport accross the box. The Nusselt number is defined as the ratio of total heat flux H to the conductive heat flux. Thus we have:

$$N = H/(\kappa\Delta/L) \tag{2}$$

There are have been many experimental studies of the dependence of N on the Rayleigh

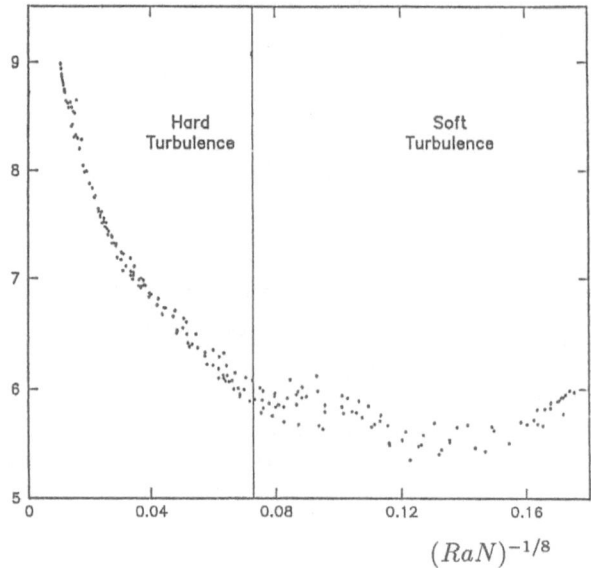

$$(RaN)^{-1/8}$$

Figure 1. The boundary layer Rayleigh number Rb in the experiment of Castaing et al. (1989) . The abcissa decreases with growing global Rayleigh number Ra.

number. The theory of Malkus (1954), which we shall describe in a simplified way below, predicts that the mean flow is marginally stable with respect to fluctuations. It results in the prediction of a relationship of the form

$$N \sim Ra^{1/3} \tag{3}$$

For instance the experiments of Goldstein and Tokuda (1980) in water yield exponents close to 1/3 for the $N - Ra$ correlation while experiments in air (Goldstein and Chu 1971) have produced smaller exponents. It has been suggested that remaining differences are due to higher order corrections such as those suggested by Long (1976). Recently however, experiments in helium gas by Libchaber and his coworkers (Castaing et al. 1989) have produced results which are clearly inconsistent with a 1/3 exponent.

To discuss the experimental results as well as the theory it is interesting to consider the thermal boundary layer near the top and bottom plates of the cell. It is likely that most of the temperature drop occurs accross a thin layer of thickness λ near the top and botttom walls. It is then possible to estimate the heat flux as:

$$H = \kappa \Delta / 2\lambda, \tag{4}$$

and the Nusselt number is:

$$N = L/2\lambda. \tag{5}$$

It is of course assumed that the motions of the fluid in the interior do not penetrate the boundary layer. Order of magnitude estimates by Kraichnan (1962) and Shraiman and Siggia (preprint) show that this will not happen at the Rayleigh numbers of most experiments. However, as argued by Howard (1966), if the boundary layer thickness λ is too large the buoyancy of the fluid may break it into plumes. To estimate when this will happen, one may construct a boundary layer Rayleigh number:

$$Rb = \frac{g\alpha\Delta\lambda^3}{2\kappa\nu} = \frac{Ra}{16N^3}. \tag{6}$$

It is likely that the boundary layer will become unstable when this number overcomes some critical number Rb_c. As the "top" of the boundary layer is not a rigid plate but another fluid region it is somewhat analogous to a free-slip boundary. Assuming that Rb_c is somewhere between the critical Rayleigh number for two rigid boundaries and the critical number for two free-slip bondaries we get:

$$Rb_c \sim 10^3. \tag{7}$$

(This probably still overestimates Rb_c since convection would overshoot above the top of the boundary layer). The boundary layer is then stable if:

$$Rb < Rb_c. \tag{8}$$

From (5), (6), (7) and (8) one gets:

$$N > \frac{1}{2^{1/3}20} Ra^{1/3}. \tag{9}$$

Marginal stability theory claims that the system realizes this lower bound. On the other hand, the experiments of Libchaber as well as the theories in Castaing et al. (1989) and Siggia and Shraiman (1990) suggest a different scaling:

$$N \sim Ra^{2/7}. \tag{10}$$

An interesting way to compare the experimental data to the predictions of stability theory (equation (8)) is to estimate Rb from the experiment. From (5), (6) one has:

$$Rb = Ra/(16N). \tag{11}$$

The values of Rb estimated using the above expression are shown on figure 1. It is seen that Rb grows without limit, but does not reach yet the estimated value of Rb_c. While there is a large uncertainty on the value of this upper bound for Rb, it is still possible that the 2/7 regime is a transient regime.

In order to predict what will happen to the boundary layer when Ra increases, it is necessary to revise the above stability theory. Indeed in discussing the stability of the boundary layer, we neglected the effect of the flow in the interior. However this flow is responsible for an average shear in the vicinity of the wall. In the absence of time dependence of the flow near the wall, one has for $z < \lambda$:

$$u(z) \simeq z/\tau \tag{12}$$

This will be true for instance if λ is much smaller than the thickness of a viscous sublayer in a wall jet. The presence of this shear modifies considerably the stability properties of the flow. Indeed it was known for a long time, since the experiments performed by Bénard and Avsec (1938) that a shear flow would damp rolls perpendicular to the direction of shear. For Couette flow, inviscid linear theory was investigated by Kuo (1963) and the finite viscosity case was done by Gallagher and Mercer (1965), and Ingersoll (1966). Linear stability theory was also investigated independently by the author (Castaing et al. 1989). It confirms the experimental results and shows that the critical Rayleigh number for the development of convection is raised as a function of the shear. On the other hand, recent experiments by Solomon and Gollub (1990) see no damping effect of shear. Since these experiments are performed in a turbulent cavity with an oscillating shear they are in a somewhat different category than the linear theory and the plane Couette flow experiments.

Let Re be the Reynolds number for the shear:

$$Re = \lambda^2 \tau / \nu \tag{13}$$

Then the Richardson number J is defined as:

$$\overline{J} = -Ra/(Pr Re^2) = -g\alpha\Delta\tau^2/\lambda \tag{14}$$

This number may be seen as the ratio of the shear frequency to the oscillation frequency of a stratified flow. At very high Re and Ra the inviscid theory of Kuo predicts that the Couette flow is unstable for $\overline{J} < 0.75$.

The nonlinear properties of a sheared flow heated from below are known only through experiment and numerical simulations. Lipps (1971) performed 2d numerical simulations up to $Ra = 35840$ and for various values of $\overline{J}$. They showed that the flow was indeed stable for large enough $\overline{J}$.

3. Numerical simulations at moderate Rayleigh numbers

In what follows the results of our numerical solution of the Boussinesq equations are described. Dimensionless units classical in the literature shall be used. In these units the equations read:

$$Pr^{-1}(\mathbf{u}_t + \mathbf{u}.\nabla\mathbf{u}) = \nabla^2\mathbf{u} - \nabla p + Ra\hat{\mathbf{z}}, \tag{15}$$

$$T_t + \mathbf{u}.\nabla T = \kappa\nabla^2 T, \tag{16}$$

$$\nabla.\mathbf{u} = 0. \tag{17}$$

The boundary conditions on the bottom plate are:

$$\mathbf{u} = 0, \qquad T = 0.5 \qquad \text{at} \qquad z = 0. \tag{18}$$

On the top plate:

$$\mathbf{u} = RePr \qquad T = -0.5 \qquad \text{at} \qquad z = 1. \tag{19}$$

where Re is an arbitrary Reynolds number. Horizontal boundary conditions are periodic with period l. This is intended to mimic a region of a boundary layer of horizontal extent much larger than its thickness. Initial conditions are of the form

$$T(x, z, 0) = a_1 sin(2\pi x/l)sin(\pi z) + z - 0.5, \tag{20}$$

$$\mathbf{u}(x, z, 0) = 0. \tag{21}$$

Thus the size of the box fixes the initial wave number $k = 2\pi/l$. It is interesting to investigate the dependency of the Nusselt number on wavelength. Thus we repeated all experiments with several values of the wavelength l. Finally, all experiments were made with $Pr = 0.7$, in order to be close to conditions in air and helium.

Numerical solution of the equations was obtained by a finite difference code which is described in the appendix. The code was checked by comparing the results for an increasing number of finite difference grid points. Define the number of points n in the boundary layer using definition (5) of the boundary layer thickness. At $R = 10^6$ the experiments with 64 points in the vertical have $n = 4$ points in the boundary layer and the error on N is on the order of 1%. In all numerical experiments care was taken to retain at least this level of precision. To determine the effect of shear, two series of runs were made for $Ra = 5\ 10^4$ and $Ra = 10^6$, both with various values of the wave number k. The results are summarized in figures 2 and 3. At small Reynolds number little effect on N is seen.

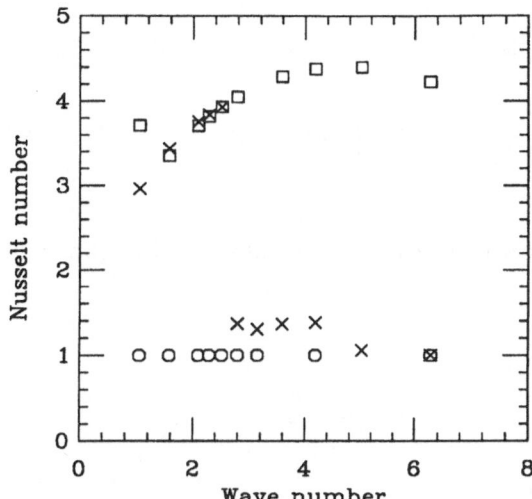

Figure 2. The Nusselt number is plotted on this figure against the wavenumber $k = 2\pi/l$ for a variety of numerical experiments. $Ra = 5 \cdot 10^4$ and $Pr = 0.7$ for all the experiments. $\square$: $Re = 0$, $\times$: $Re = 200$, $\circ$: $Re = 400$.

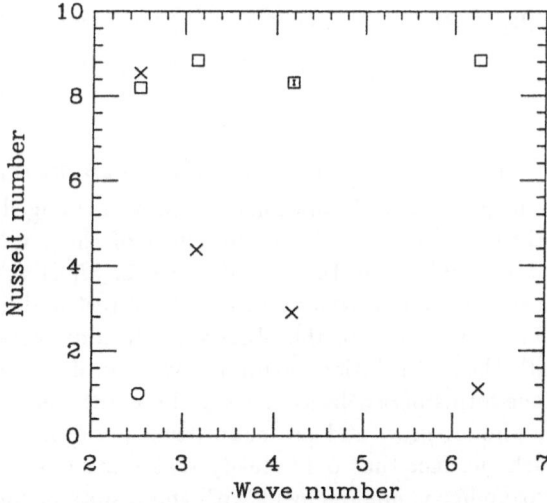

Figure 3. Same as figure 2 but for $Ra = 10^6$. $\square$: $Re = 0..$ $\times$: $Re = 894.$, $\circ$: $Re = 2000..$ Dynamical states are labelled as in figure 2, with also T: turbulent as described in text.

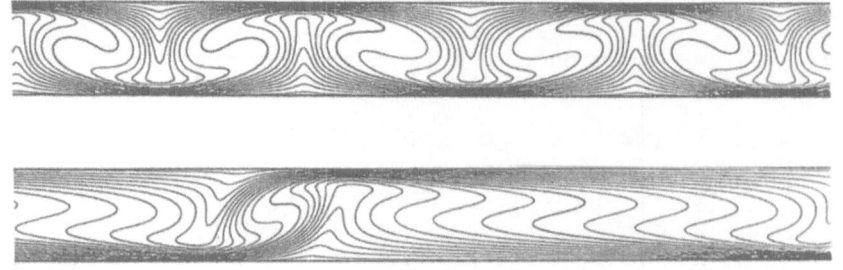

Figure 4. Solutions without shear (top) and with shear (bottom) ($Ra = 510^4$).

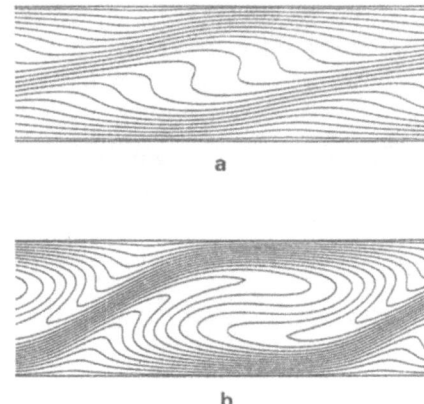

a

b

Figure 5. The temperature field at two instants, for $Ra = 510^4$, $Re = 200$. The plume is suppressed by the wind just after the state shown in (b) has been reached.

At large Re all solutions relax to a steady state. At intermediate Re the effect varies with k. The heat transport in long wavelength solutions is almost unchanged by shear. Typical temperature fields are shown in figure 4. The only effect of shear is to bind the rising and falling plumes into a pair which travels at speed $RePr/2$. Short wavelength solutions are most affected by shear: the system returns to a state of rest and heat is transported by conduction alone. The complexity of the observed solutions increases dramatically with Ra. At $Ra = 5.10^4$, sheared solutions exhibit a variety of dynamical states, from steady solutions to various forms of oscillations. Near the point were the travelling wave solution is destabilized, plumes grow and are suppressed periodically (figure 5). This should be contrasted with the fact that only steady states are observed with no shear. The dependence on wavenumber is in agreement with the results of Lipps, who observed that large wavelength solutions were selected in large boxes $l >> 1$.

4.1. Shear induced thermal turbulence at high Rayleigh numbers

At $Ra = 10^6$, convection between plates at rest shows only steady states and periodic

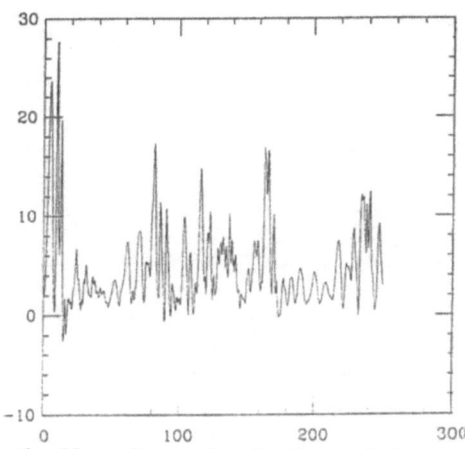

Figure 6. Time series for the Nusselt number in the turbulent state described in the text, at $Ra = 10^6$, $k = \pi$, $Re = 894$.

or quasiperiodic oscillations. This is similar to the observations of Lennie et al. (1988) who also found only regular oscillations. Chaotic states were indeed found by Goldhirsch et. al.(1989), but with rigid lateral walls, which seem to be responsible for recirculating flows in the corners and a more complex dynamics. When shear is added no surprising results are obtained for high Reynolds numbers or long wavelengths. Travelling pairs of falling and rising plumes are still observed. However, when shear is added at moderate wavelengths, a state of interesting dynamical complexity is seen. Figure 6 shows the Nusselt number time series for this state, while figures 7 and 8 show time series and snapshots of the temperature field. Clearly, this state is irregular both in space and time. This strongly suggests that the number of non linear degrees of freedom is large, although no measurement of this number has been made yet. Figure 7 shows plumes attached to the boundary layers that may stay quasi steady for a long time and burst intermittently towards the opposite plate. All fields on figure 7 have internal boundary layers as pronounced as the ones near the walls. Plume bursting occurs intermittently and is responsible for the spikes in figure 6. Temperature time series were taken in the two boundary layers and at mid-height, and at several points at each level (figure 8). Histograms of temperature values from these series yield an approximation to the probability distribution of temperature. These probability distributions are interesting from many points of view, not the least because they yield a clear signature of the transition from soft to hard turbulence in the Chicago experiment. Indeed the original motivation for the present work was to mimic the essential features of hard thermal turbulence. However histograms of the center probes yield no evidence of an exponential distribution (figure 9).

5. Turbulent states with no shear

At Rayleigh numbers of 10^7 with no shear ($Re = 0$) and an aspect ratio of two ($l = 2$), the flow configurations with no shear are seen to drastically change. The system does not reach a stationary solution. Instead a quasi-steady large scale flow with two counter-circulating eddies is observed, together with small amplitude fluctuations in the form of detached plumes rotating with the large scale flow. The detached plumes are small scale objects and are responsible for the signal of figure 10. This signal has jumps reminiscent

Figure 7. The temperature field for the turbulent state at three instants in time. Plumes are seen to burst on the top figure, while a more quiescent state is seen at the bottom. Same parameters as in figure 6, with 64 × 128 grid points.

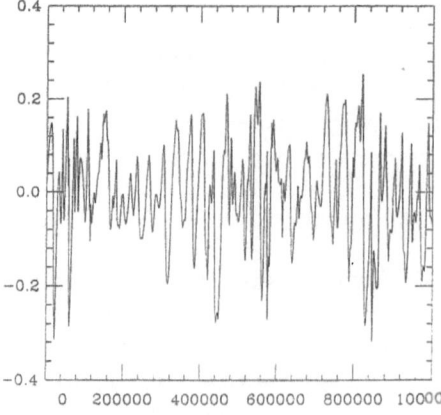

Figure 8. Temperature time series of the center probe in the turbulent state of figure 6 and 7.

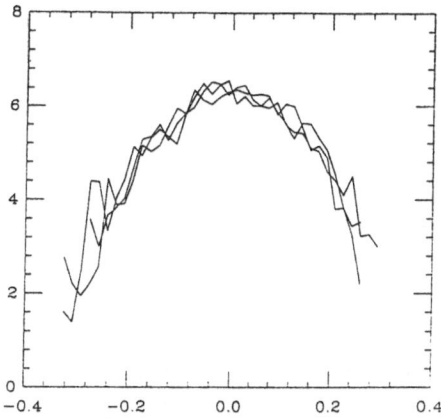

Figure 9. Temperature histogram for the turbulent state of figures 6-8.

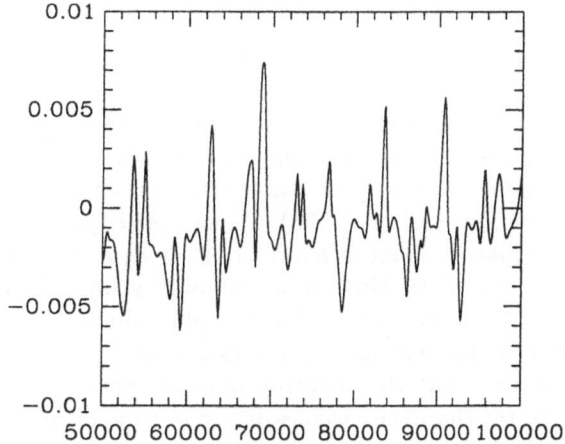

Figure 10. Temperature time series at $R = 10^7$ with no shear ($Re = 0$) and aspect ratio $l = 2$. Same time units as on figure 8.

of those seen in the hard thermal turbulence experiments. The measured distribution of temperature, although quite noisy, is somewhat reminiscent of an exponential (figure 11).

This state is not the only one observed at 10^7. Another state was found with an oscillating large scale flow in which the two main plumes vacillate between two oblique positions. It is likely that such states would not exist with rigid lateral walls.

The motivation for the experiments with no shear is somewhat different from those reported in the previous sections. It is meant to reproduce the results in the entire box of a large Rayleigh number experiment, and not only the boundary layer. It is interesting to notice that exponential distributions and even time dependence are not seen at $Ra = 10^6$ in contrast with the results obtained by De Luca et al. (1990) at $Pr = 7.$.

6. Conclusion

We have been able to show that in the range of Rayleigh numbers investigated, a shear flow could at a high enough Reynolds number suppress convection. In many instances were convection is not suppressed the Nusselt number is strongly reduced. At long wavelengths, however, the shear may have little effect and may even enhance slightly the heat flux. It

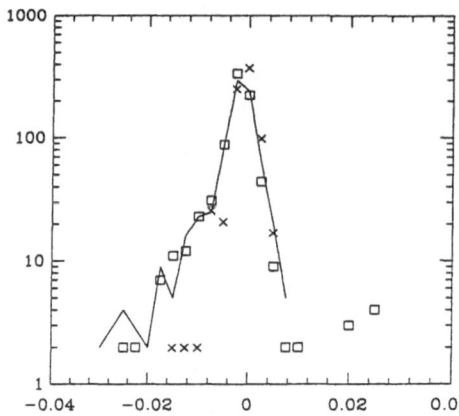

Figure 11. The histogram of temperatures for the time series at $R = 10^7$, in lin-log coordinates. The various □, × and the solid line correspond to 3 different probes.

is interesting to remark that the effect of shear, and specially the opposition between long and short wavelengths, is similar to the one that was observed in Spiegel and Zaleski (1984) for reaction diffusion equations. The stabilisation effect is apparently inconsistent with the results of Solomon and Gollub (1990). A possible explanation is that the latter experiments involve an oscillating shear in a large box. A shear that oscillates too rapidly will be faster than the characteristic time of the boundary layer λ^2/κ, and will have no significant effect. On the other hand a shear that would oscillate more slowly would create recirculation in the box, thus forcing directly a larger heat flux. To avoid this effect the shear must be fast enough so that the shearing motions do not penetrate the interior, which imposes a period faster than the viscous diffusion time accross the boundary layer λ^2/ν. Thus for Pr of order 1 the stabilisation effect seems difficult to observe through a forcing of the boundary layer.

The numerical experiments also confirm that the most interesting region in the Ra, Re, k parameter space is the one of marginal stability of convection with shear. In this region the flow shows complex features at relatively low Rayleigh numbers. The state observed numerically differs however from real experiments on hard turbulence in the probability distributions. It may still be of interest as a simple, 2d numerical model of instable flow with Gaussian statistics. Indeed, interesting effect such as internal boundary layers are seen. It is interesting to not that these layers are not responsible for the characteristic hard turbulence signal.

Experiment in the absence of shear, which attempt to model the entire convection experiment, yield realistic distributions. The transition to turbulence occurs at $R = 10^7$, close to the point where the transitions from soft to hard turbulence appears in real 3D flows.

Finally, it is clear that the modeling of the bottom region of a thermal turbulence experiment could be improved over the present Couette flow model. Instead of periodic boundaries in the horizontal specific inflow and outflow conditions could be used. The top of the model could be merged into a core region simulated numerically. The core region itself should be simulated using a turbulence model, since the number of grid points would otherwise be enormous.

Acknowledgments

The computational resources for this work were provided by the Scientific Committee of the Centre de Calcul Vectoriel pour la Recherche. The author thanks J. Gollub, T. Solomon, B. Shraiman, E. Siggia and J. Werne for interesting and stimulating conversations as well as for providing preprints of their work prior to publication.

Appendix

The numerical method is inspired by the artificial compressibility method of Chorin (1967). Let the field $\mathbf{u}$ have components (u_1, u_2). The Boussinesq equation may be written:

$$(\frac{\partial u_i}{\partial t} + u_j\frac{\partial u_i}{\partial x_j})Pr^{-1} = -\frac{1}{\rho}\frac{\partial p}{\partial x_i} + Ra\delta_{i2} + \frac{\partial^2 u_i}{\partial x_j \partial x_j}. \tag{22}$$

Summation over indices i, j is implied. The temperature equation is unchanged:

$$\frac{\partial T}{\partial t} + \frac{\partial u_i T}{\partial x_i} = \frac{\partial^2 T}{\partial x_j \partial x_j} \tag{23}$$

The principal numerical difficulty comes from the incompressibility equation. It is useful to replace it by

$$\frac{\partial \rho}{\partial t} = -\frac{\partial \rho u_i}{\partial x_i}, \tag{24}$$

where the pressure is given by

$$p = c^2\rho \tag{25}$$

where c is an adjustable sound speed in the new compressible equations. Let u_m be the maximum velocity at time t. The solutions converge to the incompressible solutions when the Mach number $M = u_m/c$ goes to 0. We use the artificial incompressibility method to find the time dependent solution whereas Chorin and other authors suggested its use for the search of time independent solutions only. The above equations are discretized by a finite difference method: velocity, temperature and pressure variables are represented on a staggered mesh as in the marker and cell method. Time derivatives are approximated by first order in time finite differences. Stability conditions for this method are derived in Peyret and Taylor (1983).

The method may seem to waste some computer time because of artificial compressibility, although this is not really the case. Indeed many more time steps might be needed than for a "truly incompressible" method, in which instead of (24) the condition

$$\nabla.\mathbf{u} = 0 \tag{26}$$

is satisfied exactly and the pressure is found by inverting the Poisson equation. In such a method a necessary condition fpr stability is $u_m\tau/h < 1$. However it is a generally true rule of thumb that in a first order in time method the time step should verify $v\tau/h < 1/5$. Thus one may reach Mach numbers $M \simeq 0.2$ without wasting computer time. On the other hand M should not be much larger, because the method would become inaccurate. M should not be smaller either because the first order method would become too expensive for the achieved accuracy. A truly incompressible, second order method would be more efficient. The present method is thus cost effective for medium accuracy calculations. The simulations at $R = 10^7$ with no shear cost about 30 mn of CPU time on a single processor of a CRAY2.

References

Bénard, H., and D. Avsec 1938: Travaux récents sur les tourbillons cellulaires. J. Phys. Radium **9**, 486-500 (1938).

Chorin A. J. 1967: A numerical Method fo Solving Incompressible Viscous Flow Problems J. Comp. Phys. , **2**, 12.

Castaing B., G. Gunaratne, F. Heslot, L. Kadanoff, A. Libchaber, S. Thomae, X.-Z. Wu, S. Zaleski and G. Zanetti 1989: Scaling of Hard Thermal Turbulence in Rayleigh Bénard Convection, J. Fluid Mech.**204** 1-30.

De Luca E.E., J. Werne, R. Rosner and F. Cattaneo 1990: Numerical Simulations of Soft and Hard Turbulence: Preliminary Results for Two Dimensional Convection, Phys. Rev. Lett. **64**, 2370-2373.

Domaradzki J.A. and R.W. Metcalfe, 1988: Direct Numerical Simulation of the Effects of Shear on Turbulent Rayleigh Bénard Convection, **193**, 499-531.

Gallagher, A.P. and A. McD. Mercer 1965: On the disturbances in plane Couette flow with a temperature gradient, Proc. Roy. Soc. London A 286, 117-128.

Goldhirsch I., R.B. Pelz and S.A. Orszag 1989: Numerical Simulations of thermal convection in a 2-d box, J. Fluid Mech. **199**, 1.

Goldstein R.J. and T.Y. Chu 1971: Turbulent convection in a horizontal layer of air, Prog. Heat Mass Transfer **2**, 55-75.

Goldstein R.J. and Tokuda S. 1980: Heat Transfer by Thermal Convection, Int J. Heat and Mass Transfer, **23**, 738-740.

Howard, L. N. 1966: Convection at high Rayleigh Number, in *Proceedings of the 11th Congr. Appl. Mech. Munich*, 1109-1115, Springer.

Ingersoll, A.P. 1966b: Convective instabilities in Plane Couette Flow, Phys. Fluids, **9**, 682-689.

Kraichnan R.H. 1962: Turbulent Thermal Convection at Arbitrary Prandtl Number, Phys. Fluids **5**, 1374-1389.

Kuo H.L. 1963: Perturbations of plane Couette Flow in Stratified Fluid and Origin of Cloud Streets, Phys. Fluids **6**, 195-211.

Lennie T.B., D.P. Mc Kenzie, D.R. Moore and N.O. Weiss 1988: The Breakdown of Steady Convection, J. Fluid Mech. **188**, 47-85.

Lipps F. B. 1971: Two-dimensional NUmerical Experiments iin Thermal Convection with Vertical Shear , J. Atm. Sci. 28, 3-19.

Long R.R. 1976: Relations Between Nusselt number and Rayleigh number in Turbulent Thermal Convection, J. Fluid Mech. **73**, 445-451.

Malkus, W.V.R. 1954a: Discrete Transitions in Turbulent Convection, Proc. Roy. Soc. (London) A225, 185-195.

Malkus, W.V.R. 1954b: The heat transport and spectrum of thermal turbulence, Proc. Roy. Soc. (London) A225, 196-212. Ingersoll, A. P. 1966: J. Fluid Mech. **25**, 209.

Peyret R. and T. D. Taylor 1983: *Computational Methods for Fluid Flow*, Springer.

Solomon T. H. and J.P. Gollub 1990: Sheared Boundary Layers in Turbulent Rayleigh Bénard Convection, Phys. Rev. Lett. **64**, 2382-2384.

Spiegel E. and S. Zaleski 1984: Shear Induced Instability in Reaction Diffusion Systems, Physics Letters 106 A, 335-338.

CLOSED FLOWS:

EXPERIMENTS

EFFECT OF NOISE ON BIFURCATIONS AND PATTERNS IN

DISSIPATIVE SYSTEMS

Guenter Ahlers

Department of Physics and
Center for Nonlinear Science
University of California
Santa Barbara CA 93106

Bifurcations and the evolution and nature of patterns in nonlinear dissipative systems are usually discussed in terms of deterministic equations of motion for the macroscopic variables [1] which neglect the "microscopic" degrees of freedom of the system. There are circumstances, however, under which the neglected degrees of freedom play an important role. Usually their influence can then be included in some average sense by the addition of a stochastic force to the deterministic equations,[2] for instance to the Navier-Stokes equation for Rayleigh-Bénard convection[1] (RBC) or to the hydrodynamic equations for electro-convection (EC) in a thin layer of a nematic liquid crystal[3] to which a voltage is applied. Simplified models with limited ranges of applicability usually have to be derived from these stochastic equations to make specific predictions feasible.[4-7]

Although it was recognized some time ago that stochastic forces associated with the microscopic degrees of freedom due to the thermal motion of the atoms or molecules of a system should in principle be taken into consideration,[2,8-12] it seemed that their effect would be unobservably small for macroscopic systems where typical dissipative energies are many orders of magnitude larger than the microscopic energy $k_B T$. Nonetheless, it has been possible recently to observe directly the fluctuations, which are the response of the system to the stochastic field, below the first bifurcation point in a hydrodynamic system, namely in EC in a nematic liquid crystal.[13] Figure 1 is a reproduction of a digitally processed shadowgraph image of EC at a control-parameter value $\epsilon = -10^{-3}$, where $\epsilon = V^2/V_c^2 - 1$ with V the amplitude of a time-periodic voltage applied to the system and V_c the critical amplitude. Particularly in the lower right and upper left portions, patches of weak convection rolls are observable. These patches are not permanent, but rather decay after a characteristic time and new ones grow again in a different location. Data like those in Fig. 1, taken repeatedly at constant time intervals, can be used to calculate spatial and temporal correlation functions. Spatial correlation functions at three values of ϵ are shown in Fig. 2.[13] They show the typical convective roll wavelength as fast oscillations, and the typical spatial extent of the fluctuating patches as a decay length of the envelope. Extrapolation of the envelope to zero spatial delay yields the mean square amplitude of the fluctuations. Figure 3 shows the fluctuation amplitudes at negative ϵ.[13] They are compared with two theoretical models. One of them, given as a dashed line, is

The Global Geometry of Turbulence
Edited by J. Jiménez, Plenum Press, New York, 1991

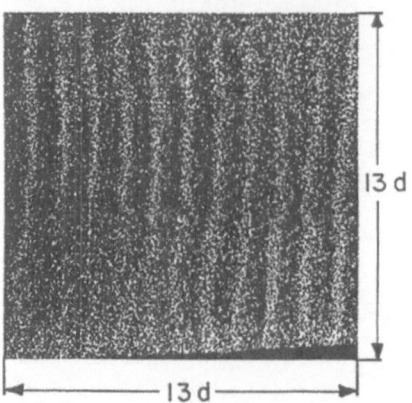

13 d

13 d

Fig. 1. A digitally enhanced shadowgraph image of fluctuations in electro-convection. The sample is a layer of thickness 14 μm of the nematic liquid crystal MBBA, and the applied voltage has an amplitude which is about 0.1 percent below the threshold value for the onset of convection. The finite spatial extent, as well as the periodicity of the convection rolls in the fluctuating patches, are noticeable. After Ref. 13

a single-mode model which is obtained by first averaging over the spatial degrees of freedom of the system.[4-7] It does not contain any explicit spatial variation, and is given by the equation

$$\tau_0 \dot{A} = \epsilon A - g A^3 + h(t) \tag{1}$$

where the dot indicates differentiation w.r.t. time, and where h(t) is the stochastic field with correlations given by

$$\langle h(t)h(t') \rangle = 2F\delta(t - t') \tag{2}$$

This model has been very successful for the discussion of fluctions near the instability point of a laser where there are no spatial degrees of freedom. [14-17] It is apparent from Fig. 3 that it is inadequate for the description of the fluctuation amplitudes below the bifurcation in EC. A model which considers the spatial degrees of freedom of the system explicitly is given by the equation

$$\tau_0 \dot{A} = \epsilon A + \xi_\parallel A_{xx} + \xi_\perp A_{yy} + h(t, x, y) \tag{3}$$

where A_{xx} and A_{yy} are second spatial derivatives in the direction parallel and perpendicular to the roll wave vector respectively.[18] The corresponding deterministic equation has been derived as the appropriate envelope equation from the hydrodynamic equations of this system.[3] The noise term $h(t, x, y)$ is delta-correlated both in time and in the two spatial directions x and y. The solid line through the data in Fig. 3 is a fit of this model to the experiment. The agreement clearly is excellent.[19] The required noise intensity is within a factor of two or so of thermal noise. Thus it seems that the microscopic degrees of freedom of this system, which are eliminated in the hydrodynamic equations, play a major role in producing the observed fluctuations.

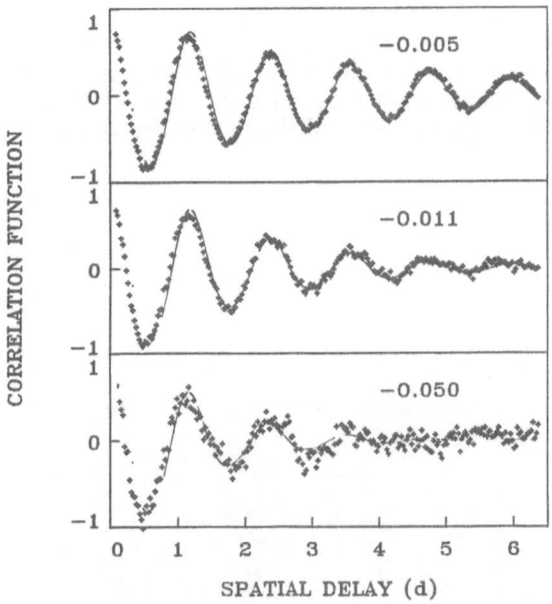

Fig. 2. Spatial correlation functions for data like those shown in Fig. 2 for the three ϵ values shown in the figure. The spatial periodicity of the rolls in the fluctuating patches of Fig. 1 is reflected in the fast oscillations of the correlation function, and the ϵ-dependent finite extent of the patches is reflected in the spatial decay of the envelope. After Ref. 13

In an earlier set of experiments, the role of fluctuations in the pattern-formation dynamics had been studied, and it had been shown that the transition from the unstable linear state above the bifurcation to the stable nonlinear states can evolve from fluctuations rather than from deterministic disturbances when the latter are made sufficiently small.[20,21] When this is the case, then the pattern which evolves is irreproducible from one experimental "run" to the next, and is determined by the particular realization of the noise which prevails during the sensitive early stages of that particular run when the amplitudes are still small. Figure 4 shows four shadowgraph images of convection patterns which evolved after $\epsilon = \Delta T / \Delta T_c - 1$

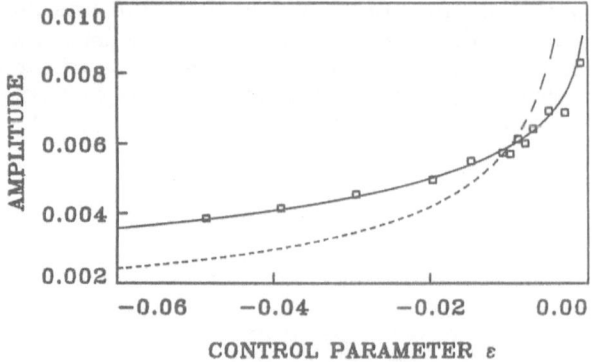

Fig. 3. Amplitudes of the fluctuations below the bifurcation. The dashed line is a fit of a single-mode stochastic model to the data at $\epsilon = -0.01$. The solid line is a fit of a stochastic model with two spatial degrees of freedom (Eq. 3) to the data at $\epsilon = -0.01$. After Ref. 13

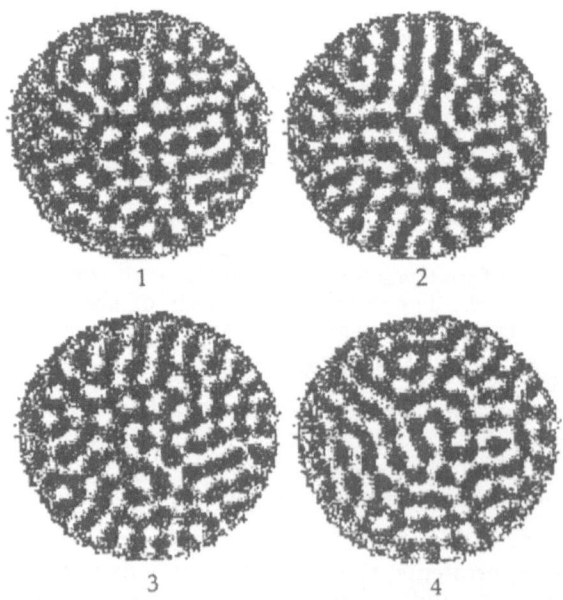

Fig. 4. Emerging patterns for four separate, nominally identical, experimental runs. It is apparent that the patterns are irreproducible from one run to the next. After Ref. 20.

was ramped slowly as a function of time from negative to positive values. The random appearance, and the irreproducibility of the patterns indicates a stochastic origin. In this case, however, the noise below the bifurcation was too feeble to be observable directly. The experimentally observed dynamics of the pattern evolution could be explained in terms of stochastic models only by invoking a noise intensity considerably larger than thermal noise.

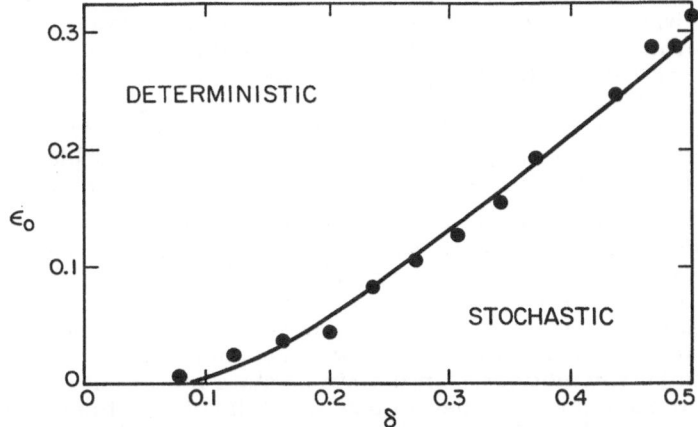

Fig. 5. The boundary between the deterministic and the stochastic region in the $\epsilon_0 - \delta$ plane for RBC with modulation at frequency $\Omega = 1$. The individual points are experimental measurements, and the solid line is the prediction (Ref. 22) based on the stochastic single-mode equation (Eq. 1 with Eq. 4) and the noise intensity measured in the ramp experiments. After Ref. 21.

In a third set of experiments, it has been shown that stochastic effects can be important not only in the initial pattern-formation process, but can also be responsible for time dependence associated with noise-induced transitions between different patterns.[20,21] Noise -induced time dependence will occur whenever the barrier between attractors corresponding to different spatial structures is sufficiently small. Quantitative information about such a case has been obtained from experiments involving time-periodic variations of the Rayleigh number in RBC which led to a periodic decay and re-growth of the pattern amplitude as a stationary process. In this experiment ϵ was varied according to

$$\epsilon = \epsilon_0 + \delta \cos(\Omega t) \tag{4}$$

When $\delta > \epsilon_0$, then ϵ becomes negative during part of the modulation cycle. If ϵ becomes sufficiently negative for a sufficiently large fraction of the cycle, then the amplitude of the convective flow grows to a macroscopic value, and then decays again to an unobservably small value once during each cycle. For a given Ω there is a region in the $\epsilon_0 - \delta$ plane where the pattern is irreproducible from one modulation cycle to the next.[20] In that region the appearance of the patterns is much like that of those in Fig. 4. For larger ϵ_0, above a remarkably sharp boundary in the $\epsilon_0 - \delta$ plane, an initial pattern is essentially reproduced from one cycle to the next. The irreproducible region corresponds to that part of the parameter space in which the flow amplitude becomes sufficiently small to be influenced significantly by the stochastic field. Experimental measurements[20] of the boundary between the deterministic and the stochastic regions are shown in Fig. 5 as solid points. As noted by Hohenberg and Swift,[22] important features of the experiment can be reproduced by the stochastic single-mode model, Eq. 1, with ϵ given by Eq. 4. For positive ϵ_0, the corresponding deterministic equation has two stable limit cycles for its solutions.[23] One corresponds to positive, and the other to negative amplitudes. When the parameters are such that the separation between the attractor basins of these two solutions becomes small, then the stochastic field may cause transitions between the positive and negative solutions. This switching between attractor basins corresponds to the irreproducibility of the experimental pattern from one cycle to the next. One can define a boundary between the stochastic and the deterministic regions in Fig. 5 by choosing a value of the average number of times for which the system remains on the same attractor as the dividing line. This, and the noise intensity inferred from the ramp experiment discussed above, yields the line shown in Fig. 5. It is an excellent representation of the experimental data.

In contrast to the experimental observations mentioned above, the deterministic equations of motion, for example the NS equation for fluid systems, predict no fluctuations below the bifurcation. They yield strictly deterministic dynamics, for instance in the sense that the system either remains in the unstable linear state above the first bifurcation, or that it evolves to a final state which is uniquely determined by the particular deterministic disturbance which is applied to initiate the dynamics and by the boundary conditions. Once a stable state is chosen, they predict that the system will remain in it regardless of how small the barrier to attractor basins of neighboring states may be. Although the experimental examples described above of exceptions to these deterministic predictions of necessity are taken from the parameter range near the first bifurcation where definitive, quantitative experiments are possible, there is no reason to believe that stochastic effects are always unimportant for instance at large Rayleigh or Reynolds numbers in fluid systems. Particularly under those conditions there is often a great multiplicity of nearby solutions to the deterministic equations, and stochastically induced transitions between neighboring solutions seem likely at least in some parameter ranges.

This work was supported by the Department of Energy under Grant No. DOE 84ER13729, and by the National Science Foundation under Grant No. NSF-DMR88-14485. The work described in this review involved a number of collaborators. I am particularly indebted to Helmut Brand, David Cannell, Pierre Hohenberg, Christopher Meyer, Ingo Rehberg, and Jack Swift.

REFERENCES

1. S. Chandrasekhar, *Hydrodynamic and Hydromagnetic Stability*, (Oxford University Press, London, 1961).

2. L.D. Landau and E.M. Lifshitz, *Fluid Mechanics* (Addison-Wesley, Reading MA, 1959).

3. E. Bodenschatz, W. Zimmermann, and L. Kramer, *J. Phys. France* **49**, 1875 (1988).

4. G. Ahlers, M.C. Cross, P.C. Hohenberg and S. Safran, *J. Fluid Mech.* **110**, 297 (1981).

5. H. van Beijeren and E.G.D. Cohen, *J. Stat. Phys.* **53**, 77 (1988).

6. H. van Beijeren and E.G.D. Cohen, *Phys. Rev. Lett.* **60**, 1208 (1988).

7. P.C. Hohenberg, J.B. Swift, and G. Ahlers, to be published.

8. V.M. Zaitsev and M.I. Shliomis, *Zh. Eksp. Teor. Fiz.* **59**, 1583 (1970) [Engl. transl.: *Sov. Phys. JETP* **32**, 866 (1971)].

9. R. Graham, *Phys. Rev. A* **10**, 1762 (1974).

10. J.B. Swift and P.C. Hohenberg, *Phys. Rev. A* **15**, 319 (1977).

11. G. Dewel, P. Borkmans, and D. Walgraef, *J. Phys. C: Solid State Phys.* **12**, L491 (1979).

12. D. Walgraef, G. Dewel, and P. Borkmans, *Phys. Rev. A* **21**, 397 (1980).

13. I. Rehberg, S. Rasenat, M. de la Torre-Juarez, W. Schöpf, F. Hörner, G. Ahlers and H.R. Brand, to be published.

14. F.T. Arecchi, G.S. Rodari, and A. Sona, *Phys. Lett. A* **25**, 59 (1967).

15. F.T. Arecchi, M. Giglio, and A. Sona, *Phys. Lett.* **25**, 341 (1967).

16. H. Risken and H.D. Vollmer, *Z. Phys.* **201**, 323 (1967).

17. H. Risken, in *Progress in Optics*, edited by E. Wolf (North-Holland, Amsterdam, 1970), Vol. VIII, pp. 239.

18. This model ignores the fact that the fluctuations which are observed actually are travelling waves of convection rolls, thus indicating that the bifurcation is of the Hopf type. However, the measured phase velocity of the waves is quite small and thus the model discussed here is likely to be a good approximation.

19. Because of the ultraviolet divergence characteristic of two-dimensional models with truncation at second-order gradient terms, a cutoff value for the wavenumber had to be used in addition to the field strength F as an adjustable parameter.[13] The required cutoff wavenumber turned out to be approximately equal to one.

20. C.W. Meyer, G. Ahlers and D.S. Cannell, *Phys. Rev. Lett.* **59**, 1577 (1987).

21. C. W. Meyer, G. Ahlers, and D. S. Cannell, to be published.

22. P.C. Hohenberg and J.B. Swift, *Phys. Rev. Lett.* **60**, 75 (1988).

23. G. Ahlers, P. C. Hohenberg and M. Lücke, *Phys. Rev. A* **32**, 3493, 3519 (1985).

HEXAGONAL CONVECTIVE CELLS

C. Pérez-García (*), S. Ciliberto (‡) & E. Pampaloni (‡)

(*) Departamento de Física, Universidad de Navarra,
    31080 Pamplona, Navarra, Spain
(‡) Istituto Nazionale di Ottica, Largo E. Fermi 6,
    50125 Arcetri-Firenze, Italia

ABSTRACT. The transition between two patterns with different symmetry is studied experimentally in thermal convection in a layer of fluid heated from below (Rayleigh-Bénard convection). The experimental results agree with theoretical calculations based on a model of three coupled Ginzburg-Landau equations. The influence of defects on the transition between a pattern of hexagons an a pattern of rolls is also analysed.

1. Introduction

One of the simplest pattern forming system where spatial regular patterns can be studied with great accuracy is thermal convection. When a horizontal layer of fluid heated from below (Rayleigh-Bénard convection) the rest (conducting) state is replaced by motions which organize themselves to form patterns. Usually these patterns are formed by rolls. However, by adding some complexity (poor conducting boundaries, a binary mixture, non-Boussinesq conditions, etc.) other symmetries are also possible [1].

At secondary thresholds a transition between different symmetries is also possible. In these nonequilibrium transitions the state with higher symmetry (hexagons, squares) is replaced by one with lower symmetry (rolls). As the properties of the fluid and the boundary conditions of the system can be regulated with great accuracy, thermal convection is a useful system to study such kind of transitions. Another advantage of this systems is that some simplified models, based on the central manifold theory [2] can by obtained from the original balance equations. Although these models are only valid near threshold they allow to make analytical calculations of the transition between different symmetries.

We have studied the finite size effects and the role of the defects on the transition between a hexagonal pattern and a pattern of rolls in convection. However it is important to point out that many of our findings are not specific of the particular system chosen in our experiments but are rather general, because they can be explained by topological arguments based on the symmetries of the system.

2. Convection under non-Boussinesq conditions

We recall briefly the main results on the Rayleigh-Bénard

convection. A horizontal layer enclosed between two plates is heated from below. The most relevant parameter of this system is the Rayleigh number $R = \alpha g \Delta T d^3 / \nu \kappa$, where α is the volumetric expansion coefficient, g the acceleration of gravity, ν the kinematic viscosity, κ the thermal diffusivity coefficient, d the layer depth and ΔT the temperature difference between the plates. When R exceeds a threshold value R_c, convective flow starts, producing a pattern of rolls with a well defined wavenumber k_c. The values of these critical parameters depend on the cell size and geometry, and on the boundary conditions. For example, in the case of an infinite layer between conducting plates they are R_c = 1708 and $k_c \approx 3.11/d$. However experiments are made in finite geometries. This fact is taken into account by means of a parameter, the aspect ratio, defined as $\Gamma = L/d$, where L is a characteristic horizontal length of the convective cell. When $\Gamma \gg 2\pi/k_c$ the system has many spatial periods an therefore many convective cells are present in the system.

This brief description is valid for simple fluid under the simplest conditions. The main approximation used is the so called Oberbeck-Boussinesq (OB) approximation: the temperature dependence of the fluid parameters is neglected, except for the thermal expansion effects responsible for buoyancy forces. Moreover, the viscous dissipation is neglected in comparison with the conductive term in the energy balance equation. There are some theoretical investigations on the effects of small departures from OB approximation ([3]). Some experiments partially confirm some of the predictions of theoretical works ([4]).

We recall briefly here the main results of theoretical analyses. The departures from the OB approximation give rise to the following consequences: i) near threshold the convective pattern is hexagonal; ii) subcritical motions are possible and iii) an hysteretic transition between hexagons and rolls is predicted, when the heating increases. Therefore the bifurcation is **transcritical**, and several threshold values can be distinguished

$$\varepsilon_a \leq \varepsilon_c \leq \varepsilon_r \leq \varepsilon_h \qquad (2.1)$$

where ε represents the supercritical heating ($\varepsilon = \Delta T - \Delta T_c / \Delta T_c$ where ΔT_c indicates the critical temperature difference across the convective layer). For $\varepsilon_a \leq \varepsilon \leq 0$ hexagons and no-convective state are possible solutions. In the interval $0 \leq \varepsilon \leq \varepsilon_r$ only hexagons are stable. Between $\varepsilon_r \leq \varepsilon \leq \varepsilon_h$, hexagons and rolls can coexist. For $\varepsilon \geq \varepsilon_h$ only rolls are stable. These thresholds and the direction of the flow depend on the departures from the OB approximations.

3. Experimental set up

In Fig 1. one can see a schematic section of a convective cell used in our experiments. The lateral walls are made in plexiglas. These can be changed to vary Γ and boundary conditions. The bottom is a copper plate (CP) whose upper surface is finished to a mirror quality. This surface is protected with a nickel film to avoid oxidation. The bottom of this plate is heated with an electrical resistor (ER). The upper plate is a sapphire window (SP) whose top is cooled by water

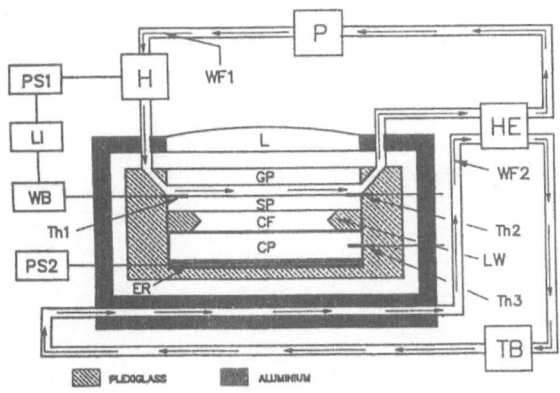

Fig. 1. Schematic diagram of the convective cell

circulation (WF1), which is confined on the other side by the glass window (GW). This device allows for optical inspection of convective motions. The whole cell is inside a thermostatic box that reduces the thermal fluctuations from outside. The temperature of the cooling water circuit (Wa) is stabilized by a thermal bath and a feedback loop that controls the temperature of the upper plate. This ensures a long term temperature stability of about $\Delta T = \mp 0.001$ K. (More details can be found in ([5])). The convective fluid is pure water.

The qualitative features of the patterns are determined by a digital enhanced shadowgraph technique. A different optical technique, based on the deflection of a laser beam that sweeps the fluid layer ([7]) allows to determine local properties (local temperature gradients) in the fluid, from which it is possible to reconstruct global quantities (temperature field, heat flow) after integration ([8]). The shadowgraphic and the laser beam deflection technique are based in the variations of the refractive index induced by temperature variations.

Two reconstructions of the temperature field on an array of 128×128 points on a square area of 4.9×4.9 cm^2 can be seen in Fig. 2.

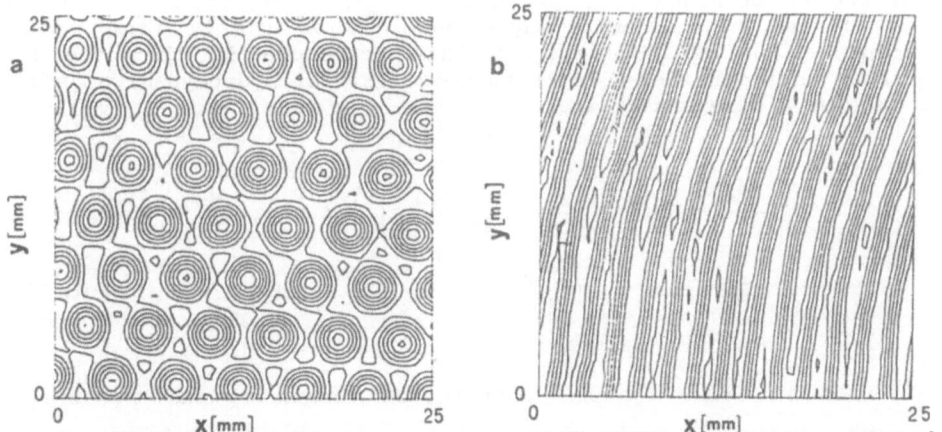

Fig. 2. Temperature field reconstructed from the laser scanning technique corresponding to a) $\varepsilon = 0.022$ and b) $\varepsilon = 0.053$.

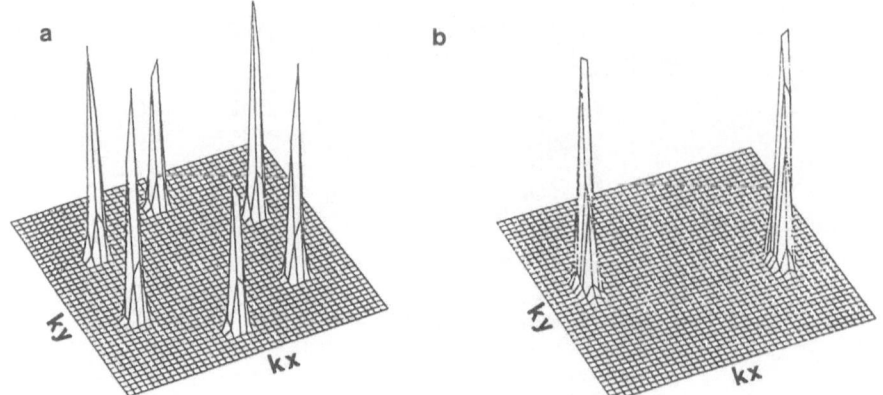

Fig 3. Spectra of the temperature fields for a) $\varepsilon = 0.022$ and
b) $\varepsilon = 0.053$

The corresponding Fourier spectrum $S(k_x, k_y)$ of the temperature field computed in the full scanning area is also shown in Fig 3

4. Convective patterns under non-Boussinesq conditions

The layer of pure water is enclosed in a cylindrical cell with $d = 0.18$ cm and $\gamma = 20$. The experiment have been made at a mean working temperature of 28.3°C. At this temperature the Prandtl number $P = \nu/\kappa$ is $P = 5.62$ and the horizontal thermal diffusion time (characteristic of horizontal motions of the convective cells) is $\tau_h = 2.45$ h ([6]). The critical temperature difference in these conditions is $\Delta T_c = 12.6$ K. This temperature difference across the cell is sufficient to induce non-Boussinesq effects.

Convective patterns can be described by means of evolution equations for the slowly varying amplitudes, (**amplitude equations**). These can be obtained by a perturbation analysis around the critical modes of the hydrodynamics equations ([2]) or simply from symmetry arguments ([8]). The advantage of this procedure is that the final nonlinear equations are simpler than the initial hydrodynamics ones. In the case of a hexagonal pattern they take the form

$$\tau_0 \frac{\partial A_i}{\partial t} = \left\{ \varepsilon + \xi_0^2 \left[\frac{\partial}{\partial x_i} - \frac{i}{2k_c} \frac{\partial^2}{\partial y_i^2} \right]^2 \right\} A_i + a\, A_{i+1}^* A_{i+2}^* -$$

$$- b \left[\sum_{i \neq j} |A_j|^2 \right] A_i - c\, |A_i|^2 A_i \quad (i = i \bmod 3) \qquad (3.1)$$

The relaxation time τ_0, the correlation length ξ_0 and the critical Rayleigh number R_c and wavenumber k_c are obtained from a linear stability analysis of the full hydrodynamics equations of the system. The superscript * denotes the complex conjugate and $\partial/\partial x_i$ and $\partial/\partial y_i$ are the spatial derivatives parallel and perpendicular to the vector $\mathbf{k}_i$, respectively. The physical meaning of the nonlinear coefficients a, b and c, will be specified later on. Stable stationary solutions are i) a pattern of rolls $|A_1| \neq 0$, $|A_2| = |A_3| = 0$ and ii) a hexagonal pattern $|A_1| = |A_2| = |A_3|$. A linear stability analysis around these solutions allows to determine that rolls are the unstable solutions for $0 \leq \varepsilon \leq \varepsilon_r$, where ε_r is given by

$$\varepsilon_r = \frac{a^2 c}{(b-c)^2} \tag{3.2}$$

Hexagons are stable in the interval $\varepsilon_a \leq \varepsilon \leq \varepsilon_h$ with

$$\varepsilon_a = - \frac{a^2 c}{4(2b+c)} \qquad\qquad \varepsilon_h = \frac{a^2 (b+2c)}{(b-c)^2} \tag{3.3}$$

The amplitude of the different modes present in the convective system can be obtained by means of the laser deflection technique briefly described above. The Fourier spectrum $S(k_x, k_y)$ (see Fig. 3) of the temperature field $T(x,y)$ present some peaks whose centers of mass are at the vertices of the vectors $\mathbf{k}_j$ and $-\mathbf{k}_j$. Each pair of peaks present corresponds to a set of rolls (see fig. 3). Once the $\mathbf{k}_j$ are determined, one shifts the Fourier transform by $-\mathbf{k}_j$ in order to have the corresponding peak at the origin. Afterwards, the contributions of other peaks are filtered with a low pass filtering function ([11]). The anti-Fourier transform gives the amplitude $A_j(x, y)$ of each set of rolls. An easy calculation allows to recover the real part R_j as well as the phase φ_j of this roll.

It is interesting to relate the local description of amplitude equations with the global heat flow measurements which are experimentally accessible. The comparison can be made taking into account that the normalized heat flow is related with the amplitudes by the relation

$$N \equiv \frac{(N-1)R}{R_c} = \frac{1}{S} \int dxdy \sum_{i=1}^{3} |A_i|^2 = \sum_{i=1}^{3} \langle |A_i|^2 \rangle_{x,y} \tag{3.4}$$

where the bracket $\langle \rangle_{x,y}$ indicates the average on the horizontal plane. For hexagons the normalized Nusselt number N gives

$$N_h = 3|A_h|^2 = \frac{3}{C} \left\{ \frac{a^2}{4C} + \varepsilon + a \sqrt{a^2 + 4C\varepsilon} \right\} \tag{3.5}$$

where $C = c+2b$. For rolls the expression is simply

$$N_r = |A_r|^2 = \varepsilon/c \tag{3.6}.$$

As a consequence the slope of $N_r(\varepsilon)$ gives a direct information about the coefficient c, while the determination of b and a from N_h requires a more delicate fitting of the dependence on ε and $\varepsilon^{1/2}$

Of course, for a comparison of (3.4)-(3.5) with heat flow measurements one must take into account the contributions of lateral effects to the heat transport. This have been in some recent works ([9]). These effects also modifies the transition thresholds (3.2) and (3.3) as shown in ref ([9]) (See also the contribution of Dr. Walgraef in these proceedings).

5. Stationary defects in hexagonal patterns

The influence of the defects on the transition threshold between different symmetries is clearly shown in experiments. However, their influence on the heat flow is very small, because this parameter is global and defects are localized in space. In this paragraph we only intend to give a qualitative view of the two types of stationary defects that appear hexagons and rolls can compete.

In hexagonal patterns the stable defects is the pentagon-heptagon pair ([10]). As suggested recently ([11]) this defect corresponds to a local transition between an hexagonal pattern and a pattern of rolls, because at the core of these pairs only one of the three systems of rolls that form the hexagonal symmetry survives, i.e., only an amplitude is different from zero. In this situation the single amplitude is bigger than the three amplitudes in the hexagonal regime. This is due to the fact that, as one can deduce from (3.5) and (3.6), rolls are more efficient to transport heat than hexagons ([6]). However, this effects in only local and it cannot be detected merely by global calorimetric measurements. Following the procedure to calculate the amplitudes described above, one can recover also the phases of the sets of rolls present in the system. In the case of a penta-hepta pair the phases of the two set of rolls that vanish in some point suffer a jump, one of $+2\pi$ and the other of -2π, while the third one does not have any singularity. This indicates that this defect can be seen as a dislocation in each of the sets of rolls whose amplitude vanishes at the core of the defect.

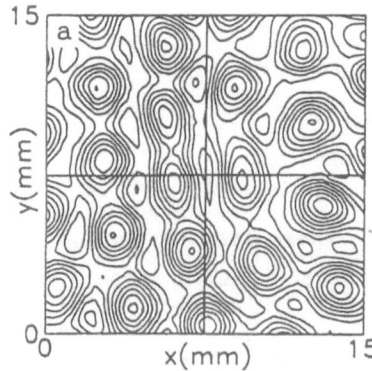

Fig. 4. A penta-hepta pair in a pattern of hexagons in water under non-Boussinesq conditions ([11]).

When the rolls are dominant the typical defect is not a pure' dislocation, as in the usual case, but a grain boundary. (This can be proved by very general topological arguments ([11])) In the core of such a grain boundary hexagonal cells are still present as one can see in the temperature field in Fig. 5. This confirms that the unstable "phase" reappears in the defects of the stable one.

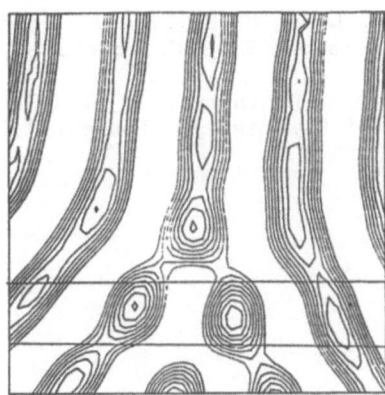

Fig. 5. Temperature field showing a grain boundary in water under non-Boussinesq conditions (^{11}).

experiment in water under non-Boussinesq conditions (^{6}). As one can deduce from (2.7), when two oblique modes are present the third one must also be present and no singularity must be observed in the phases. The grain boundary acts as a seed of nucleation of hexagons when one decreases the temperature difference above the transition threshold ε_h.

Although these results are preliminary, the main conclusion is that defects can play an important role similar to that of condensation nuclei in first order phase transitions in equilibrium a point that must be analysed in more detail in future works.

Acknowledgements

We acknowledge the EEC twinning project for finantial support under contract SC1-0035-C and of the Spain-Italy integrated action n°49 (1990). We also thank Prof. P. Coullet and Dr. J. Lega (Université de Nice) for helpful discussions

REFERENCES

(1) Wesfreid J.E. & Zaleski S., eds. 1983 **Cellular Structures in Instabilities** , Springer, Berlin
(2) Manneville P, **Dissipative Structures and Weak Turbulence** 1990, Academic, Boston
(3) Palm E., 1960 J. Fluid Mech. **8**, 183
 Busse F.H., 1967 J. Fluid Mech. **30**, 625
(4) Hoard C.Q., Robertson C.R. & Acrivos A. 1970 Int. J. Heat Mass Transfer **13**, 849
 Dubois M., Bergé P. & Wesfreid J.E. 1978 J. Physique 39, 1253
 Ahlers, G. 1980 J. Fluid Mech. **98**, 137
(5) Ciliberto S., Pampaloni E. & Pérez-García C. 1988 Phys. Rev. Lett. **61**, 1198
(6) Pampaloni E., Pérez-García C., Albavetti L. & Ciliberto S. J. Fluid Mech (submitted)
(7) Ciliberto S., Francini F. & Simonelli, F. 1985 Opt. Commun. **54**, 381
 Rubio M.A., Ciliberto S & Albavetti L., J. Fluid Mech (to appear)

(8) Caroli B., Caroli C. & Roulet B. 1984 J. Crystal Growth **68**, 677

(9) Pérez-García C., Pampaloni E. & Ciliberto S. 1990 Europhys. Lett. **12**, 51

(10) Pantaloni J. & Cerisier P. in ref. 1, p. 197

(11) Ciliberto,S., Coullet P., Lega J., Pampaloni E. & Pérez-García C., preprint

THEORETICAL MODELS:

THE NAVIER-STOKES AND RELATED EQUATIONS

VORTEX DYNAMICS AND TURBULENCE

P.G. Saffman

Applied Mathematics 217-50
California Institute of Technology
Pasadena, California, 91125

INTRODUCTION

We shall define Vortex Dynamics to be the study of laminar (i.e. non turbulent) solutions of the equations

$$\frac{\partial \boldsymbol{\omega}}{\partial t} + \mathbf{u} \cdot \nabla \boldsymbol{\omega} = -\boldsymbol{\omega} \cdot \nabla \mathbf{u} + (\nu \nabla^2 \boldsymbol{\omega}), \tag{1}$$

$$\mathbf{u} = \mathrm{curl}^{-1} \boldsymbol{\omega}. \tag{2}$$

for which $\boldsymbol{\omega}$ is compact. It is also implied that usually $\nu \ll 1$.

The role of vortex dynamics in turbulence is based on the idea that turbulence is a random, chaotic, disorganized superposition of basically organized quasi-laminar coherent structures, which are characterized by being compact distributions of vorticity and can be called vortices. Such a picture implies that phase correlations are important in the dynamical processes by which the turbulence evolves, and that spectra alone will not give a complete representation of the turbulence properties. This approach is certainly useful from a kinematic or qualitative standpoint, and clearly has proved valuable in the interpretation of the modern flow visualization diagnostics, but despite much effort it has yet to prove itself as a quantitative or dynamical approach.

Earlier work of a kinematic nature was carried out by Synge & Lin (1943), Townsend (1951), Corrsin (1962), Saffman (1966) and Tennekes (1968). Synge & Lin calculated the correlation functions produced by a random superposition of Hill spherical vortices, and Townsend found the form of the energy spectrum of a superposition of two-dimensional and axisymmetric Burgers vortices (exact steady solutions of the Navier-Stokes equations for uni-directional vorticity in a uniform irrotational straining field). Neither approach led to a Kolmogorov type inertial range or modeled satisfactorily properties of the turbulence.

Corrsin considered assemblies of vortex sheets of thickness proportional to the Kolmogorov microscale. The work of Saffman was an attempt, based on a random superposition of vortex sheets, to provide a physical explanation for the occurrence of the Kolmogorov scale and account for

The Global Geometry of Turbulence
Edited by J. Jiménez, Plenum Press, New York, 1991

the effects of intermittency on the qualitative dependence of high order structure functions on the Reynolds number. A similar approach by
Tennekes employed vortex tubes. The predictions of these physical models do not seem to agree with the results of either experiments or recent
relatively high Reynolds number numerical simulations (e.g. Vincent &
Meneguzzi 1990)). Moffatt (1984) employs spiral vortices to obtain arbitrary power laws.

Lundgren (1982) appears to have been the first to use dynamically
evolving vortices. He calculated unsteady non-axisymmetric Burgers type
solutions of the Navier-Stokes equations consisting of a central viscous smoothed core surrounded by rolling up vortex sheets and obtained
the spectrum induced by a random superposition of these structures. The
remarkable feature of this approach is that it predicted (with apparently
no disposable constants) the $k^{-5/3}$ inertial range dependence of the energy
spectrum on the wave number k. However, the coefficient is arbitrary, and
although physical assumptions can be made to produce the Kolmogorov $\epsilon^{2/3}$
dependence, this does not appear automatically from the model. Flatness
factors or skewness coefficients were not calculated.

It does not seem unfair to say that up to the present the role of vortex dynamics in providing quantitative predictions for the structure of
homogeneous turbulence has not been particularly successful. However,
various new dynamical phenomena in vortex dynamics have been uncovered in
the last two decades (due principally but not entirely to the application
of the techniques of scientific computing), which may shed some light on
the qualitative evolution of turbulent flows, both homogeneous and inhomogeneous. These have not yet been incorporated into physical models
(apart from models of the turbulent mixing layer, e.g. Saffman 1981), and
the purpose of this review is to describe briefly some the recent developments in vortex dynamics and comment on a possible role in turbulence.

DYNAMICAL PHENOMENA

Six basic phenomena can be identified.

I   Fusion and fission of vortex patches.

II   Waves and instabilities.

III Singularity formation.

IV   Breaking and reconnection of vortex tubes.

V   Filamentation of vortex patches.

VI   Core bursting.

We now comment on these.

Fusion and fission of vortex patches

Fusion was first demonstrated numerically by Roberts & Christiansen
(1972). They considered a pair of equal two-dimensional co-rotating vortices, and showed that the vortices merge if too close together. Fission
was demonstrated theoretically by Moore & Saffman (1971) and confirmed
numerically (Moore & Saffman 1975). It occurs when a two-dimensional vortex patch is placed in a uniform straining field. When the strain exceeds
a number times the vorticity in the patch, equilibrium solutions are not
possible and the patch disintegrates. Moore & Saffman (1975) and Saffman

(1981) used this process to predict the growth of the coherent structures in the turbulent mixing layer by the process called tearing, in which a vortex patch grows in size by entrainment of irrotational fluid and then disintegrates when its vorticity is reduced to a level below that required for equilibrium in the strain caused by the other vortices, the fragments then being absorbed by its neighbours. The opposite process of pairing results from the fission when neighboring patches come too close together as they rotate as predicted by the pairing instability mode of a row of line vortices.

These two phenomena provide mechanisms whereby the scales can change in turbulent flow, fusion increasing the scale and fission decreasing it. The question is whether an equilibrium is reached, or if one or the other becomes dominant. Studies of two- and three-dimensional turbulence indicate that the answer to this question may depend upon the spatial dimension. There are also doubts about the relevance of studies of uniform vortex patches, since variations of vorticity inside the vortices may be important. See Melander et al. (1987) for calculations of the evolution of a non-uniform vortex patch.

Waves and instabilities

The studies of wave propagation and instabilities in vortices constitute a huge literature. The work starts with Kelvin (1880), and new phenomena are still being discovered (e.g. the short wave instability discovered by Pierrehumbert (1986) and analyzed further by Bayley (1986) and Landman & Saffman (1987)). Moreover, important open questions still remain. Instabilities can be broadly characterized in two ways, as individual or cooperative. The individual type depends upon the internal structure of the vortex. For instance, the Krutzsch (1939) instability of vortex rings and the Widnall, Bliss & Tsai (1974) instability of weakly strained vortices. The cooperative instabilities are those that depend upon the velocity fields induced according to the Biot-Savart law by vortex filaments, and depend only weakly on the internal structure; for example, the instability of vortex arrays such as the Karman vortex street to two-dimensional and three-dimensional instabilities (see Robinson & Saffman 1982).

These studies give qualitative insights about the type of structures that might occur in turbulent flow. Robinson & Saffman claim that the cooperative instabilities suggest why large scale organized structures are seen in the mixing layer and to a lesser extent in the wake, but hardly at all in the boundary layer. The turbulent spot is an individual type of disturbance, more related to transition than to fully developed turbulence.

One of the more important open questions is the stability to three-dimensional disturbances of the stretched Burgers vortices. (It is convenient to call straining the effect of a two-dimensional uniform irrotational field in a plane perpendicular to the vorticity, while stretching refers to the deformation produced by an irrotational velocity field with a principal rate of strain parallel to the vorticity.) Robinson & Saffman (1984) studied the stability to two-dimensional disturbances and concluded that they were stable. (They also calculated non-axisymmetric equilibrium solutions).

Singularity formation

The possible existence of singularity formation in solutions of the incompressible unsteady Euler equations with smooth initial data is one of the more fascinating questions in fluid dynamics. The evidence in three-dimensional flow is uncertain, and some years ago the majority opinion was inclined towards the existence of a finite time singularity. However, recent studies are tending to suggest that no singularity forms in a finite time; see for instance the calculations by Pumir & Siggia (1990) of colliding vortices in a straining field. The primary reason for not believing in the singularity is the result that vorticity does not stretch itself, as can be seen from writing the vorticity equation in the form

$$\frac{\partial \boldsymbol{\omega}}{\partial t} = \mathbf{e} \cdot \boldsymbol{\omega}, \tag{3}$$

where e is the rate of strain tensor whose value at a point is independent of the vorticity at the point (although it depends of course on the distribution of vorticity elsewhere). The non-existence of a singularity is also supported by the Buntine & Pullin (1989) calculation of colliding straight vortices, although this model does not allow for any amplification of the stretching component of the straining field.

Both calculations suggest that vortex tubes are transformed into vortex sheets, and this is consistent with the measurements in homogeneous turbulence of the principal rates of strain α, β, γ (see Townsend 1951) which give $\alpha\beta\gamma < 0$. Remembering that $\alpha + \beta + \gamma = 0$, by virtue of the incompressibility of the fluid, it follows that on average distributions of vorticity should be pulled out into sheets rather than tubes.

On the other hand, there is now very strong (almost rigorous) evidence that vortex sheets in two-dimensional incompressible flow develop finite time singularities as predicted by Moore (1979). Moore showed, and this has subsequently been confirmed by other workers, that a vortex sheet with initial shape given parametrically by

$$Z(\Gamma, 0) = \Gamma + i\varepsilon \sin \Gamma \tag{4}$$

'blows up' in a time $t_c \sim \log(1/\varepsilon)$. The form of the singularity is the creation of infinite curvature while the slope of the sheet remains finite. It is not known how the sheet develops after the critical time t_c, and there is no reason to think that 'desingularization' leads to a unique solution. Pullin (1989) has suggested that the sheet rolls up into pairs of vortex tubes after the singularity has formed. Perhaps this is the cascade process: large scale tubes interact by mutual and external straining, and form smaller scale sheets which produce via the singularity yet smaller scale tubes. It is significant that Lundgren (1982) demonstrated that the statistics of the flow field produced by rolling up spiral vortex sheets would have a Kolmogorov $k^{-5/3}$ inertial range energy spectrum.

<u>Reconnection</u>

This double cascade resembles the model proposed by Saffman (1966), except in that model Görtler instability of curved flow was suggested for the secondary cascade producing the smallest scales. The present ideas are more appealing, although still lacking hard quantitative support, and moreover the primary stage may have some bearing on the mechanisms of macroscopic mixing which is one of the prime features of turbulent flow. The primary stage in which vortex tubes collide and form new tubes, and thin sheets when the Reynolds number is large, is the reconnection process associated with the breaking and rejoining of vortex tubes. This was seen by Scorer & Davenport (1970) in aircraft trailing vortices, and by Kambe & Takao (1971) in colliding vortex rings. Schatzle (1987) has carried out the most extensive quantitive measurements to date, and there has been an extensive series of numerical simulations by many workers in which unsteady solutions of the incompressible Euler and Navier-Stokes equations have been found.

204

Saffman (1990) has proposed a physical model with the object of trying to isolate the physics and uncover the scaling laws, as well as provide a framework for analysis of experimental and numerical data and suggest experiments and computations. The idea is that there are four stages in the breaking and reconnection process.

(i) As the vortex filaments are brought together by strain in the plane perpendicular to the vorticity, there is a cancellation of vorticity by viscous diffusion.

(ii) This leads to a pressure rise in the core as the centrifugal force is weakened.

(iii) An axial pressure gradient is produced by the variations of axial pressure.

(iv) The axial flow convects vorticity away from the region of collision, and rejoining takes place essentially kinematically as vortex lines cannot end inside the fluid.

Notice that external axial strain is not required.

Saffman writes down equations to model these physical processes. Unfortunately they are not sufficiently simple to allow analytical solution, and numerical methods are necessary. The equations were solved with values of the parameters corresponding to Schatzle's data, and reasonable agreement found between observed and predicted values of physical variables. For example, Saffman predicts the time T_B for the vortex line breaking to occur should scale like

$$T_B \sim \frac{5\pi R^2}{\Gamma_0} \left(\frac{a}{R}\right)^{5/4} \sqrt{\log\left(\frac{\Gamma_0 (a/R)^{1/2}}{2\pi^2 \nu}\right)}, \tag{5}$$

and the time T_R for the reconnection to occur scales like

$$T_R \sim \frac{2\pi R^2}{\Gamma_0} \left(\frac{a}{R}\right)^{1/2}. \tag{6}$$

In the inviscid limit, the equations take the form

$$\frac{\partial \delta}{\partial t} = -(\alpha + \lambda E)\delta \tag{7}$$

$$\frac{\partial A}{\partial t} = -E A, \tag{8}$$

$$\frac{\partial w}{\partial t} = -w\frac{\partial w}{\partial z} - \frac{\partial p}{\partial z}. \tag{9}$$

Here, δ is the distance between the centres of the vortex cores, $A = bh$ is proportional to the area of the vortex cores, b and h being the linear dimensions of the cores, w is the axial velocity inside the cores and $E = \partial w/\partial z$. α denotes the strain which is convecting the cores towards each other, and λ is a constant $\frac{1}{2} \leq \lambda \leq 1$ is a constant which measures the amount of asymmetry in the straining field (Saffman took $\lambda = 1$). p is the pressure in the core, which depends on the core circulation Γ and the deformation or aspect ratio of the core, $\theta = h/b$. The formula used, based on the Moore & Saffman (1971) results for an elliptical vortex in a straining field is

$$p = \frac{\gamma^2 f(\theta)}{2\pi^2 bh}, \quad f = \frac{\theta}{1+\theta^2}, \quad \theta = 1 + 4\left(\frac{\pi\alpha bh}{\Gamma} + \frac{bh}{\delta^2}\right). \tag{10}$$

These equations still have to be solved numerically. The numerics indicates that the axial strain E saturates and eventually decays, there is no sign of any 'blowup', and the equations predict no singularity for the Euler equations. The core deformation depends sensitively on λ, but even for $\lambda = \frac{1}{2}$, for which the deformation is least, the aspect ratio $h/b \to \infty$, indicating that the collision tends to produce sheets.

Meiron & Shelley (private communication) have pointed out that the inviscid model has a simple asymptotic solution when $t \to \infty$. In this limit, $w \sim \beta(t)\,z$, while all other variables become independent of z and are just functions of t. Then asymptotically

$$\frac{d\beta}{dt} = -\beta^2, \quad \frac{dA}{dt} = -\beta A, \quad \frac{d\delta}{dt} = -(\alpha + \lambda\beta)\delta, \tag{11}$$

where $A = bh$. The solution of these equations is

$$\beta = \frac{1}{t + t_0}, \quad A \propto \frac{1}{t + t_0}, \quad \delta \propto \frac{e^{-\alpha t}}{(t + t_0)^\lambda}. \tag{12}$$

Note that $A/\delta^2 \to \infty$, so that tubes are deformed into sheets.

Filamentation

This phenomenon, in which thin braids of vorticity are ejected from the boundaries of vortex patches, appears to have been first discovered by Roberts & Christiansen (1972) in a study of the merging of co-rotating vortex patches. Deem & Zabusky (1978) observed the phenomena for the oscillations of a perturbed circular vortex, and a thorough study has been carried out by Dritschel (1988) for this geometry. In addition to this form which can be called extrusive filamentation, Pullin (1981) observed intrusive filamentation of the boundary of vortex layers, in which filaments of irrotational fluid are ingested by the vortex layer. A remarkable feature of filamentation is that it occurs when the patch is linearly stable, and even when non-linearly stable in a low order norm. For instance, a circular vortex patch is stable in the L^0, L^1 and L^2 norms. The first and third are consequences of the conservation of vorticity and angular impulse, the second follows from the Schwarz inequality

$$\int \omega r \, dS \le \sqrt{\int \omega r^2 \, dS \int \omega \, dS}. \tag{13}$$

Pullin et al. (1990) proposed that there are three basic mechanisms of filamentation.

(i) Linear instability

(ii) Non-linear instability

(iii) Wave breaking (Dritschel (1988))

The first was demonstrated by Polvani et al. (1988), in a study of the evolution of a linearly unstable elliptical vortex patch (with $a/b > 3$). The second mechanism (Pullin et al. 1989) attributes filamentation to a secondary instability. Waves of finite amplitude are supposed to be generated on a stable interface, and the instability of these nonlinear waves produces filamentation by generation of hyperbolic stagnation points in a frame of reference moving with disturbance extrema. The third is perhaps the most interesting, as it occurs without any growth in amplitude of the the disturbance and is the steepening (like shock wave formation) of a bump on the interface and there seems to be no minimum amplitude condition. Dritschel's (1988) estimates of the time to breaking and the nature of the filamentation process have been confirmed by Pullin & Moore (1990). If l denotes the width of the bump, a its height, and R the radius of the patch, the numerical data suggests the expression for the time to breaking t_B (Pullin, private communication)

$$t_B = \frac{1}{\omega}\left(1.5 + 6.5\frac{l}{R}\right)\frac{l^2}{a^2}, \tag{14}$$

where ω is the magnitude of the vorticity in the patch. Meiron (private communication) has found that for the case in which the disturbance is periodic with the same amplitude a and wavelength l, the time to breaking is somewhat larger, i.e. a single bump breaks faster than a series of bumps.

For weakly non-linear disturbances to a circular vortex patch, a Korteweg-DeVries type equation can be constructed (Dritschel 1988). It has the form

$$\phi_t + \tfrac{1}{2}\omega_0(\phi_\theta + \mathcal{H}(\phi)) - \frac{\omega_0}{2R^2}\phi\phi_\theta$$
$$= \omega_0 \frac{\partial}{\partial\theta}\left(-\frac{1}{3R^4}\phi^3 + \frac{1}{24\pi R^4}\int_0^{2\pi}\frac{(\phi(\theta) - \phi(\theta'))^3}{1 - \cos(\theta - \theta')}\,d\theta'\right) + O(\phi^4), \tag{15}$$

where $\phi = \tfrac{1}{2}(r^2 - R^2)$. For disturbances with the same value of l/R, it is claimed by Dritschel that this equation also gives the same proportionality of t_B on a^{-2} that is found for the Euler equations. Since a vortex patch is a Hamiltonian system, it would be of interest to study the evolution of (15) using a symplectic integrator, but this is a task for the future.

Core bursting

A vortex tube with axial flow can experience a type of transition called vortex breakdown or more descriptively core bursting, which has been studied intensively in the last 30 years. It appears that under some conditions, a flow in which the flow varies slowly along the tube cannot be sustained, and variations with a scale comparable to the core radius appear. The radius of the core can also increase and significant non-axisymmetric components can appear. If vortex tubes constitute an important part of the turbulent velocity field, then it can be expected that core bursting will also play a role in the structure and evolution of the turbulence, but as yet no attempt to incorporate the phenomena into vortex models has been carried out.

The theory of core bursting for laminar vortices is still uncertain. There are probably at least three possible mechanisms. First, external pressure gradients may decelerate the flow causing a stagnation point to appear on the axis and the formation of a bubble of reverse flow; the core bursting is then akin to boundary layer separation. Second, the vortex may become unstable to non-axisymmetric disturbances. Third, downstream boundary conditions may become incompatible with the flow in the tube, so that the functions describing the dependence of flow properties on axial distance develop limit points or folds. These would be associated with the existence of standing waves of long wavelength, and could be interpreted in terms of a transition between subcritical and supercritical flow.

ACKNOWLEDGMENT

The support during the writing of this review by the Department of Energy, Applied Mathematical Sciences (DE-AS-03-76ER72012 KC-07-01-01) is gratefully acknowledged.

REFERENCES

Bayly, B.J., 1986, Three-dimensional instability of elliptical flow, Phys. Rev. Lett., 57, 2160:2163.

Buntine, J.D., and Pullin, D.I., 1989, Merger and cancellation of strained vortices, J. Fluid Mech., 205, 263:295.

Corrsin, S., 1962, Turbulent dissipation fluctuations, Phys. Fluids 5, 1301:1302.

Deem, G.S. and Zabusky, N.J., 1978, Vortex waves: stationary V states, interactions, recurrence, and breaking, Phys. Rev. Lett., 40, 859:862.

Dritschel, D.G., 1988, The repeated filamentation of two-dimensional vorticity interfaces, J. Fluid Mech., 194, 511:547.

Kambe, T. and Takao, T., 1971, Motion of distorted vortex rings, J. Phys. Soc. Japan, 31, 591:599.

Kelvin, Lord, 1880, Vibrations of a columnar vortex, Phil. Mag., (5) 10, 155:168.

Krutzsch, C.H., 1939, Über eine experimentell beobachtete Erscheinung an Wirbelringen bei ihrer translatorischen Bewegung in wirklichen Flussigkeiten, Ann. Phys., 35, 497:523.

Landman, M.J. and Saffman, P.G., 1987 On the 3-dimensional instability of vortices in a viscous fluid, Phys. Fluids, 30, 2339:2342.

Lundgren, T. S., 1982, Strained spiral vortex model for turbulent fine structure, Phys. Fluids 25, 2193:2203.

Melander, M.V., McWilliams, J.C. and Zabusky, N.J., 1987, Axisymmetrization and vorticity-gradient intensification of an isolated two-dimensional vortex through filamentation. J. Fluid Mech. 178, 137:159.

Moffatt, H.K., 1984, Simple topological aspects of turbulent vorticity dynamics. Turbulence and chaotic phenomena in fluids (Ed T. Tatsumi), 223:230, Elsevier.

Moore, D.W., 1979, The spontaneous appearance of a singularity in the shape of an evolving vortex sheet, Proc. Roy. Soc. A, 365, 105:119.

Moore, D.W. and Saffman, P.G., 1971, Structure of a line vortex in an imposed strain. Aircraft wake turbulence and its detection, 339:354, Plenum press..

Moore, D.W. and Saffman, P.G., 1975, The density of organized vortices in a turbulent mixing layer, J. Fluid Mech., 69, 465:473.

Pierrehumbert, R.T., 1986, A universal shortwave instability of two-dimensional eddies in an inviscid fluid, Phys. Rev. Lett., 57, 2157:2159.

Polvani, L.M., Flierl, G.R., and Zabusky, N.J., 1989, Filamentation of unstable vortex structures via separatrix crossing: A quantitative estimate of onset time, Phys. Fluids, A1, 181:184.

Pullin, D.I., 1981, The nonlinear behavior of a constant vorticity interface at a wall, J. Fluid Mech., 108, 401:421.

Pullin, D.I., 1989, On similarity flows containing two-branched vortex sheets. Mathematical aspects of vortex dynamics (Ed R Caflisch), SIAM 97:106.

Pullin, D.I., Jacobs, P.A., Grimshaw, R.H.J. and Saffman, P.G., 1989, Instability and filamentation of finite-amplitude waves on vortex layers of finite thickness. J. Fluid Mech, 209, 359:384.

Pullin, D.I. and Moore, D. W., 1990, Remark on a result of D. Dritschel. Phys. Fluids, to appear.

Pumir, A. and Siggia, E. 1990, Collapsing solutions to the 3-D Euler equations, Phys. Fluids, A2, 220:241.

Roberts, K.V. and Christiansen, J.P., 1972, Topics in computational fluid mechanics, Comput. Phys. Commun, 3(supp), 14:32.

Robinson, A.C. and Saffman, P.G., 1982, Three-dimensional stability of vortex arrays, J. Fluid Mech., 125, 411:427.

Robinson, A.C. and Saffman, P.G., 1984, Stability and structure of stretched vortices, Stud. App. Math., 70, 163:181.

Saffman, P.G., 1966, Lectures on homogeneous turbulence, Topics in nonlinear physics (Ed N. Zabusky), 485:614. Springer Verlag.

Saffman, P.G., 1981, Vortex interactions and coherent structures in turbulence, Transition and Turbulence (Ed R.E. Meyer), 149:166. Academic Press.

Saffman, P.G., 1990, A model of vortex reconnection, J. Fluid Mech., 212, 395:402.

Schatzle, P.R., 1987, An experimental study of fusion of vortex rings, Ph. D. thesis. Caltech.

Scorer, R.S. and Davenport L.J., 1970, Contrails and aircraft downwash. J. Fluid Mech., 43, 451:464.

Synge, J.L., and Lin, C.C. 1943, On a statistical model of isotropic turbulence, Trans. Roy. Soc. Canada, 37, 45:79.

Tennekes, H., 1968, Simple model for the small-scale structure of turbulence, Phys. Fluids, 11, 669:670.

Townsend, A.A., 1951, On the fine scale structure of turbulence, Proc. Roy. Soc. A, 208, 534:542.

Vincent, A. and Meneguzzi, M., 1990, The spatial structure and statistical properties of homogeneous turbulence, submitted, J. Fluid Mech.

Widnall, S.E., Bliss, D.B. and Tsai, C-Y. 1974 The instability of short waves on a vortex ring. J. Fluid Mech. 66, 35:47.

CONTROL OF THE TURBULENT BOUNDARY LAYER AND DYNAMICAL SYSTEMS THEORY: AN UPDATE[*]

Gal Berkooz, Philip Holmes, John Lumley

Cornell University
Ithaca, NY 14853, USA

ABSTRACT

We expand the velocity field in the vicinity of the wall in empirical eigenfunctions obtained from experiment. Truncating our system, and using Galerkin projection, we obtain a closed set of non-linear ordinary differential equations with ten degrees of freedom. We find a rich dynamical behavior, including in particular a heteroclinic attracting orbit giving rise to intermittency. The intermittent jump from one attracting point to the other resembles in many respects the bursts observed in experiments. Specifically, the time between jumps, and the duration of the jumps, is approximately that observed in a burst; the jump begins with the formation of a narrowed and intensified updraft, like the ejection phase of a burst, and is followed by a gentle, diffuse downdraft, like the sweep phase of a burst. The magnitude of the Reynolds stress spike produced during a burst is limited by our truncation. The behavior is quite robust, much of it being due to the symmetries present (Aubry's group has examined dimensions up to 128 with persistence of the global behavior). We have examined eigenvalues and coefficients obtained from experiment, and from exact simulation, which differ in magnitude. Similar behavior is obtained in both cases; in the latter case, the heteroclinic orbits connect limit cycles instead of fixed points, corresponding to cross-stream waving of the streamwise rolls. The bifurcation diagram remains structurally similar, but somewhat distorted. The role of the pressure term is made clear - it triggers the intermittent jumps, which otherwise would occur at longer and longer intervals, as the system trajectory is attracted closer and closer to the heteroclinic cycle. The pressure term results in the jumps occurring at essentially random times, and the magnitude of the signal determines the average timing. Stretching of the wall region shows that the model is consistent with observations of polymer drag reduction. Change of the third order coefficients, corresponding to acceleration or deceleration of the mean flow, changes the heteroclinic cycles from attracting to repelling, increasing or decreasing the stability, in agreement with observations. The existence of fixed points is an artifact introduced by the projection; however, a decoupled model still displays the rich dynamics. Numerous assumptions made in Aubry et al. (1988) can now be proved exactly. Feeding back eigenfunctions with the proper phase can delay the bursting, (the heteroclinic jump to the other fixed point), decreasing the drag. It is also possible to speed up the bursting, increasing mixing to control separation. Our approach is optimal for short time tracking in control.

THE PROPER ORTHOGONAL DECOMPOSITION

Lumley (1967) proposed a method of identification of coherent structures in a random turbulent flow. This uses what Loève (1955) called the Proper Orthogonal Decomposition, and which is often called the Karhunen-Loève expansion. An advantage of the method is its objectivity and lack of bias. Given a realization of an inhomogeneous, energy integrable velocity field, it consists of projecting the random field on a candidate structure, and selecting the structure which maximizes the projection in quadratic mean. The calculus of variations reduces this problem to a Fredholm integral equation of the first kind whose symmetric kernel is the autocorrelation matrix. The properties of this integral equation are given by Hilbert Schmidt theory. There is a denumerable set of eigenfunctions (structures). The eigenfunctions form a complete orthogonal set, which means that the random field can be reconstructed. The coefficients are uncorrelated and

---

[*] Prepared for presentation at NATO Advanced Research Workshop *The Global Geometry of Turbulence*: *Impact of Nonlinear Dynamics*, July 8-14, 1990, Rota, Spain. Supported in part by: the U. S. Air Force Office of Scientific Research, The U. S. Office of Naval Research (Mechanics Branch and Physical Oceanography Program), The U. S. National Science Foundation (programs in Applied Mathematics, Fluid Mechanics, Meteorology and Mechanics, Structures & Materials) and the NASA Langley Research Center.

their mean square values are the eigenvalues themselves. The Kernel can be expanded in a uniformly and absolutely convergent series of the eigenfunctions and the turbulent kinetic energy is the sum of the eigenvalues.

The most significant point of the decomposition is perhaps the fact that the convergence of the representation is optimally fast since the coefficients of the expansion have been maximized in a mean square sense. Berkooz *et al.* (1990) have shown that the n terms of this decomposition contain at least as much energy as n terms of any other decomposition.

Application of the Proper Orthogonal Decomposition to the Shear Flow of the Wall Region

The flow of interest here is three dimensional, approximately homogeneous in the streamwise direction (x_1) and spanwise direction (x_3), approximately stationary in time (t), inhomogeneous and of integrable energy in the normal direction (x_2). In the homogeneous directions the spectrum of the eigenvalues becomes continuous, and the eigenfunctions become Fourier modes, so that the proper orthogonal decomposition reduces to the harmonic orthogonal decomposition in those directions. See Lumley (1967, 1970, 1981) for more details.

We want a three dimensional decomposition which can be substituted in the Navier-Stokes equations in order to recover the phase information carried by the coefficients. We measure the two velocities at the same time and determine $<u_i(x_1,x_2,x_3,t)\, u_j(x'_1,x'_2,x'_3,t)> = R_{ij}$. From R_{ij} we will determine the eigenfunctions. Since the flow is quasistationary, R_{ij} does not depend on time, nor do the eigenvalues and eigenfunctions. The information in time is carried by the coefficients $a^{(n)}$ which are still "stochastic", but now evolve under the constraint of the equations of motion. We also change the Fourier integral into a Fourier series, assuming that the flow is periodic in the x_1 and x_3 directions. The periods L_1, L_3 are determined by the first non-zero wave numbers chosen. Finally, each component of the velocity field can be expanded as the triple sum

(1) $$u_i(x_3,x_2,x_3,t) = \frac{1}{\sqrt{L_1 L_3}} \sum_{k_1 k_3 n} e^{2\pi i(k_1 x_1 + k_3 x_3)}\, a^{(n)}_{k_1 k_3}(t)\phi^{(n)}_{i(k_1 k_3)}$$

where

(2) $$\int \Phi_{ij}(x_2, x'_2)\phi_j^{(n)}(x'_2)dx'_2 = \lambda^{(n)}\phi_i^{(n)}(x_2),$$

and we have to solve equation (2) for each pair of wave numbers (k_1,k_3). Φ_{ij} now denotes the Fourier transform of R_{ij} in the x_1, x_3 directions.

EXPERIMENTAL RESULTS

The candidate flow we are investigating is the wall region (which reaches $x_2^+ = 40$; x_2^+ is the distance from the wall normalized by kinematic viscosity and friction velocity) of a pipe flow with almost pure glycerine (98%) as the working fluid Herzog (1986). From this data the autocorrelation tensor R_{ij} was obtained and the spatial eigenfunctions were extracted by numerical solution of the eigenvalue problem. The results show that approximately 60% of the total kinetic energy is contained in the first eigenmode (figure 1) and that the first three eigenmodes capture essentially the entire flow field as far as these statistics are concerned.

THE DYNAMICAL EQUATIONS

We decompose the velocity—or the pressure—into the mean (defined using a spatial average) and fluctuation in the usual way. We substitute this decomposition into the Navier-Stokes equations. Taking the spatial average of these equations we obtain, in the quasi stationary case, an approximate relation between the divergence of the Reynolds stress and the mean pressure and velocity.

(3) $<u_{i,j}u_j> = -1/\rho\, P_{,i} + \nu\, U_{,jj}\delta_{i1}$.

(where $u_{i,j}$ indicates the derivative with respect to x_j of u_i, and similarly for the other terms; repeated indices are summed). Equation (4) may be solved to give the mean velocity in terms of the Reynolds stress in a parallel flow. This reduces the slope of the mean velocity as the structures become stronger, stabilizing the system. (This depends on the sign of the Reynolds stress, which is certainly positive for the first structure, though not necessarily for the higher modes).

After taking the Fourier transform of the Navier Stokes equations and introducing the truncated expansion, we apply Galerkin projection by multiplying the equations by each successive eigenfunction in turn, and integrating over the domain.

212

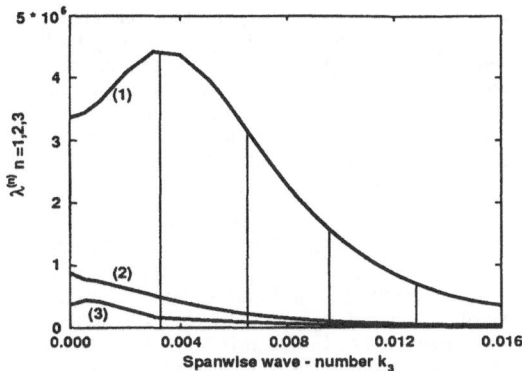

Figure 1. Convergence of the proper orthogonal decomposition in the near-wall region ($x_2^+ = 40$) of a pipe flow according to experimental data. Turbulent kinetic energy in the first three eigenmodes. $\lambda^{(n)}$ (n = 1, 2, 3) function of the spanwise wavenumber (from Herzog, 1986).

By use of the continuity equation and the boundary conditions (vanishing of the normal component at the wall, and at infinity) it can be seen by integration by parts that the pressure term would disappear if the domain of integration covered the entire flow volume. Since this is not the case (rather the domain is limited to $X_2^+ = 40$, where X_2^+ indicates the value of x_2^+ at the upper edge of the integration domain), there remains the value of the pressure term at X_2^+, which represents an external perturbation coming from the outer flow.

ENERGY TRANSFER MODEL

The exact form of the equations obtained from the decomposition, truncated at some cut-off point (k_{1c}, k_{3c}, n_c), does not account for the energy transfer between the resolved (included) modes and the unresolved smaller scales. The influence of the missing scales will be parameterized by a simple generalization of the Heisenberg spectral model in homogeneous turbulence. Such a model is fairly crude, but we feel that its details will have little influence on the behavior of the energy-containing scales, just as the details of a sub-grid scale model have relatively little influence on the behavior of the resolved scales in a large eddy simulation. This is a sort of St. Venant's principle, admittedly unproved here, but amply demonstrated experimentally by the universal nature of the energy containing scales in turbulence in diverse media having different fine structures and dissipation mechanisms (see Lumley (1972) for a fuller discussion). The only important parameter is the amount of energy absorbed.

We will refer to α_1 as a Heisenberg parameter. We will adjust α_1 upward and downward to simulate greater and smaller energy loss to the unresolved modes, corresponding to the presence of a greater or smaller intensity of smaller scale turbulence in the neighborhood of the wall. This might correspond, for example, to the environment just before or just after a bursting event, which produces a large burst of small scale turbulence, which is then diffused to the outer part of the layer.

A term representing the energy fluctuation in the unresolved field due to the resolved field appears in the equation for the resolved field, and can be combined with the pressure term. We assume that the deviation (on the resolved scale) in the kinetic energy of the unresolved scales is proportional to the rate of loss of energy by the resolved scales to the unresolved scales. This term gives some quadratic feed-back. For generality we call this parameter α_2, although in all work presented in this paper, we have set $\alpha_1 = \alpha_2$.

Thus the Heisenberg model introduces two parameters in the system of equations, one, α_1, in the linear term, the other one, α_2, in the quadratic term. The equations therefore have the following form:

(4) $\qquad da_{k_1 k_3}^{(n)} /dt = L + (\nu + \alpha_1 \nu_T)L' + Q + \alpha_2 Q' + C$

where L and L' represent the linear terms, Q the direct quadratic terms, Q' the quadratic pseudo-pressure term and C the cubic terms arising from the Reynolds stress.

IMPLICATIONS FOR THE FLOW IN THE WALL REGION

Numerical integrations of 3, 4, 5 and 6 mode models have been carried out, but we shall only report in detail on the 6 mode (5 active mode) simulations here. Note that the (0, 0) mode is uncoupled and inactive.

There is a rich dynamical behavior, but we focus here on the behavior for $1.37 < \alpha < 1.61$, when a family of globally attracting double homoclinic cycles G exists, connecting pairs of saddle points which are π out of phase with respect to their second (x_2, y_2) components. The system spirals away from one saddle (the laminar phase) until it is far enough to leap to the other, and then repeats the process, to return to the first. The existence of the cycles G implies that, after a relatively brief and possibly chaotic transient, almost all solutions enter a tubular neighborhood of G and thereafter follow it more and more closely. As they approach G, the duration of the "laminar" phase of behavior increases while the bursts remain short. In an ideal, unperturbed system, the laminar duration would grow without bound, but small numerical perturbations, such as truncation errors, prevent this occurring in our numerical simulations. More significantly, the pressure perturbation will limit the growth of the laminar periods. Thus there is an effective maximum duration of events, which is reduced as α is decreased from the critical value $\alpha_b \sim 1.61$.

In Figure 2 we show the time histories of the modal coefficients for $\alpha = 1.45$. A description of the motion of the eddies during a burst is given in Figure 3 for $\alpha = 1.4$ by plotting u_2 and u_3 at 14 different times during one of the transitions shown in figure 2. Before and after the event, two pairs of streamwise vortices are present in the periodic box. However, pictures 1 and 14 are shifted in the spanwise direction by π. Moreover it is possible to adjust the value of the Heisenburg parameter ($\alpha \sim 1.5$) so that the bursting period is 100 wall units as experimentally observed (Kline *et al.*, 1967). It is found that, in this case, the "burst" lasts 10 wall units which is also the right order of magnitude. During one of these events there is a sudden increase in Reynolds stress, though smaller than observed. An event consists of a sudden intensification and sharpening of the updraft between eddies (5, 6 & 7, fig.3), followed by a drawing apart of the eddies, and the establishment of a gentle downdraft between them (9, 10 & 11, fig. 3): these are similar respectively to the ejection and sweep events that are observed.

PHYSICAL INTERPRETATION

Keith Moffatt points out that non-trivial solutions to the Navier Stokes equations, having no streamwise variation and driven by a streamwise (mean) velocity dependent only upon distance from the wall, should ultimately decay. This is easily seen from a simplified model with a single cross-stream Fourier mode for each velocity component and a fixed linear mean velocity profile. The streamwise velocity component (u_1) is fed from the mean velocity gradient by the component normal to the wall (u_2). However, neither u_2 nor u_3 has a source of energy. Both u_2 and u_3 decay exponentially from their initial values, with u_1 at first rising, but ultimately decaying exponentially also. The ratio of the Reynolds stress to the energy at first rises, but ultimately decays to zero algebraically.

In our ten-dimensional model, however, the ratio of Reynolds stress to energy does not decay, but is bounded away from zero, as is easily proved (Berkooz *et al.*, 1990), providing the energy source which makes the non-trivial fixed points and heteroclinic cycles possible. Berkooz has also shown (op cit) that, since the contributions of some of the higher modes to the Reynolds stress are of opposite sign to that of the first mode, the Reynolds stress for higher-order approximations will not be bounded away from zero. Thus we expect an "accurate" model lacking streamwise variations, but including many spanwise modes and several eigenfunctions, to exhibit the appropriate decay properties, the trivial solution $u = 0$ being a stable fixed point.

The proximal cause for the non-zero Reynolds stress/energy ratio when only the first eigenfunction is included, therefore, is the fact that the vector eigenfunctions have scalar coefficients. Hence, the u_1 and u_2 components in each mode are held in a non-evolving ratio. The eddies which occur in the real boundary layer, of course, have streamwise variation, and temporal variation. They each go through a life cycle, growing to a maximum and decaying. Only in a statistical sense is the ensemble stationary. The stationary behavior of the model reflects the stationary behavior of the ensemble, rather than the non-stationary behavior of the members. The Reynolds stress of the model (relative to the energy) is endowed by the empirical eigenfunctions with the value measured in the real boundary layer. In this way the cross-stream velocity components can extract energy from the mean flow. Hence, the empirical eigenfunctions are, in a sense, a closure approximation that embodies the effects of streamwise structure and unsteadiness in the value of the Reynolds stress represented by the relative sizes of their components. In this sense the model only appears to belong to the subspace of fields without streamwise variation.

In the present context, the vital question is whether the complex and apparently physically significant dynamical behavior of the ten-dimensional model is an artifact of the projection, like the fixed points. Happily we can give strong assurance that this is not so. We have constructed a decoupled model (Berkooz *et al.*, 1990) in which the streamwise component and those normal to the streamwise direction have separate coefficients. Solutions of this model decay properly, as described in the first paragraph (figure 4). The Reynolds stress (relative to the energy) decays to zero. The "fixed points" now drift slowly toward the origin. They are still connected by "ghosts" of heteroclinic cycles, so that the same bursting phenomenon occurs, but the bursts are now modulated by the slow decay. The bursts only occur while the cross-stream components are non-zero. There is a relatively long period after the cross-stream components have decayed during which only the streamwise component remains, no bursting occurs, and the streamwise component decays slowly to zero. We feel that this is probably the explanation for the common observation that the sublayer consists primarily of "streaks" - the streamwise remnants of eddies whose cross-stream components have decayed. The fraction of time during which there is cross stream activity (u_2, u_3 and bursting) is relatively short, and most of the time the scene would be dominated by the streak left behind.

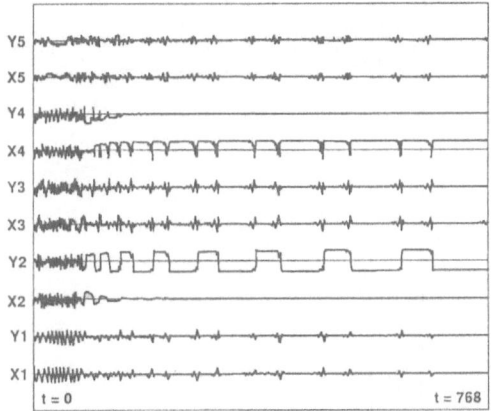

Figure 2. Time histories of the real (x_i) and imaginary (y_i) parts of the coefficients for a value of the Heisenberg parameter of $\alpha = 1.45$.

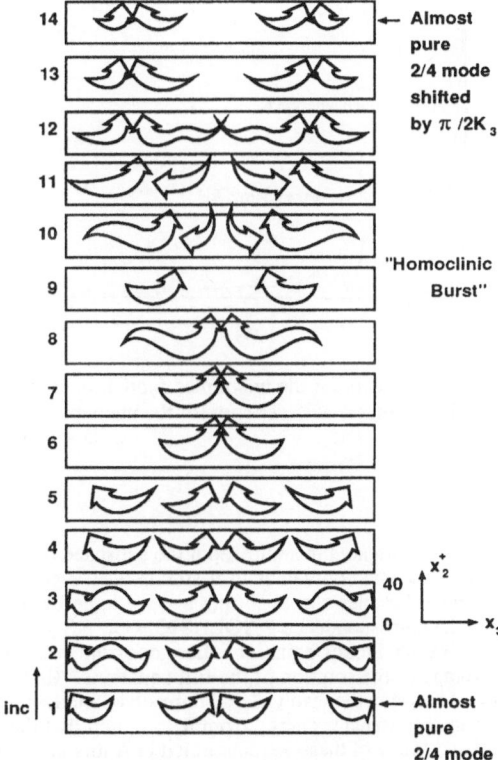

Figure 3. Intermittent solution (corresponding to an Heisenberg parameter a = 1.4) during a burst, times equally speced from the bottom. Each snapshot is a cross section of the flow (normal to the streamwise firection) from the wall (bottom) to $x_2^+ = 40$ (top), of width $Dx_3^+ = 333$.

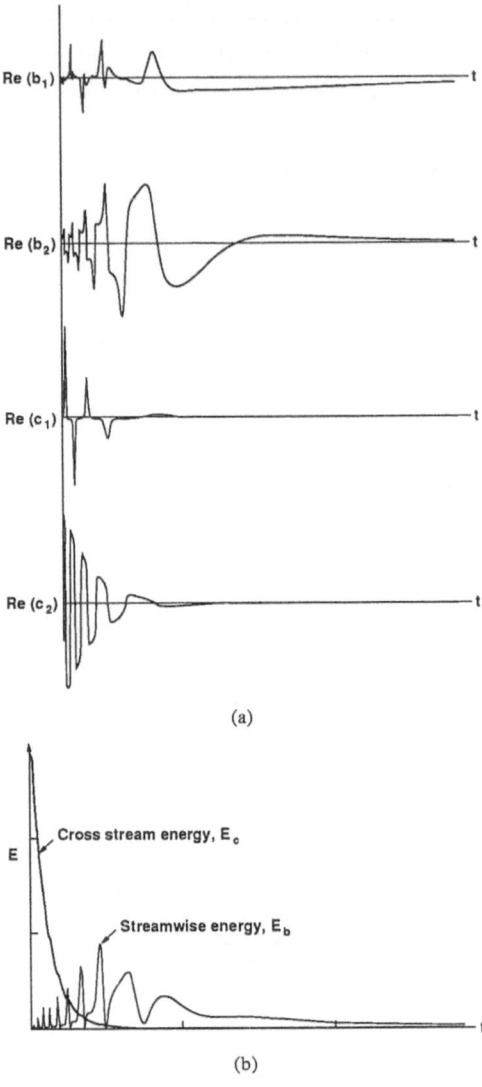

Re (b₁)

Re (b₂)

Re (c₁)

Re (c₂)

(a)

E

Cross stream energy, E_c

Streamwise energy, E_b

(b)

Figure 4. (a) Evolution of modal components for the uncoupled model: b_j are the streamwise and c_j the cross stream components respectively. (b) Evolution of the energy for the uncoupled model.

Holmes *et al.* (1990) have investigated in some depth the subspace of no streamwise variation. If we let P( ) be a projection operator, which is equivalent to a streamwise average, we can split the field into resolved modes $\mathbf{r} \in R$ and unresolved modes $\mathbf{s} \in S$, so that $\mathbf{u} = \mathbf{r} + \mathbf{s}$, and $P(\mathbf{s}) = 0$, $P(\mathbf{u}) = \mathbf{r}$, then Berkooz (Berkooz *et al.* 1990) has shown the correspondence P(SOLUTION OF NS) to the SOLUTION P(NS). That is, if we streamwise average the Navier Stokes equations, how does the solution of the averaged equations correspond to the streamwise average of the solution of the full equations. He has shown in addition: that the Leonard stresses (the cross stresses between the resolved and unresolved modes) vanish on the average; that the perturbation Reynolds stresses can only transfer energy from R to S; and that the energy loss from R to S can be represented by an eddy viscosity. Many of these were assumed in Aubry *et al.* (1988).

We have truncated in our ten-dimensional system the mechanism that represents the production of higher wavenumber energy when an intense updraft is formed, presumably as a result of a secondary instability. Thus, although our eddies are capable of exhibiting the basic bursting and ejection process, the labor is in vain. A contribution is made only to the low wavenumber part of the streamwise fluctuating velocity and the Reynolds stress. Recently however, Aubry & Sanghi (1989) have extended the model to include 1, 2 and 3

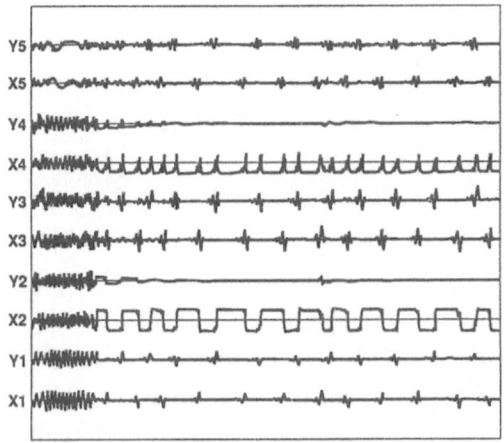

Figure 5. Similar to figure 2, but with the pressure term operative. Note that the inter-burst period is randomized, and on average, stabilized.

streamwise Fourier components, going up to studies of 38 complex (78 real) differential equations (very recently extended to 128 real). Addition of the streamwise components does not change the basic behavior of the system. For the most part the streamwise components are relatively quiescent; following a burst, however, they are excited, contributing to the Reynolds stress.

Initially we did not exercise the pressure term, which appeared due to the finite domain of integration. The order of magnitude that we estimated for this term was small, and for that reason we at first neglected it. It has, however, an important effect, while not changing the qualitative nature of the solution.

The term has the form of a random function of time, with a small amplitude. This slightly perturbs the solution trajectory constantly; away from the fixed points this has little effect, but when the solution trajectory is very close to these points, the perturbation has the effect of throwing the solution away from the fixed point, so that it need not wait long to spiral outward. This results in a thorough randomization of the transition time from one solution to the other, while having little effect on the structure of the solution during a burst. While in the absence of the pressure term (and round-off error), the interburst time tends to lengthen as the solution trajectory is attracted closer and closer to the heteroclinic cycle, with the pressure term, the mean time stabilizes.

One of the important findings of this work is the suggestion of the etiology of the bursting phenomenon. That is, presuming that the abrupt transitions from one fixed point to the other can be identified with a burst, these bursts appear to be produced autonomously by the wall region, but to be triggered by pressure signals from the outer layer. Whether the bursting period scales with inner or outer variables has been a controversy in the turbulence literature for a number of years. The matter has been obscured by the fact that the experimental evidence has been measured in boundary layers with fairly low Reynolds numbers lying in a narrow range, so that it is not really possible to distinguish between the two types of scaling. The turbulent polymer drag reduction literature is particularly instructive, however, since the sizes of the large eddies, and the bursting period, all change scale with the introduction of the polymer (Kubo & Lumley,1980; Lumley & Kubo, 1984). The present work indicates clearly that the wall region is capable of producing bursts autonomously, but the timing is determined by trigger signals from the outer layer. This suggests that events during a burst should scale unambiguously with wall variables. Time between bursts will have a more complex scaling, since it is dependent on the first occurrence of a *large* enough pressure signal *long* enough after a previous burst; "long enough" is determined by wall variables, but the pressure signal should scale with outer variables.

FURTHER CONSEQUENCES

We are, of course, concerned about the robustness of our findings. We have tried eigenfunctions generated from exact numerical simulations of channel flow, by Moser and Moin at the Center for Turbulence Research (Stanford/NASA Ames). These eigenfunctions are superficially similar to those from Herzog's data, but result in changes of the order of 20% in the values of the coefficients in the equations. The bifurcation diagram is similar, but the fixed points are replaced by limit cycles. Physically, this means that the eddies are wiggling from side to side instead of sitting still. This makes no essential difference, and is even more realistic physically. The intermittent behavior remains.

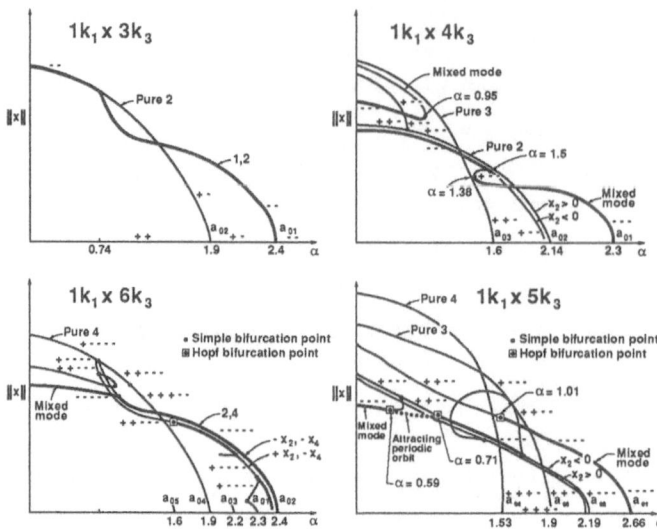

Figure 6. Bifurcation diagrams for models with varying numbers of cross-stream modes (from Stone, 1989). Note in the six-mode model that modes 3 and 5 have been suppressed for clarity. Note the similarity of the basic structure.

In her thesis, Stone (Stone & Holmes, 1990) investigated models with various numbers of cross-stream modes: 3, 4, 5 and 6. She found that the bifurcation diagrams had a backbone common to all of these systems, and were all structurally similar. In particular, the intermittent behavior was common to all. This is illustrated in figure 6.

In addition, Stone (Stone & Holmes, 1990) found that a small change in the value of the coefficients of the third order terms could change the heteroclinic cycles from attracting to repelling. This is illustrated in Figure 7 (lower), where one can see that the system begins on a traveling wave, but is gradually attracted to the heteroclinic cycle. In Figure 7 (upper) we see the opposite - the system starts on the heteroclinic cycle, but is repelled by it, and ends on a traveling wave. (We show only the values of the first two transverse Fourier modes - the others are quiescent). This would be a dynamical systems curiosity, if we could not relate it to the physics. However, if we consider the case $\partial_t U_1 = kU_1$ (an exponential increase or decrease of the mean velocity) we find that this results in a change in the real part of the cubic terms for $k_1 = 0$. When k changes sign the addition to the real part of the cubic term changes sign. This phenomenon is related to the destabilization and stabilization known to be induced by deceleration and acceleration of the flow (as by an adverse or favorable pressure gradient). Although we have discussed here the effect of temporal acceleration and deceleration, the same qualitative effect is obtained from a spatial acceleration and deceleration. Making the heteroclinic cycle more attractive would increase the time between bursts, stabilizing the flow, and vice versa.

Stone (Stone & Holmes, 1990) also predicted and measured histograms of the bursting period (Figure 8a). These look reasonably similar to measurements of the same by Kline *et al.* (1967), (Figure 8b).

Bloch and Marsden (1989) have shown that it is possible to stabilize this system by feedback, in the absence of noise. That is, if an eigenfunction is fed back with the proper phase, the system can be held in the vicinity of a fixed point for all time. In the presence of noise, however, (such as the pressure perturbation from the outer layer) the system cannot be stabilized completely; however, it can be held in a neighborhood of the fixed point for a longer time. When the system finally wanders so far from the fixed point as to make it uneconomical to recapture it, it is allowed to leave. The same procedure is carried out at the other fixed point. The effect is to increase the mean time between bursts, and hence to reduce the drag. Of course, the system can be made to work the other way, also, kicking the system away from the fixed point whenever it comes too close, resulting in a decrease in the mean time between bursts, and an increase in drag. This would be useful in avoiding separation or improving mixing, for instance. In recent work, Berkooz (1990) introduced the notion of short term tracking time, which measures the time over which a dynamical system model tracks the true dynamics accurately; for control, it must be of the order of the wall region time scales. Berkooz (1990) showed that dynamical systems based on the Proper Orthogonal Decomposition have, on the average, the best short term tracking time for a given number of modes.

In drag reduction by polymer additives, one of the accepted mechanisms (Kubo & Lumley, 1980; Lumley & Kubo, 1984) is the stabilization of the large eddies in the turbulent part of the flow, allowing the eddies to grow bigger and farther apart, as observed. Aubry *et al* (1989) tried stretching the eddy structure in the wall region, producing drag reduction, and found the bifurcation diagrams morphologically unchanged, except that the bifurcations occurred for larger and larger values of the Heisenberg parameter. This suggests that the motions giving rise to the bifurcations are more and more unstable, the more the region is stretched, requiring a larger and larger value of the Heisenberg parameter to stabilize them. Now, the Heisenberg parameter represents the loss of energy to the unresolved modes. However, the crudeness of the model is such, that it cannot distinguish between loss to the unresolved modes and loss to any other dissipation mechanism, such as viscosity or extensional viscosity. As far as the large scales are concerned, all losses are the same. Hence the findings of Aubry *et al* (1989) are completely consistent with the idea of the larger eddies being less stable, and able to grow to this larger, less stable size due to the stabilizing effect of the polymer.

Finally, Bloch and Marsden (1989) have shown that, within the assumptions of the scenario above, polymer drag reduction is equivalent to control of the wall region. That, is they showed that an increase of the Heisenberg parameter was equivalent to control by feeding back eigenfunctions, and would lead to a reduction in the bursting rate, and hence to a decrease in the drag. According to the scenario above, this would lead to a stabilization, and result in a growth of the eigenfunctions. This has the important consequence that a controlled boundary layer would be very similar to a polymer drag-reduced boundary layer. From experience with the polymer-drag-reduced boundary layer, we know that it would be a robust layer, still turbulent though with a reduced bursting rate, relatively insensitive to roughness and external disturbance. This is important from the standpoint of applications. Other drag reduction schemes connected with stabilization of the laminar layer are not robust in this sense, and are very sensitive to external disturbances and surface roughness.

From a practical point of view, the feed back could be implemented by an array of hot film sensors to detect the presence, location and strength of an eigenfunction. Piezoelectrically raised welts could produce a negative eigenfunction (see figure 9) by overturning the vorticity in the boundary layer.

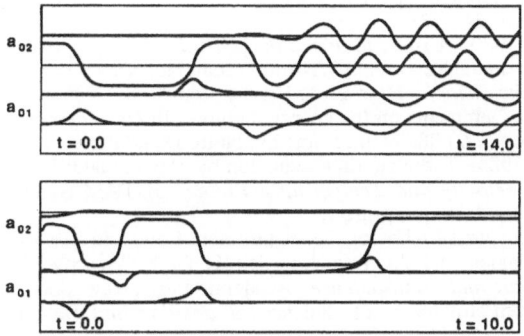

Figure 7. A model system with two streamwise modes (one active) and three cross-stream modes (two active). We show only the 01 and 02 modes (the others are unexcited). In the upper figure one of the third order coefficients is -2.69, and the heteroclinic cycle is repelling; in the lower, it is -3.00, and it is attracting (Stone, 1989).

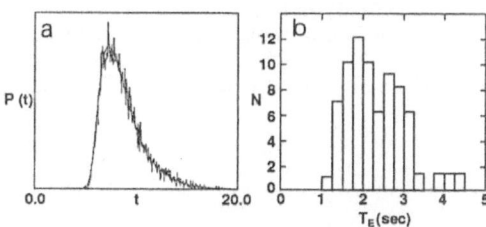

Figure 8. (a) The predicted and measured histogram of the bursting period from our model (Stone, 1989). (b) The measured histogram of the bursting period in the turbulent boundary layer, from Kline *et al.* (1967).

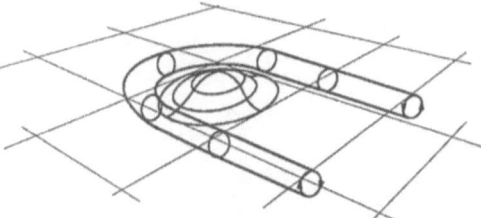

Figure 9. Schematic of a piezoelectrically produced welt on the surface, overturning the vorticity in the boundary layer and producing a negative eigenfunction.

BIBLIOGRAPHY

Armbruster, D., Guckenheimer, J. and Holmes, P. 1987. Heteroclinic cycles and modulated traveling waves in systems with O(2) symmetry. *Physica D* (to appear).

Aubry, N., Holmes, P., Lumley, J. L. and Stone, E. 1988. The dynamics of coherent structures in the wall region of a turbulent boundary layer. *J. Fluid Mech.* **192**: 115-173.

Aubry, N., Lumley and Holmes, P. , J. L. 1989 The effect of drag reduction on the wall region. *Theoretical and Computational Fluid Dynamics.* **1**, 229-248.

Aubry, N. and Sanghi, S. 1989. Streamwise and cross-stream dynamics of the turbulent wall layer. Proceedings, July meeting of ASME, New York. ed Ghia.

Berkooz, G. (1990). In preparation.

Berkooz, G., Holmes, P. J. and Lumley, J. L. (1990). Intermittent dynamics in simple models of the turbulent wall layer. *J. Fluid Mech.* (submitted for publication) .

Bloch, A. M. and Marsden, J. E. 1989. Controlling Homoclinic Orbits. *Theoretical and Computational Fluid Dynamics.* **1**(3):179-190.

Corino, E. R., and Brodkey, R.S. 1969. A visual investigation of the wall region in turbulent flow. *J. Fluid Mech.* **37**(1):1-30.

Golubitsky, M. & Guckenheimer, J. 1986. (eds). *Multiparameter Bifurcation Theory.* A.M.S. Contemporary Mathematics Series, No. 56. American Mathematical Society, Providence, R.I.

Herzog, S. 1986. *The large scale structure in the near-wall region of turbulent pipe flow.* Ph.D. thesis, Cornell University.

Holmes, P. J., Berkooz, G., and Lumley, J. L. 1990. Turbulence, dynamical systems and the unreasonable effectiveness of empirical eigenfunctions. *ICM '90 Proceedings*, Kyoto, Japan. In Press.

Kline, S.J., Reynolds, W.C., Schraub, F.A. and Rundstadler, P.W. 1967. The structure of turbulent boundary layers. *J. Fluid Mech.* **30**(4): 741-773.

Kubo, I. and Lumley, J. L. 1980. *A study to assess the potential for using long chain polymers dissolved in water to study turbulence.* Annual Report, NASA-Ames Grant No. NSG-2382. Ithaca, NY: Cornell.

Loève, M. 1955. *Probability Theory .* New York: Van Nostrand .

Lumley, J. L. 1971. Some Comments on the energy method. In *Developments in Mechanics 6*, eds. L. H. N. Lee and A. H. Szewczyk, pp. 63-88. Notre Dame IN: Notre Dame Press.

Lumley, J. L. and Kubo, I. 1984. Turbulent drag reduction by polymer additives: a survey. In *The Influence of Polymer Additives on Velocity and Temperature Fields.* IUTAM Symposium Essen 1984. Ed. B. Gampert. pp. 3-21. Berlin/Heidelberg: Springer.

Lumley, J.L. 1967. The structure of inhomogeneous turbulent flows. In *Atmospheric Turbulence and Radio Wave Propagation*, A.M. Yaglom and V.I. Tatarski:, eds.: 166-178. Moscow: Nauka.

Lumley, J.L. 1970. *Stochastic tools in turbulence.* Academic Press, New York.

Lumley, J.L. 1981. Coherent structures in turbulence. *Transition and Turbulence,* edited by R.E. Meyer, Academic Press, New York: 215-242.

Moffat, H. K. 1989 Fixed points of turbulent dynamical systems and suppression of non-linearity. In *Whither Turbulence*, ed. J. L. Lumley. Heidelberg: Springer. In press.

Moin, P. 1984. Probing turbulence via large eddy simulation. AIAA 22nd Aerospace Sciences Meeting.

Smith, C.R. and Schwarz, S.P. 1983. Observation of streamwise rotation in the near-wall region of a turbulent boundary layer. *Phys. Fluids* **26**(3) 641-652.

Stone, E. and Holmes, P. J. 1990. Random perturbations of heteroclinic attractors. *SIAM J. Appl. Math.* **50**: 726-743.

Tennekes, H. and Lumley, J.L. 1972. *A first course in turbulence.* Cambridge, MA: M.I.T. Press.

KOLMOGOROV SPECTRA AND INTERMITTENCY IN TURBULENCE

M.H. Jensen

NORDITA, DK-2100 Copenhagen Ø. Denmark

G. Paladin and A. Vulpiani

Dipartimento di Fisica Università de L'Aquila
I-67010 Coppito L'Aquila, Italy

ABSTRACT

We study a shell-model of three dimensional fully developed turbulence. By calculating the structure functions, we find corrections to the classical Kolmogorov theory due to an intermittent structure of energy dissipation. The intermittency can be described in terms of a multifractal scaling. The information dimension of the fractal measure given by the energy dissipation is found to be $D_1 \simeq 2.92$. We also study Lyapunov exponents and eigenvectors and relate the spatial intermittency to these dynamical quantities.

1. Introduction

The understanding of intermittent structures in fully developed turbulence is one of the main topics in turbulence research [1]. In the celebrated Kolmogorov cascade scenario of energy transfer, it is assumed that the energy dissipation at the Kolmogorov inner scale and the transfer between successive turbulent eddies are uniform in space leaving out intermittency phenomena. However, in several experimental measurements and numerical simulations of Navier-Stokes equations it has been observed that the large majority of the energy dissipation does not fill the space uniformly but concentrates rather on some fractal-like structures [2]. Notable examples are the velocity measurements in a channel by Sreenivasan and Meneveau [3] and the simulations by Siggia [4]. In both cases very "spiky" structures of energy dissipation have been observed giving some hints that the assumptions about uniformity in space might not be valid.

One should therefore like to test the Kolmogorov theory in details both experimentally and by simulations. Indeed, such measurements have been performed quite extensively over the last ten years. In the experiments on channel flows by Anselmet et al [5] the velocity structure functions were measured and a strong deviation from the predictions of the Kolmogorov theory is evident at high moments. These data can be fitted by the random β model [6] indicating that the dissipation takes place on a multifractal set [7]. Recently, an extensive numerical simulation of the Navier-Stokes equations also revealed velocity structure functions that indicated a similar underlying multifractal structure of the energy dissipation [8]. However, both in the experiments and in the direct numerical simulations the main difficulty arises when the scaling exponents of the structure functions are extracted. Typically, due to size limitations of the systems, it is only possible to identify scaling behavior over around half a decade in length scales. It is thus hard to be completely confident in the extrapolated exponent.

Our approach has been to investigate an approximate model of the Navier-Stokes equations where we could obtain a scaling behavior for the structure functions over several decades but where the basic requirements of fully developed turbulence, such as energy transfer and conservation (in the inertial range), are still present [9]. The appropriate model we choose is a shell model in Fourier space of the Navier-Stokes equations, introduced by Gledzer [10]. Surprisingly, this model shows corrections to Kolmogorov theory due to intermittency which are in quantitative agreement with experimental results. We calculate the fractal set of dissipation and try also to relate the corrections to the dynamics of the flow as determined by the Lyapunov exponents and eigenvectors. Besides intermittency effects, the model have an interesting feature: there are many Lyapunov exponents which are almost zero and just a few are positive. We speculate that such marginality might also be a characteristic property in fully developed turbulence. To understand the relation between the dynamics and the fractal structure of dissipation we introduce an "instantaneous" Lyapunov exponent. Spikes in this exponent appear to be strongly correlated to bursts in the energy dissipation and a localization of the corresponding eigenvector at the Kolmogorov scale. However, between the bursts, the eigenvector spreads over the whole inertial range indicating that the intermittency corrections is a complex phenomena involving energy transfer at all length scales.

2. The shell model

Various versions of shell models of the Navier-Stokes equations have been studied extensively in the last ten years for instance by Siggia et al [11] and Grappin et al [12]. Quite recently a particular system, the Gledzer model, was investigated by Yamada and Ohkitani to study the energy cascade and the Kolmogorov scaling [13,14]. In the model, the Fourier space is divided into wave vector shells $k_n = 2^n k_0$. For each shell k_n, there is a corresponding typical velocity difference u_n (over a length scale $\sim 1/k_n$). The velocity is here taken to be a complex number. Assuming that each shell interacts only with its neighbors and nearest-neighbors [15] we obtain the following set of N coupled differential equations

$$(\frac{d}{dt} + \nu k_n^2)\, u_n \; = i(a_n u_{n+1}^* u_{n+2}^* \; + \; b_n u_{n-1}^* u_{n+1}^* \; + \; c_n u_{n-1}^* u_{n-2}^*) + f\delta_{n,4} \qquad (1)$$

Here ν is the viscosity and the system is forced with an amplitude f in the fourth mode. The energy is thus injected into the system uniformly and deterministically at a large length scale. We shall use the notation:

$$\frac{du_n^R}{dt} = F_{2n-1}(\mathbf{u}) \qquad \frac{du_n^I}{dt} = F_{2n}(\mathbf{u}),$$

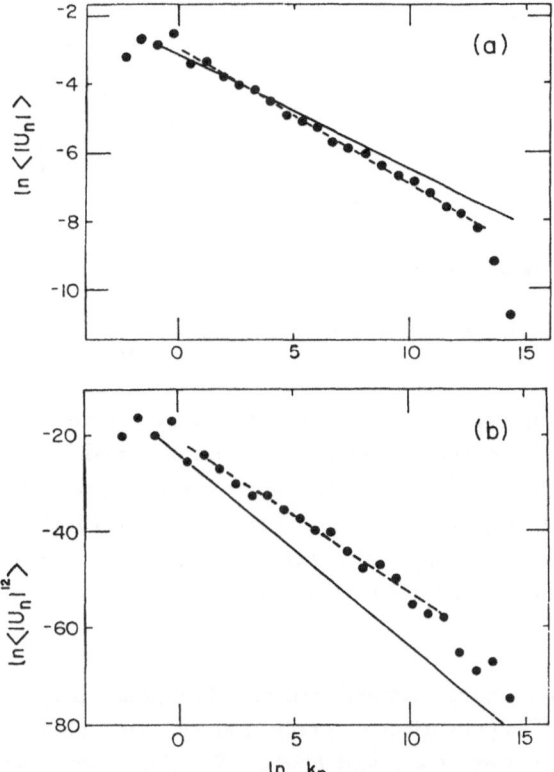

Fig.1. Plot of the structure functions, $\ln< |u_n|^Q >$ versus $\ln k_n$ for a): $Q=1$ and b): $Q=12$ found from simulations with $N=27$ over 10^4 time units. The full lines have slopes respectively -1/3 and -4 from Kolmogorov scaling. The dashes lines have slopes -0.39 and -3.18, respectively.

where u_n^R (u_n^I) denotes the real (imaginary) part of u_n. By demanding energy conservation in the invicid limit without forcing ($\nu = f = 0$) we obtain the coefficients in Eq.(1) as

$$a_n = k_n, \qquad b_n = \frac{-k_{n-1}}{2} \qquad c_n = \frac{-k_{n-2}}{2} \tag{2}$$

$$b_1 = b_N = c_1 = c_2 = a_{N-1} = a_N = 0$$

Notice that the Kolmogorov cascade law $u_n \propto k_n^{-\frac{1}{3}}$ is an unstable fixed point of Eq.(1) when $\nu = f = 0$.

To calculate the structure functions, we consider the scaling of the velocity differences over small length scales (but larger than the Kolmogorov inner length): $\delta u(\ell) \equiv |\mathbf{u}(\mathbf{x} + \ell) - \mathbf{u}(\mathbf{x})| \sim \ell^h$. From dimensional analysis one obtains the standard Kolmogorov result $h = \frac{1}{3}$ for the singularity exponent such that the higher moments scale as

$$< \delta u(\ell)^Q > \propto \ell^{\varsigma_Q} , \qquad \text{with} \qquad \zeta_Q = Q/3 \tag{3}$$

where $< \cdot >$ is a spatial average. In the shell model the velocity differences are instead

223

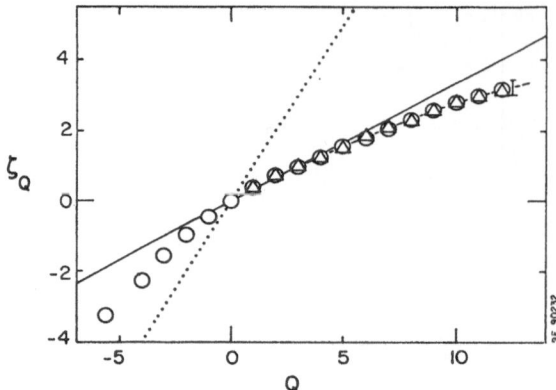

Fig.2. The structure function exponent ζ_Q plotted versus Q for values between -6 and 12. The squares are obtained from integration of Eq.(1) with N=27 shells, the crosses with N=19 shells. The full line shows the Kolmogorov result $\zeta_Q = Q/3$; the dashed line is the random β model fit with $x = 0.12$. The slope of ζ_Q seems to approach 1 as Q becomes negative (indicated by the dotted line).

averaged over time and we have obtained structure functions up to the order $Q = 12$, by a numerical integration of Eq.(1) with N=19 and 27 shells. Fig.1 shows a plot of $\ln < |u_n|^Q >$ versus $\ln k_n$ for the two cases, Q=1 and Q=12. We find a good scaling over about five decades between the shell of the energy input and the shell of the Kolmogorov inner scale which defines the inertial range. The slope is $-\zeta_Q$ and is different from the Kolmogorov result $< |u_n|^Q > \propto k_n^{-Q/3}$ which is indicated by the full drawn line. Fig.2 collects the results for the structure function exponents for two different cases, one with N=19 shells and $\nu = 10^{-6}$, the other with N=27 and $\nu = 10^{-9}$. We note that the exponent ζ_Q is not linear in Q. The results are in very good agreement with the experimental results by Anselmet et al obtained from velocity measurements in a channel flow [5]. Our data suggest an underlying multifractal structure of the energy dissipation since ζ_Q can be obtained by the random β model [6,7], as shown by the dotted line in Fig.2.

To understand the underlying multifractal structure in more details we introduce the generalized dimensions D_Q [16] of the energy dissipation. The "measure" of dissipation in a ball Λ of radius ℓ centered in a given fluid point x is $\epsilon(\ell) = \int_\Lambda d^3x \, \epsilon^*(x)$, where $\epsilon^*(x)$ is the density of energy dissipation . In the inertial range the moments scale as

$$< \epsilon(\ell)^Q > \propto \ell^{(Q-1)\,D_Q\,+\,3} \tag{4}$$

By dimensional counting, one sees $\epsilon(\ell) = \ell^3 \, \delta u^3(\ell)/\ell$ so that the relation between dimensions and structure function exponents is

$$D_Q \;=\; \frac{\zeta_{3Q} + 2Q - 3}{Q - 1} \tag{5}$$

D_0 is the fractal dimension of the set where the energy dissipation concentrates. We numerically find that $\lim_{Q \to 0} \zeta_Q = 0$, i.e. $D_0 = 3$, in agreement with some experiments [3] and some recent numerical integrations of the Navier Stokes equation [8]. This result is not in

224

contradiction with the fractal nature of turbulence. Indeed, the multifractal approach considers a hierarchy of singularities h and related fractal sets $S(h)$ of fluid points $\mathbf{x}$, such that $|\mathbf{u}(\mathbf{x}+\boldsymbol{\ell})-\mathbf{u}(\mathbf{x})| \sim \ell^h$. The fractal dimensions $D(h)$ of these sets are related to the exponents ζ_Q by the Legendre transformation [7]

$$\zeta_Q = \min_h [h\,Q \ - \ D(h) \ + \ 3] \tag{6}$$

If there are fluid regions (with fractal dimension 3) where the velocity gradients are not singular, that is

$$|\mathbf{u}(\mathbf{x}+\boldsymbol{\ell})-\mathbf{u}(\mathbf{x})| \sim \ell, \tag{7}$$

Eq. (6) implies that $\zeta_Q = Q$ for Q small enough. On the contrary, a non zero fractal codimension (i.e. $\zeta_0 \neq 0$) is obtained assuming $|\mathbf{u}(\mathbf{x}+\boldsymbol{\ell})-\mathbf{u}(\mathbf{x})| = 0$ in the non-active 'laminar' regions instead of (7).

Nevertheless, the relevant dimensionality is the fractal dimension of the probability measure (information dimension) D_1 rather than the dimension of its support. D_1 is given by the derivative of ζ_Q around $Q = 3$. In fact, it can be shown [7] that the most probable behavior of the velocity gradients is given by the singularity $\bar{h} = (D_1 - 2)/3$. This means that the probability of finding a singularity $h \neq \bar{h}$ vanishes when $Re \to \infty$. We have estimated $3 - D_1 = 0.08 \pm 0.02$, which can be compared with the value $3 - D_1 = 0.13$ obtained by different experiments [3] in real turbulence.

We have also computed the structure functions for negative Q, and $d\zeta_Q/dQ$ seems to approach the value 1 corresponding to a non-singular velocity gradients, that is $h = 1$ (see Fig.2). We should add that the calculations for negative Q's are very uncertain and extensive simulation of these phenomena will be reported elsewhere.

3. Intermittency and Lyapunov Exponents

In order to get more insight into the intermittency corrections described in the previous paragraph we have investigated how the dynamical behavior of the model is correlated to intermittency effects. An appropriate way to characterize the dynamics is by means of the Lyapunov exponents $\lambda_1 \geq \lambda_2 \geq ... \geq \lambda_{2N}$. They have been calculated by Yamada and Ohkitani [13] and Fig.3 shows the Lyapunov spectrum, $\frac{1}{H}\sum_{i=1}^{j} \lambda_i$ versus $\frac{j}{D}$, where H is the Kolmogorov entropy and D is the Kaplan-Yorke dimension. The spectrum is plotted for three different values of the viscosity indicating a similarity scaling structure. We note a very interesting characteristic of the $N \to \infty$ limit of the spectrum: There is a small fraction of positive exponents, then a large fraction of marginal (i.e. zero) exponents, followed by the negative ones of order $-\nu k_i^2$ which correspond to the viscous damping and are responsible for the strongest contraction rates. The appearance of many almost zero Lyapunov exponents might be connected to the existence of spatial power law scaling for the structure functions. In fact, it seems difficult to maintain the spatial power laws if there were many exponentially diverging directions in the $2N$ phase space for orbits [17]. A finite density of zero Lyapunov exponents, a sort of "weak chaos", has been obtained by Ruelle for the discrete spectrum of operators that linearizes the Navier-Stokes equations [18] and appears also in Hamiltonian systems near integrability [19].

In Fig.4 we show the corresponding real parts of the eigenvalues, $\gamma_i + \imath\phi_i$, of the jacobian matrix of Eqs.(1) computed at the Kolmogorov fixed point given by $|u_n| = k_n^{-1/3}$. The

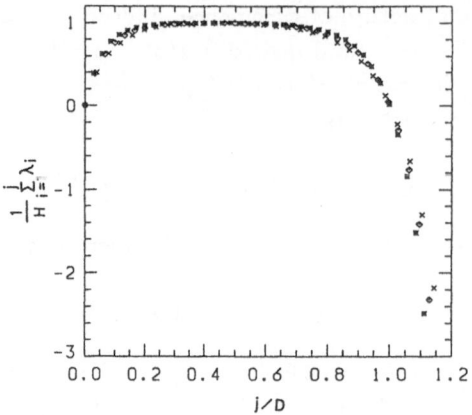

Fig.3. The Lyapunov spectrum from Ref. [13] for $\nu = 10^{-7}$, $10^{-8}, 10^{-9}$. Plotted is $\frac{1}{H}\sum_{i=1}^{j} \lambda_i$ versus $\frac{j}{D}$, where H is the Kolmogorov entropy and D is the Kaplan-Yorke dimension. Note the similarity of the spectrum for different values of ν. A small fraction of the spectrum are positive exponents but the dominating part consists of marginal exponents.

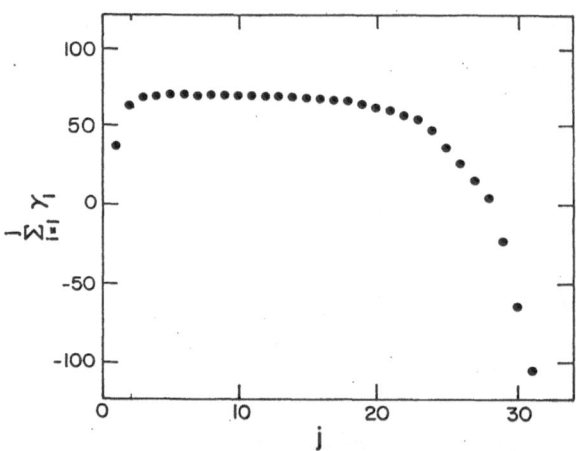

Fig.4. Real parts of the eigenvalues of the Jacobian matrix of Eqs.(1) at the Kolmogorov fixed point. $\sum_{i=1}^{j} \gamma_i$ versus j, where the eigenvalues are ordered such that $\gamma_1 \geq \gamma_2 \geq .... \geq \gamma_{2N}$.

behavior is qualitatively similar to that of the Lyapunov exponents. Again, there are a large number of almost zero (although slightly negative) γ_i (note that the sum of the real parts of the eigenvalues are plotted to allow the connection with Fig.3). There are only few γ_i which are very large in absolute value. The spectrum of the dynamics on the attractor, Fig.3, seems therefore to be closely related to the Kolmogorov fixed point, since there are only few directions which are strongly contracting or expanding. We thus expect that the system spends a large fraction of the time in the neighborhood of the fixed point.

Whereas the Lyapunov exponents give information about the relevant time scales the corresponding Lyapunov tangent vectors can resolve the interplay between various length

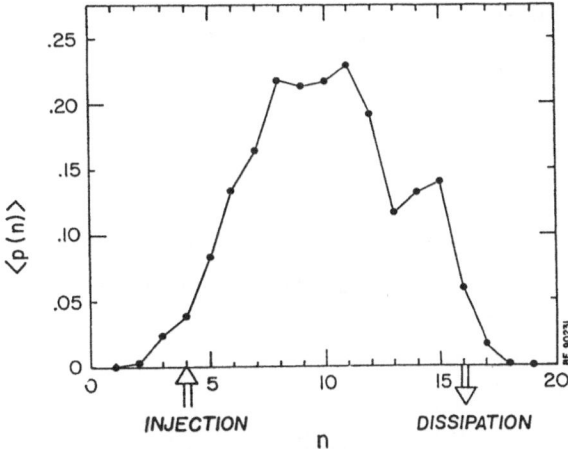

Fig.5. $< p(n) > \equiv < |e_1(2n-1)|^2 + |e_1(2n)|^2 >$, averaged over 3500 time units, plotted versus n for $N = 19$. Note that $\sum_n p(n) = 1$

scales during time evolution. Actually, as shown by Ohkitani and Yamada [13,14] the time average of the i'th Lyapunov vector gives a measure of the localization on the various shells related to the instability of the i'th Lyapunov exponent.

In particular, there is a strong correspondence between Lyapunov eigenvectors of the last negative Lyapunov exponents and dissipative modes following the end of the inertial scaling range, since the viscous damping is responsible of the strongest contraction rates, so that $\lambda_{2i} \approx \lambda_{2i-1} \propto -\nu k_i^2$ for $i \approx N$. On the other hand, the eigenvector corresponding to the small Lyapunov exponents are concentrated, although in a less sharp way, in the inertial wavenumbers. This weak correspondence, as well as the existence of a large number of small negative γ_i of the jacobian matrix at the kolmogorov fixed point, suggest that the power scaling law are related to the presence of a large number of (almost) zero Lyapunov exponents. In this picture, the largest Lyapunov exponent - and its eigenvector - should be responsible of the intermittency.

We have therefore integrated the linearized equation

$$\frac{dz_i}{dt} = A_{i,j} \cdot z_j \qquad i, j = 1, ..., 2N \tag{8}$$

for the time evolution of an infinitesimal increment $\mathbf{z} = \delta \mathbf{U}$, where $A_{n,j} \equiv \partial F_n / \partial U_j$ is the jacobian matrix of eqs (1), and $\mathbf{U} = (u_1^R, u_1^I, ..., u_N^R, u_N^I)$. The solution for the tangent vector $\mathbf{z}$ can thus be formally written as $\mathbf{z}(t_2) = M(t_1, t_2) \cdot \mathbf{z}(t_1)$, with $M = \exp \int_{t_1}^{t_2} A(\tau) d\tau$. We introduce a stability basis e_i given by the eigenvectors of the matrix M so that a generic tangent vector $\mathbf{z}(t)$ is projected by the evolution along e_1 (i.e. $\mathbf{z}(t) = c \, exp(\lambda_1 t) \, e_1$) a part corrections $O(exp - |\lambda_1 - \lambda_2|t)$ (for a discussion of Lyapunov eigenvectors versus Lyapunov stability vectors, as discussed here, se Ref.[20]). The time average of the first eigenvector is shown in Fig.5. There is no correspondence with the large scale, and the average spreads in the whole inertial range.

It is thus interesting to compute the time evolution of the stability eigenvector $e_1(t)$. In fact, we observe that its projection on the different vector shells varies quite violently in time. We estimate the fraction of the stability vector in the shell k_n as $p(n) \equiv |e_1(k_n)|^2 / \sum_j |e_1(k_j)|^2$, where $|e_1(k_n)|^2 = |e_1(2n-1)|^2 + |e_1(2n)|^2$. In particular, we

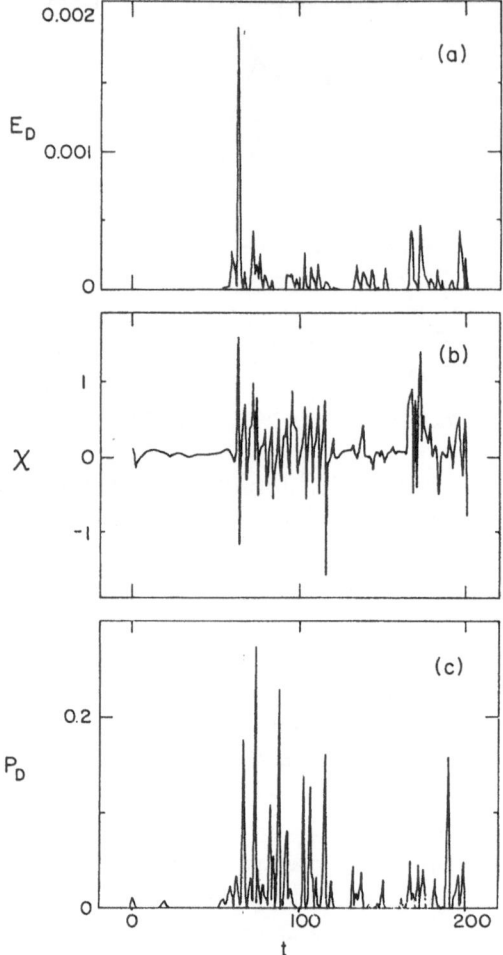

Fig.6. Temporal sequences (200 time units) of E_D (Fig.3a), the instantaneous Lyapunov exponents χ (Fig.3b), and P_D (Fig.3c) for $N = 19$. The sequences show that the energy bursts are in correspondence with large deviations of χ and localization of its eigenvector on the dissipative modes. Note the laminar phase (very small E_D values) during the first 50 time units corresponding to almost vanishing p_D and χ.

Note that $\lambda_1 = \lim_{T \to \infty} (1/T) \int_0^T dt\, \chi(t)$. The value of χ is an indication of the global chaoticity of the system at a given instant. We now focus in particular on the dissipation scale which for the case of N=19 and $\nu = 10^{-6}$ is at the wave number $k_D = k_{15}$. In Fig.6 see that during an intermittent burst there appears to be a concentration on the components related to the dissipation (Kolmogorov) scale. To make this idea more precise, we have computed the response after a time τ to an infinitesimal perturbation, defining an instantaneous maximum Lyapunov exponent as

$$\chi_\tau(t) \equiv \frac{1}{\tau} \ln\, |\frac{\mathbf{z}(t + \tau)}{\mathbf{z}(t)}| \qquad (9)$$

we show time series of for three quantities: a). The energy dissipation E_D estimated by $|u_{15}|^2$; b). The instantaneous Lyapunov exponent χ; c). The eigenvector fraction on k_D, $p_D \equiv p(k_{15})$. First of all, we note the strongly intermittent nature of energy dissipation. Furthermore, the spikes in E_D are intimately followed by spikes in as well χ as p_D. Also, for the quiet region of E_D (the first 50 time steps), the other variables are vanishingly small. The apparent correlation can be quantified calculating time correlation functions between each two of the quantities. We find that the correlation is very strong at zero time difference but is fastly decaying with increasing time difference. Actually, the time correlation function between E_D and χ even becomes negative after one time step [9]. This effect can be directly seen at the time series in Fig.6b. If the value of χ suddenly undergoes a big positive burst, it is immediately followed by a big negative burst. This suggests the following picture: By a big chaoticity burst, the system is suddenly "kicked" away from the normal energy cascade given a motion in the phase space close to the Kolmogorov fixed point of Eq.(1). However, the system immediately tries to correct from the excursion and contracts back towards the Kolmogorov fixed point, so that one finds negative χ (with large absolute value).

The implication of the calculation shown in Fig.6 is that strong energy intermittency is triggered by strong instantaneous chaoticity and a subsequent localization on the dissipation wave numbers and we might be tempted to argue that the intermittency corrections have to do with violent motion (given by the largest Lyapunov exponents) at scales close to the Kolmogorov length. This is however too simplified. The maximal Lyapunov exponent λ_1 is a global quantity and when averaging the corresponding eigenvector we find that it spreds all over the inertial range. This has also been observed in Ref.[13,14]. The intermittency is a complex phenomenon involving the transfer in the whole inertial range. Actually, what

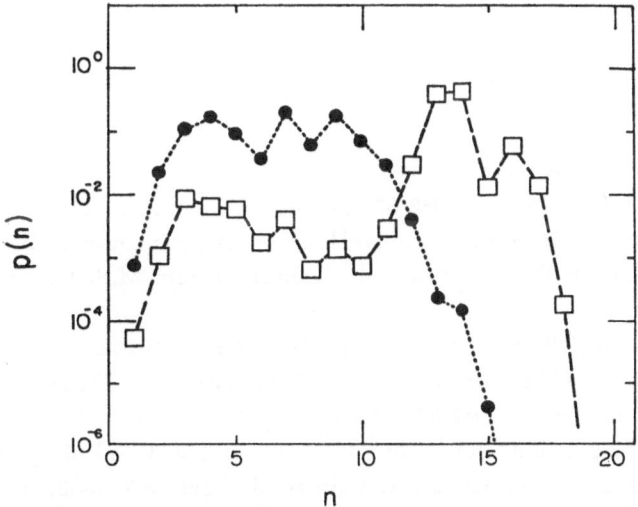

Fig.7. Distribution of the global chaoticity among the shells during a laminar and a chaotic phase. $\ln p(n)$ is plotted versus n for $N = 19$. The crossed circles (linked by a dotted line) correspond to a 'laminar' instant ($E_D = 4.2 \cdot 10^{-6}$, $\chi = 0.12$) and the squares (linked by a dashed line) correspond to a burst ($E_D = 0.39$, $\chi = 1.35$). Note that the average values are $< E_D >= (2.09 \pm 0.02) \cdot 10^{-2}$ and $\lambda_1 =< \chi >= 0.169 \pm 0.003$.

appears to happen is that the distribution in the eigenvector switches from the dissipation wave number during intermittent bursts to the inertial range during more laminar periods, as we show in Fig.7.

4. The Fractal Structure of Intermittency

We finally briefly discuss the fractal structure of the dissipation. From the calculation of the structure functions the information dimension was found to $D_1 \simeq 2.92$. But in the shell model, the spatial structure is eliminated so what do we mean by spatial intermittency ? To answer this, let us use the socalled Taylor hypothesis. Assuming a large scale mean flow velocity (which is roughly given by the velocity in the large scale shells) we may identify a time sequence with a spatial segment. In other words the time series at the Kolmogorov scale, Fig.6a, is considered as a spatial segment of energy dissipation. Clearly, the structure is very non-uniform defining a Cantor set of dissipation consisting of points whose values of E_D is larger than some gate u_g^2 (see Ref.[9] for details). With $N=19$ shells we have used a time of 10^4 time units with a gate $u_g^2 = 10^{-4}$ and obtain a Cantor set consisting of 1181 points. The corresponding dimension, obtained via the correlation integral, is found to be 0.92 ± 0.02 and assuming that the two transversal directions are smooth, the dimension of the fractal structure of the energy dissipation is $D \simeq 2.92$. Since the Cantor set consists of points where the energy is concentrated (i.e. where the "dissipation measure" is concentrated) this dimensionality is the information dimension D_1 . It is in good agreement with the value of D_1 obtained from the structure function calculation. One might also obtain the complete multifractal scaling by introducing a probability measure given by the value of the energy dissipation at each point in the set. This however goes beyond the scope of this paper.

REFERENCES

[1] P.G. Saffman, this conference

[2] B. B. Mandelbrot J. Fluid Mech. **62** 331 (1974)

[3] C.M. Meneveau and K.R. Sreenivasan Nucl. Phys. B Proc. Suppl. **2** 49 (1987);
C.M. Meneveau and K.R. Sreenivasan, P. Kailasnath and M.S. Fan Phys. Rev. **A41** 894 (1990)
C.M. Meneveau and K.R. Sreenivasan Phys. Rev. Lett. **59** 1424 (1987)
K.R. Sreenivasan and C.M. Meneveau Phys. Rev. **A38** 6287 (1988)

[4] E. D. Siggia J. Fluid Mech. **107** 375 (1981)

[5] F. Anselmet, Y. Gagne, E.J. Hopfinger and R. Antonia J. Fluid Mech. **140** 63 (1984)

[6] R. Benzi, G. Paladin, G. Parisi and A. Vulpiani J. Phys. **A17** 3521 (1984)

[7] G. Paladin and A. Vulpiani Phys. Rep. **156** 147 (1987) and references therein

[8] I. Hosokawa and K. Yamamoto J. Phys. Soc. of Japan **59** 401 (1990)

[9] M.H. Jensen, G. Paladin and A. Vulpiani Phys. Rev. **A** in press (1990)

[10] E.B. Gledzer Sov. Phys. Dokl. **18** 216 (1973)

[11] E.D. Siggia Phys. Rev. **A15** 1730 (1977)
E.D. Siggia Phys. Rev. **A17** 1166 (1978)
R.M. Kerr and E.D. Siggia J. Stat. Phys. **19** 543 (1978)

[12] R. Grappin, J. Leorat and A. Pouquet J. Physique **47** 1127 (1986)

[13] M. Yamada and K. Ohkitani J. Phys. Soc. of Japan **56** 4210 (1987)

    M. Yamada and K. Ohkitani Progr. Theo. Phys. **79** 1265 (1988)

    M. Yamada and K. Ohkitani Phys. Rev. Lett. **60** 983 (1988)

[14] K. Ohkitani and M. Yamada Progr. Theo. Phys. **81** 329 (1989)

[15] H.A. Rose and P.L. Sulem J. Physique **39** 441 (1978)

[16] H.G.E. Hentschel and I. Procaccia Physica **D8** 435 (1983)

    T. C. Halsey, M.H. Jensen, L.P. Kadanoff, I. Procaccia and B. Shraiman Phys. Rev. **A33** 1141 (1986)

[17] P. Bak private communication;

    K. Chen, P. Bak and M.H. Jensen Phys. Lett. **A** in press

[18] D. Ruelle Comm. Math. Phys. **87** 287 (1982)

[19] R. Livi, M. Pettini, S. Ruffo and A. Vulpiani J. Stat. Phys. **48** 530 (1987)

[20] S.A. Orszag, P.L. Sulem and I. Goldirsch Physica **27D** 311 (1987)

AN APPRAISAL OF THE RUELLE-TAKENS ROUTE TO TURBULENCE

R.S. MacKay

Nonlinear Systems Laboratory
Mathematics Institute
University of Warwick
Coventry CV4 7AL
ENGLAND

1. INTRODUCTION

During the 60s and 70s, the idea was put forward that turbulence corresponds to hyperbolicity (i.e. exponential separation of most nearby orbits) for the governing equations [A1,L,RT].

In particular, Ruelle and Takens [RT] proposed that one could have a sequence of transitions generating successively an attracting $(n+1)$-torus from an attracting n-torus, $n = 0,1,2,...$, and proved (under extension with Newhouse [NRT]) occurrence of flows with a strange attractor, arbitrarily close to any uniform flow on $\mathbb{T}^n$, $n \geq 3$. This gave rise to the so-called *Ruelle-Takens route to turbulence*: equilibrium $\rightarrow$ periodic state $\rightarrow$ 2-torus $\rightarrow$ 3-torus with a strange attractor, which has entered the experimentalist's repertoire (e.g. [GoL]).

In this paper, I assess the status of the Ruelle–Takens route and its relevance to turbulence, and outline recent research of [BGKM] on what happens for typical families of flows on a 3-torus, close to the uniform flows.

A lot has been written already on this subject, and I fear I will not do justice to everyone's ideas; but I trust that this paper will provide a fair appraisal and a useful indication of how one could proceed.

2. THE DYNAMICAL SYSTEMS VIEW OF FLUID FLOWS

The equations of motion for a fluid flow in a three–dimensional region Ω are assumed to give rise to a (mathematical) flow $\varphi : \mathbb{R} \times M \to M$ on a space M of fields v on Ω, specifying velocity, pressure etc.; given an initial field v, $\varphi_t(v)$ is the resulting field at time t. The set $\{\varphi_t(v): t \in \mathbb{R}\}$ (or better, the map $\varphi(v): \mathbb{R} \to M$) is called the *orbit* of v. The dynamical systems viewpoint is to study features of φ which are independent of coordinate system on M (e.g. basis for function space), for instance existence of an attracting equilibrium or periodic orbit.

The concept of strange attractor is continually evolving, but here is one definition, based on ideas of Ruelle [Ru] and Rand (private communication and [Z]). We suppose we are given a metric on M, and a differentiable structure; write the derivative as D and the norm of a tangent vector u as ‖u‖. We also suppose a measure ν on M, representing the measure class of the probability distribution of initial conditions (though there appears to be no unique way to do this when M is infinite–dimensional); so $\nu(U)$ is a notion of the size of a subset $U \subset M$.

Definition: A *strange attractor* for a flow $\varphi : \mathbb{R} \times M \to M$ is an ergodic invariant probability distribution μ on M such that:
(i) for a set S of initial conditions v, with $\nu(S) > 0$, the time average of any continuous function ψ along the orbit $\varphi(v)$, $0 \le t \le T$, converges to $\int \psi \, d\mu$, as $T \to \infty$,
(ii) for most close pairs of initial conditions in S, the orbits diverge exponentially: more formally, the *Lyapunov exponent* $\lambda(v) = \limsup_{t \to \infty} t^{-1} \log\|D\varphi_t(v)\|$ is positive μ–almost everywhere (in fact, by ergodicity of μ it is the same μ–almost everywhere).

Note that an invariant probability measure μ for a flow φ is said to be *ergodic* if for every measurable invariant subset A, $\mu(A) = 0$ or 1.

A similar definition has been proposed by Sinai [Bu]:

Definition: A *stochastic attractor* is a closed invariant set A with a neighbourhood U such that: (i) $\varphi_t(U) \subset U \; \forall \, t > 0$, and $\bigcap_{t > 0} \varphi_t(U) = A$,
(ii) for all absolutely continuous measures m with support in U, $\varphi_t{}^*m$ converges weakly to an invariant measure μ, independent of m,
(iii) the system (A, μ, φ) is mixing.

There is a comprehensive general theory of "hyperbolic attractors", where all the above properties hold (see [Ru,Pe] for overviews). Many examples of hyperbolic attractors are known, and candidates for non–hyperbolic strange attractors (see e.g. [Z]).

Many people, including myself, feel that the strange/stochastic attractor concept has the right properties for describing a turbulent state of a fluid in a bounded domain, namely, sensitive dependence on initial conditions, and yet time–averaged quantities independent of starting condition (or almost so) within its *basin* (the set of v such that (i) holds for a strange attractor, or the set of points whose orbits fall into U for a stochastic attractor).

3. THE RUELLE-TAKENS ROUTE

It is well-known (e.g. [A2,GH]) that as parameters are varied, a stable equilibrium point of a dynamical system can give birth to an attracting periodic orbit, when a complex conjugate pair of eigenvalues cross the imaginary axis (Poincaré–Andronov–Hopf bifurcation). Similarly, an attracting periodic orbit can give birth to an attracting invariant 2-torus when a complex conjugate pair of Floquet multipliers cross the unit circle (Neimark–Sacker bifurcation).

Ruelle and Takens [RT] proposed that further transitions from an n-torus to an (n+1)-torus, $n \geq 2$, are possible. In the "nearly split" situations which they study, a sequence of parameter intervals I_n, n = 0, ... N, occurs in which there is an attracting n-torus, but for $n \geq 2$, I_n does not necessarily touch I_{n+1}. This was investigated further by [CI,Sell1,Sell2], but to my knowledge, there are still many interesting open questions about what happens in these transitions.

In the nearly split situation, the flows on the resulting tori are close to uniform translations. Ruelle and Takens [RT] showed that arbitrarily close (in the C^{n-1} topology) to any uniform translation on an n-torus, $n \geq 4$, is an open set of flows with a strange attractor, indeed an Axiom–A strange attractor. This result was extended with Newhouse [NRT] to the C^∞ topology and to n = 3 in the C^2 topology.

Their work led them to conclude that "when three pairs of complex conjugate eigenvalues [for an equilibrium] have crossed [the imaginary axis], a motion asymptotic to a non–trivial Axiom–A attractor may appear" [NRT].

4. EXPERIMENTAL OBSERVATIONS

Abstracting somewhat, Ruelle and Takens' prediction was that the onset of turbulence may sometimes be identified with low-dimensional chaos of some sort. This prediction was verified in a beautiful experiment of Gollub and Swinney [GS] on the flow between rotating concentric cylinders, where they saw the introduction of two frequencies, which locked and then generation of a broadband spectrum, suggesting breakup of a 2-torus into a low-dimensional strange attractor (though the issue is complicated by the circular symmetry of the system). In subsequent experiments, estimates of dimension and Lyapunov exponent have been performed, confirming this interpretation.

As presented in [RT] and [NRT], however, the prediction was of formation of a 3- or higher-dimensional torus with a strange attractor on it. There are a few fluid experiments which have been claimed to show aspects of this.

Figure 1 shows a power spectrum from an experiment on convection in a small cell of water, by Gollub and Benson [GB]. They claim that there are three independent frequencies, and hence the motion is quasiperiodic on a 3-torus. This is probably correct, though a better way nowadays would be to do a phase portrait construction and estimate the dimension (e.g.

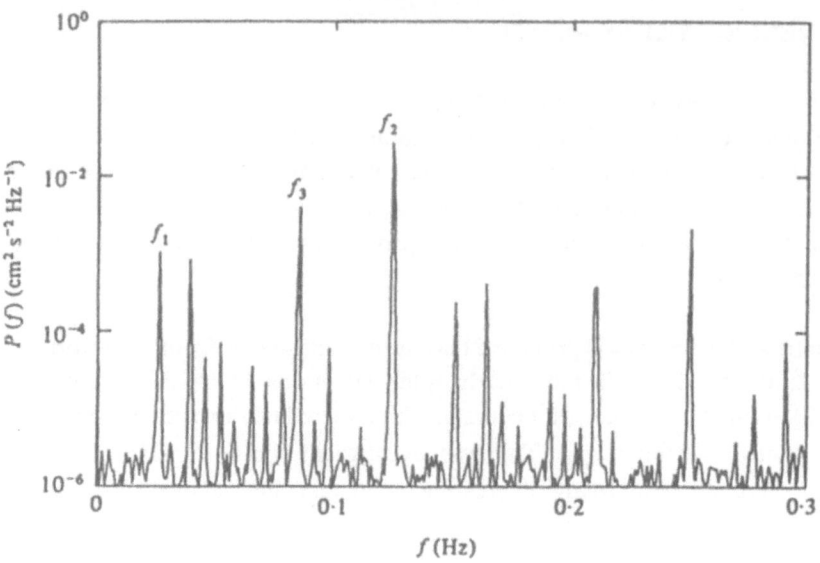

Figure 1. Power spectrum showing the presence of 3 incommensurate frequencies in convection in water (from [GB]).

[BK,BJK]). Muldoon and I are developing an automatic way of extracting topology from a time series [MM].

Figure 2 shows a power spectrum from a convection experiment of Maurer and Libchaber [ML] in liquid helium. By varying parameters it was found that there are again three frequencies, though in this picture they have locked. What is interesting is that there coexists a broad-band spectrum. Could this represent a strange attractor on a 3-torus? Again, one really needs to do phase portrait constructions in order to decide this.

[WKPS] claimed to see quasiperiodic convection with four and five incommensurate

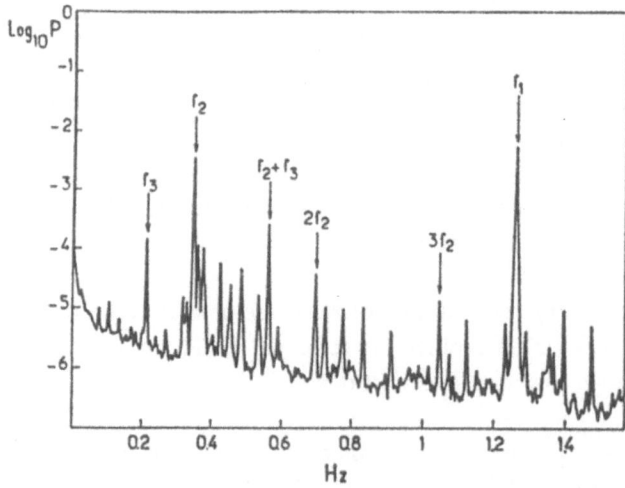

Figure 2. Fourier spectrum for a liquid Helium convection cell near the onset of turbulence, with three frequencies present, and the locking state $f_1 - 3f_2 = f_3$ (from [ML]).

236

frequencies, though with such a large number of frequencies, I think it is hard to say whether they are incommensurate or not.

The type of strange attractor which [RT] and [NRT] constructed has the property that the n modes of oscillation have a well-defined rational frequency ratio, that is, the average rates of advance of the phases of the n modes are in rational ratio. To my knowledge this has not been seen (nor looked for) in experiments.

While Maurer and Libchaber's experiment does seem to show that chaos on a three-torus can occur, it appears to be rare. What is seen more commonly in such "small" systems is transition to chaos by period doubling of by formation of two frequencies which lock, followed by a broadband component (breakup of a 2-torus?). For reviews, see [GoL,GL,Sw].

5. CRITICISMS

I have several general criticisms of Ruelle and Takens' approach and several detailed criticisms. Let us begin with the general criticisms.

5.1 General Criticisms

Firstly, is turbulence, in the sense that engineers use it, describable by a low-dimensional attractor? When a system gives the impression of being composed of many weakly interacting parts, it is unlikely to correspond to a low-dimensional attractor. Newell and others draw a distinction between "macho" and "wimpy" turbulence, based on the dimension of the attractor compared to some measure of the size of the system. So far, nonlinear dynamics can contribute only to wimpy turbulence.

Secondly, although one might argue that the onset of turbulence should be describable by low-dimensional attractors, the onset could well be by a jump to a high-dimensional attractor, e.g. by a crisis [GOY2], catastrophe or Ω-explosion [Z]. This is quite likely for systems in which the first bifurcation is subcritical.

Thirdly, how appropriate is the dynamical systems viewpoint for open systems (or even large bounded systems)? For infinite spatially homogeneous systems, the concept of space-time chaos (e.g. [BS]) may be useful, but it is not clear how to tackle the more relevant problem of semi-infinite systems, such as mixing layers. Ideas of [BR] on convective instability could help.

5.2 Detailed criticisms

Firstly, as Ruelle and Takens acknowledge [RT], the analysis of the transition from an n-torus to an (n+1)-torus is tricky for $n \geq 2$. A satisfactory story has been produced by [CI,Sell1,Sell2], and other mechanisms for obtaining n-tori with $n \geq 3$ by [Lang,IL], but to my mind there are still many questions. What can happen when an invariant torus loses normal hyperbolicity, even a 2-torus (cf. [Kan,BGOY])?

Secondly, as pointed out by many people (e.g. [Z,GOY]), chaos is rare for small perturbations of uniform translations. KAM theory shows that for most parameter values in a family of perturbations of the uniform flows on an n-torus ($n \geq 2$), the motion remains smoothly conjugate to an incommensurate uniform flow (e.g. [BMS,H]).

Thirdly, the flows they construct with a strange attractor are somewhat contrived. Although they are structurally stable, it is not clear whether any of these examples will occur in practice. Furthermore, they have the special property of a rational frequency ratio between the n modes of oscillation, which does not seem to me to be the most relevant form of chaos on a torus. It is to this third point that I have made a contribution, with Baesens, Guckenheimer, Kim and Llibre, and I wish to summarise in the next section the progress which we have made.

6. WHAT REALLY HAPPENS FOR CLOSE TO UNIFORM FLOWS ON A 3-TORUS

The results described here are a summary of work presented in [KMG, LM, BGKM1, BGKM2]. For a pedagogical introduction, see [M].

We have investigated what happens for typical perturbations:

$$\dot{x} = \Omega + g(x,\Omega,\varepsilon), \quad x \in \mathbb{T}^3, \Omega \in \mathbb{R}^3, \varepsilon \in \mathbb{R}^p, g(x,\Omega,0) = 0.$$

of the family of uniform flows on the 3-torus. Throughout, we assume small enough perturbation ε so that the flow has a cross-section.

One of the most useful concepts is the *winding ratio* of an orbit, defined to be

$$W = \lim_{t \to \infty} \frac{x(t) - x(0)}{|x(t) - x(0)|} \in S^2$$

if the limit exists, which represents the direction of the average velocity $\langle \dot{x} \rangle / |\langle \dot{x} \rangle|$. In case the limit does not exist, we define the *set of winding ratios* for the orbit to be the set of limit points of this ratio, on the 2-sphere. The *set of winding ratios for a flow* can be defined to be the union of the sets of winding ratios for its orbits (though there are other definitions).

We distinguish the following features in the parameter space $\mathbb{R}^3 \times \mathbb{R}^p$. For ease of representation and with no great loss of generality, we restrict attention to a two-dimensional slice of the parameter space, with the only condition being that it is transverse to the "rays" $L_{\Omega'} = \{s\Omega': s \in \mathbb{R}\}, \Omega' \in S^2$. An example is shown in Figure 3.

1. The KAM set This is the set of points (totally disconnected) for which the motion is conjugate to an incommensurate uniform flow. It exhausts most of the measure in parameter space for weak coupling.

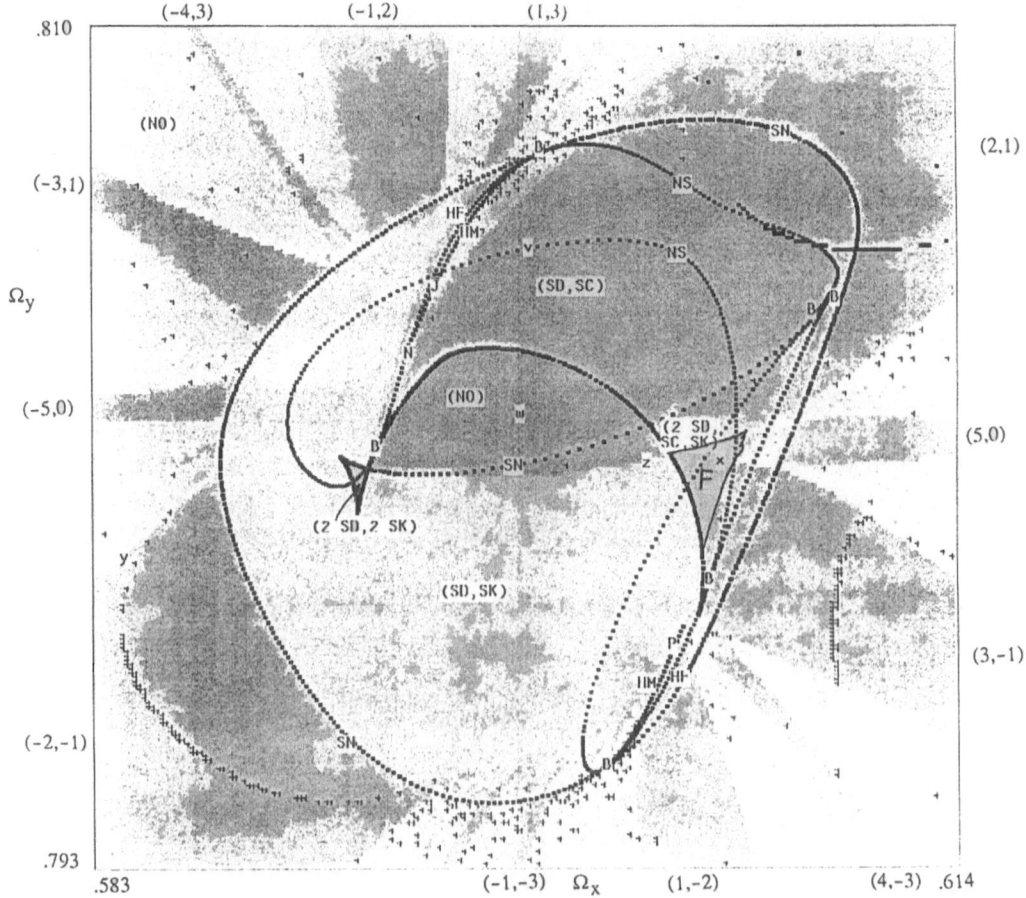

Figure 3. A parameter scan for the family of torus maps $(x,y) \mapsto (x+\Omega_x - a'\sin 2\pi y, y+\Omega_y - a'\sin 2\pi x)$, with $a' = 0.7/2\pi$ and Ω near $(3,4)/5$, indicating the $(3,4)/5$ resonance region, partial mode-locking strips of various types, a region of full mode-locking and other features (from [BGKM1]).

2. <u>Mode-locking strips</u> These are the sets where some integer combination $m.(x(t)-x(0))$ of the angles remains bounded, so the relation $m.W = 0$ is satisfied by the set of winding ratios.

3. <u>Resonance regions</u> These are the regions where there is a periodic orbit. The frequency vector of the periodic orbit has the form $\omega = k/T$, some $k \in \mathbb{Z}^3$, where T is the period of the orbit, and the resonance regions can be labelled by the value of k. The winding ratio is $k/|k|$. We prove that for weak coupling, the shape of a primitive resonance region (i.e., when k has no common factor) is often a projection of a 2-torus [KMG].

4. <u>Fully mode-locked regions</u> These are regions where two independent integer combinations $m^1.x, m^2.x$ of the angles remain bounded for all orbits. They can be labelled by the smallest non-zero $k \in \mathbb{Z}^3$ (up to change of sign) such that $m^1.k = m^2.k = 0$. Then all orbits have winding ratio $k/|k|$. The k-fully mode-locked region is contained in the k-resonance region.

5. <u>Partially mode-locked strips</u> These are the parts of the mode-locking strips which are not fully mode-locked.

6. <u>Boundary of partial mode-locking under the flow approximation</u> For weak coupling and close to a rational uniform flow ($\Omega = k/T$, some $k \in \mathbf{Z}^3$), we can describe up to a very good approximation the generic ways in which one can leave a partial mode-locking strip. There is a cross-section to $\mathbb{T}^3$ for which the return map of the uniform flow with $\Omega = k/T$ is the identity. So for flows close to this, the return map to this cross-section is close to the identity. The cross-section is a 2-torus. Smooth maps close to the identity can be approximated to arbitrary order by the time-1 map of a flow. Hence we can approximate the return map by the time-1 map of some flow on the 2-torus. The codimension-1 ways of leaving a partial mode-locking strip for flows on a 2-torus are: saddle-node of periodic orbits, rotational homoclinic cycle and saddle-node of equilibria [BGKM1].

7. <u>Complicated interactions between mode-locking strips</u> There are topological obstructions to completing the bifurcation diagram with only codimension-one bifurcations. There must also be codimension-two points to generate all the partial mode-locking strips. We have discovered a variety of such possibilities for the flow approximation. Some are indicated in Figures 4 and 5, which are conjectured bifurcation diagrams for the flow approximation to the system shown in Figure 3. We have conjectured a minimal structure that must occur in a resonance region; this is presented in [BGKM1] and justified in outline in [BGKM2].

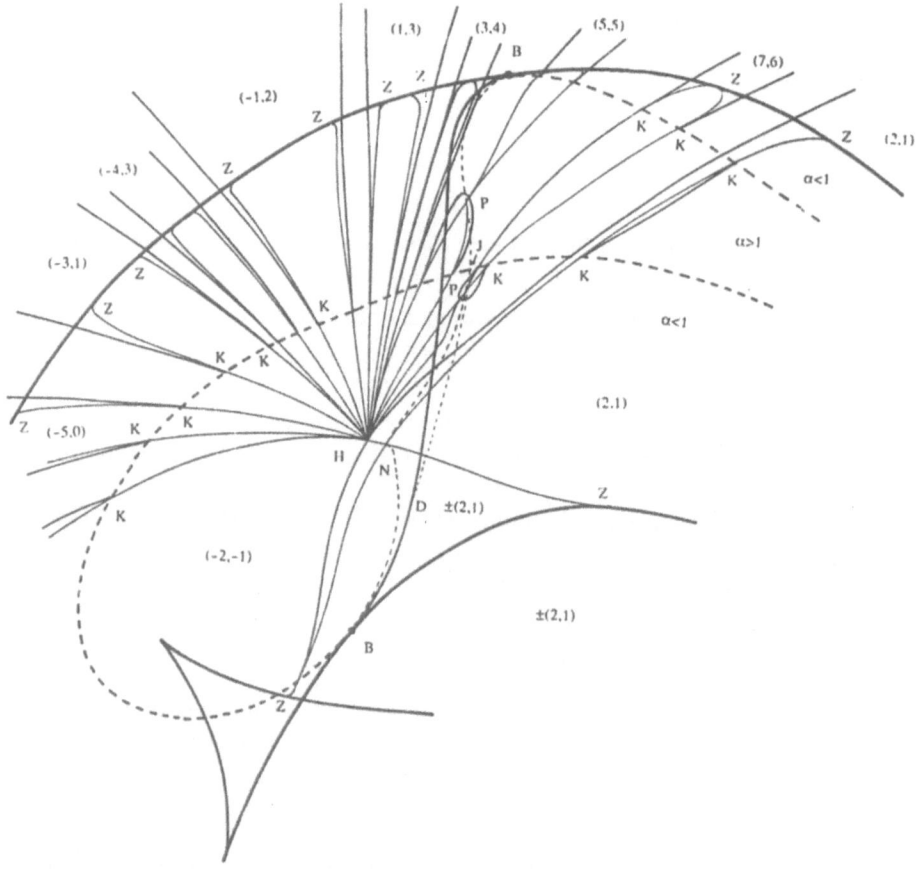

Figure 4. Conjectured bifurcation diagram for the top left of Figure 3 (from [BGKM1]).

8. Effects missed by the flow approximation Once one puts back the remainder term in the flow approximation, the transitions for flows on a 2-torus become much more complicated. In particular, there is typically topological chaos, in the form of transverse homoclinic orbits and horseshoes. We distinguish three levels of chaos for flows on a 3-torus (contrary to earlier statements, I do not now believe that one can distinguish between these by the power spectrum).

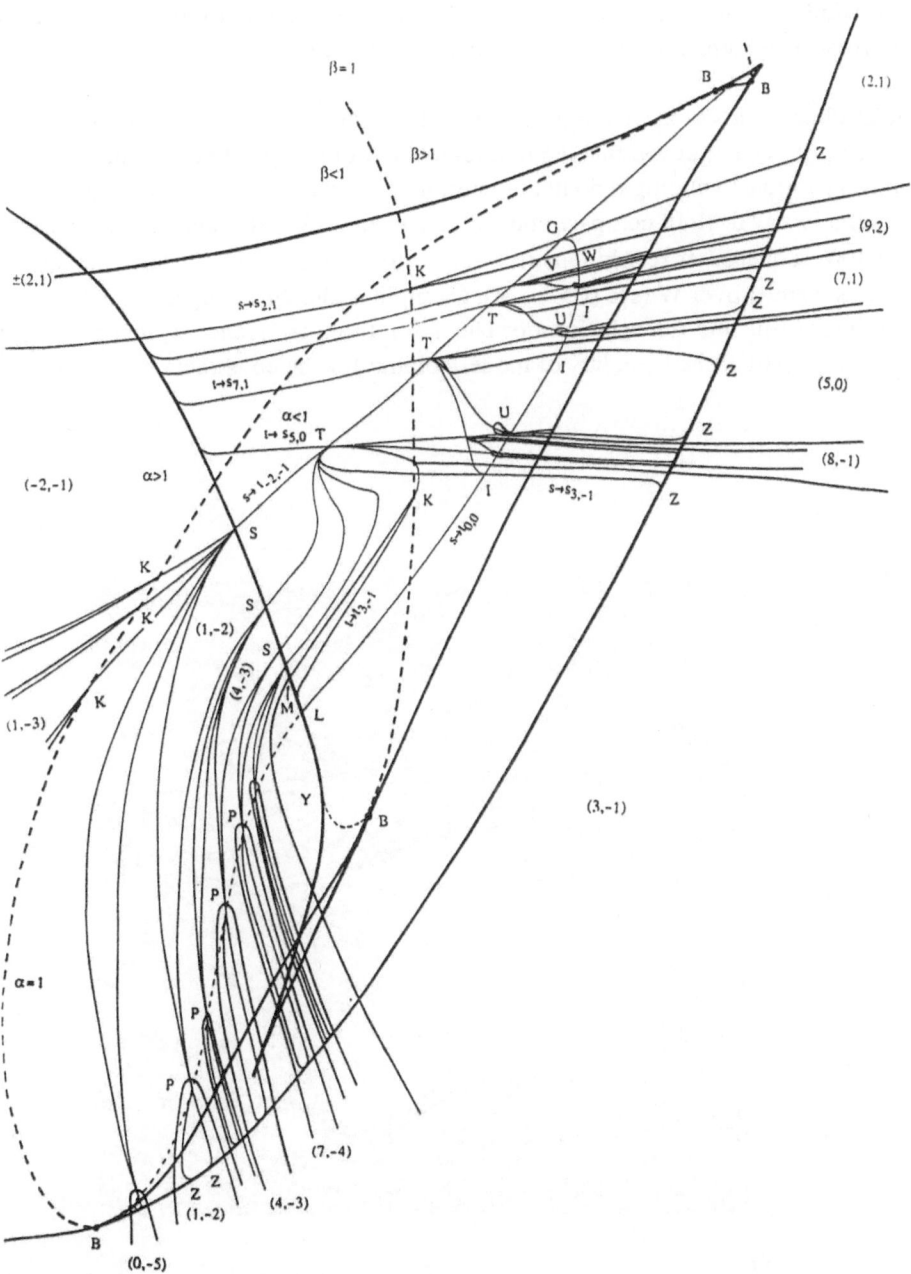

Figure 5. Conjectured bifurcation diagram for the right half of Figure 3 (from [BGKM1]).

9. <u>Contractible chaos</u> This is an invariant subset with positive topological entropy whose set of winding ratios is a single rational point. We prove that many resonance regions have "homoclinic wedges" in which it occurs [KMG]. The strange attractors of [NRT] are also examples of contractible chaos.

10. <u>Annular chaos</u> This is an invariant subset with positive topological entropy whose set W of winding ratios is contained in a commensurate line (m.W = 0, some m ∈ $\mathbb{Z}^3$) and whose orbits perform a pseudo–random sequence of revolutions in a set of directions on $\mathbb{T}^3$ such that short–term winding ratios wander over W (see [M] for a more precise definition). It occurs near every rotational homoclinic boundary of partial mode-locking.

11. <u>Toroidal chaos</u> This is an invariant set for which the set W of winding ratios has non-empty interior. It can be proved that the topological entropy is positive and there are periodic orbits for every rational winding ratio in the interior of W, there are orbits for which the set of winding ratios is any desired compact connected subset of W, and general orbits perform a pseudo-random sequence of revolutions in a set of directions round $\mathbb{T}^3$, such that short-term winding ratios wander over W (see [LM]). An example is shown in Figure 6. This, I believe is the truly interesting and relevant form of chaos for three-frequency systems. It typically occurs near every part of the boundary of mode-locking; for the construction, see Figure 7.

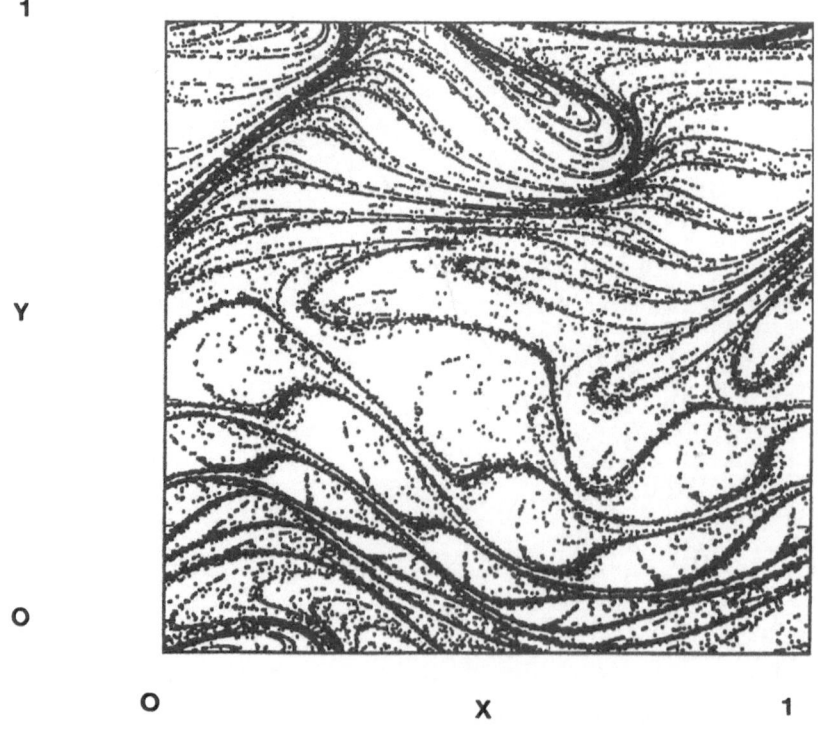

1

Y

O

O   X   1

Figure 6. An example of toroidal chaos (from [BGKM1]).

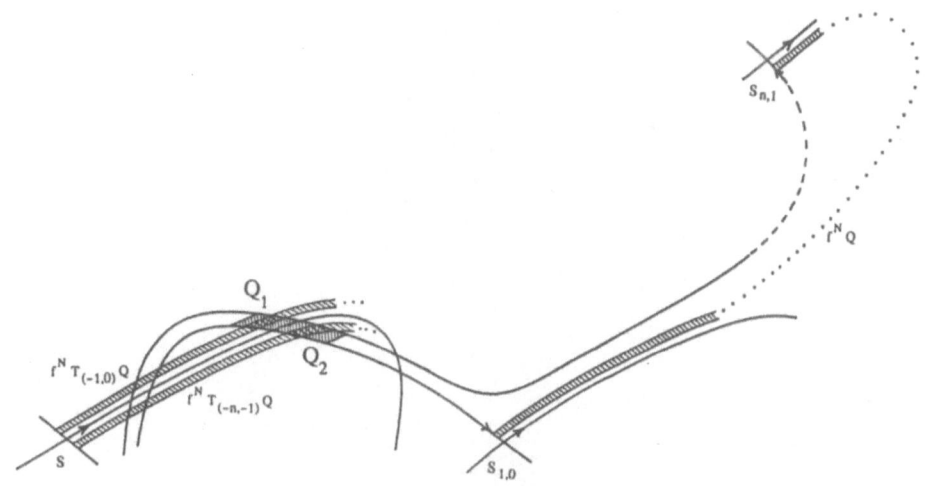

Figure 7. How toroidal chaos occurs near parts of the boundary of partial mode-locking associated with rotational homoclinic tangency (from [BGKM1]).

Our predictions have not yet been tested in fluid systems, but seem to fit well with experiments carried out on electronic circuits consisting of three coupled oscillators, by Linsay and Cumming [LC]. Figure 8 is a parameter scan taken from their paper, which we recommend the reader to compare with the bifurcation scenarios shown in Figures 4 and 5.

7. CONCLUSION

The Ruelle–Takens route is probably of limited relevance to turbulence. However, the strange attractor concept seems a good description of turbulent flows in bounded domains. What we need to do is to find the invariant measures.

This meeting is subtitled "The impact of nonlinear dynamics on turbulence". The problem of turbulence has greatly stimulated the subject of nonlinear dynamics. Unfortunately, I do not think that nonlinear dynamics has yet had much impact on the subject of turbulence. We need to examine more closely the phenomena and problems of turbulence, of which this meeting has provided a variety, and to use them to drive our mathematical theories.

I'd like to see a successful renormalisation theory developed, to capture the content of Richardson's ditty:

Big whirls have little whirls,
Which feed on their velocity.
Little whirls have lesser whirls,
And so on to infinity.

I'd like to see finite–time blowup solutions to the Euler equations, to capture the apparently random occurrence of large scale coherent events for the Navier–Stokes equations (cf. [NRR]).

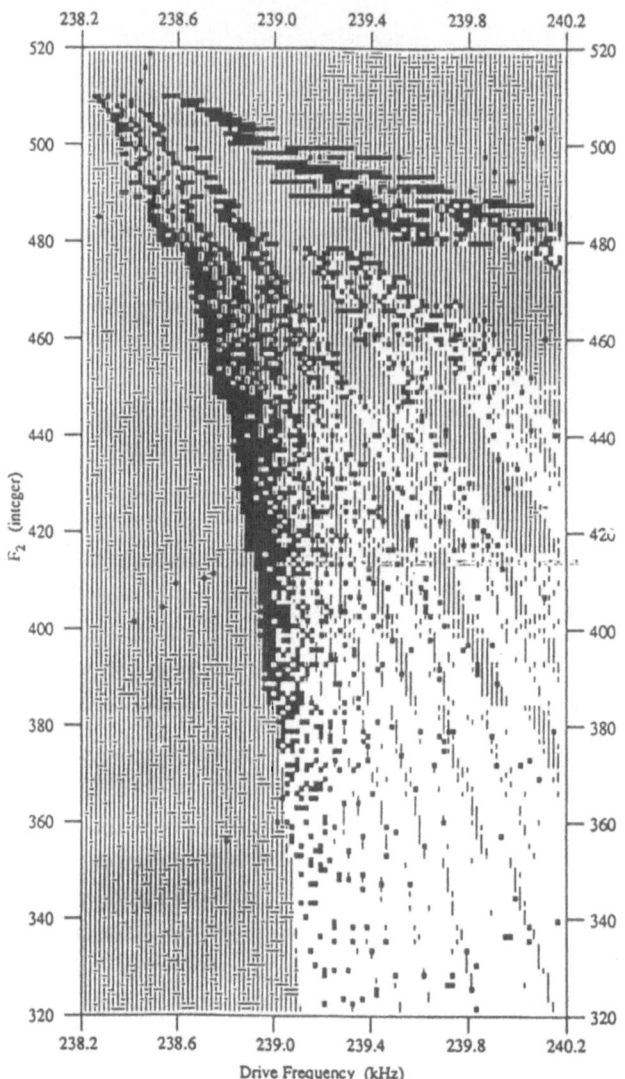

Figure 8. A parameter scan for an electronic circuit (from [LC]). Periodicity is indicated by horizontal lines, two-frequency quasiperiodicity by vertical lines, and chaos by black squares. Three-frequency quasiperiodicity is left blank.

I'd like to see existence proofs and estimates for strange attractors, and their application to compute the average mass flux down a pipe with given pressure gradient, or heat flux in a turbulent heat exchanger. I'd like to see more theory about decay of correlations for strange attractors; for hyperbolic attractors in discrete time, correlations decay exponentially [Pe], but in continuous time the issue is more complicated. This is extremely important from a physical point of view, but is only just receiving mathematical attention (e.g. Ruelle's concept of *resonances*, see [Ru] and references therein). Finally, let us develop geometric approaches to turbulence in mixing layers and other open flow systems (cf. space-time chaos), and let us produce testable predictions.

ACKNOWLEDGEMENTS

I would acknowledge fruitful collaboration with Claude Baesens, John Guckenheimer, Swan Kim and Jaume Llibre, and to thank Jerry Gollub, Albert Libchaber and Paul Linsay for permission to reproduce their figures here.

REFERENCES

[A1]   Arnol'd VI, Sur la géometrie des groupes de Lie de dimension infinie et ses applications à l'hydrodynamique des fluides parfaites, Ann Inst Fourier Grenoble 16 (1966) 319–361

[A2]   Arnol'd VI, Geometric methods in the theory of ordinary differential equations (Springer, 1983)

[BGKM1] Baesens C, Guckenheimer J, Kim S, MacKay RS, Three coupled oscillators: Mode–locking, global bifurcations and toroidal chaos, subm. to Physica D

[BGKM2] Baesens C, Guckenheimer J, Kim S, MacKay RS, Simple resonance regions for torus diffeomorphisms, an IMA Conf Proc, to appear

[BGOY]   Battelino PM, Grebogi C, Ott E, Yorke JA, Chaotic attractors on a 3–torus and torus breakup, Physica D 39 (1989) 299–314

[BMS]   Bogoljubov NN, Mitropolski Ju A, Samoilenko AM, Methods of accelerated convergence in nonlinear mechanics (Springer, 1976)

[BR]   Bohr T, Rand DA, A mechanism for the generation of turbulent spots, Physica D, to appear

[BJK]   Broomhead DS, Jones R, King GP, Topological dimension and local coordinates from time series data, J Phys A 20 (1987) L563–9

[BK]   Broomhead DS, King GP, Extracting qualitative dynamics from experimental data, Physica D 20 (1986) 217–236

[Bu]   Bunimovich LA, Strange attractors, Chapter 8.2, in Dynamical Systems II, ed Sinai Ya G (Springer, 1989)

[BS]   Bunimovich LA, Sinai Ya G, Space-time chaos in coupled map lattices, Nonlinearity 1 (1988) 491–516

[CI]   Chenciner A, Iooss G, Bifurcations de tores invariants, Arch Rat Mech Anal 69 (1979) 109–198

[GL]   Glazier JA, Libchaber A, Quasiperiodicity and dynamical systems: an experimentalist's view, IEEE Trans Circ Sys 35 (1988) 790–809

[GB]   Gollub J, Benson SV, Many routes to turbulent convection, J Fluid Mech 100 (1980) 449–470

[GoL]   Gollub JP, Libchaber A, Laboratory experiments on the transition to chaos, in Chaotic behaviour of deterministic systems, eds Iooss G, Helleman RHG, Stora R (N. Holland, 1983) 591–607

[GS]   Gollub JP, Swinney HL, Onset of turbulence in a rotating fluid, Phys Rev Lett 35 (1975) 927–930

[GOY]   Grebogi C, Ott E, Yorke JA, Are three-frequency quasiperiodic orbits to be expected in typical nonlinear dynamical systems? Phys Rev Lett 51 (1983) 339–342; and, Attractors on an n-torus: Quasiperiodicity versus chaos, Physica D 15 (1985) 354–373

[GOY2]   Grebogi C, Ott E, Yorke JA, Crises, sudden changes in chaotic attractors, and transient chaos, Physica D 7 (1983) 181-200

[GH]   Guckenheimer J, Holmes P, Nonlinear Oscillations, Dynamical Systems and Bifurcations of Vector Fields (Springer, 1983)

[H]   Herman MR, Sur la conjugaison differentiable des difféomorphismes du cercle a des rotations, Publ Math IHES 49 (1979) 5-234

[IL]   Iooss G, Los J, Quasigenericity of bifurcations to high-dimensional invariant tori for maps, Commun Math Phys 119 (1988) 453-500

[Kan]   Kaneko K, Collapse of tori and genesis of chaos (World Sci, 1985)

[Lang]   Langford WF, Periodic and steady-state mode interactions lead to tori, SIAM J Appl Math 37 (1979) 22-48

[LC]   Linsay PS, Cumming AW, Three-frequency quasiperiodicity, phase-locking and the onset of chaos, Physica D 40 (1989) 196-217

[L]   Lorenz EN, Deterministic nonperiodic flow, J Atmos Sci 20 (1963) 130-141

[M]   MacKay RS, notes written with the assitance of Oliveira A, Torus Maps, to appear in the proceedings of the Como School on Chaos, Order and Patterns, ed Cvitanovic P (Plenum)

[MM]   MacKay RS, Muldoon MR, Topology from a time series, in preparation

[ML]   Maurer J, Libchaber A, Effect of the Prandtl number on the onset of turbulence in liquid ^{4}He, J Physique Lett 41 (1980) L515-8

[NRR]   Newell AC, Rand DA, Russell D, Turbulent transport and the random occurrence of coherent events, Physica D 33 (1988) 281-303

[NRT]   Newhouse SE, Ruelle D, Takens F, Occurrence of strange Axiom-A attractors near quasiperiodic flows on $\mathbb{T}^m$, m ≥ 3, Commun Math Phys 64 (1978) 35-40

[Pe]   Pesin Ya B, General theory of smooth hyperbolic dynamical systems, Chapter 7, in Dynamical Systems II, ed Sinai Ya G (Springer, 1989)

[Ru]   Ruelle D, Chaotic evolution and Strange attractors (CUP, 1989)

[RT]   Ruelle D, Takens F, On the nature of turbulence, Commun Math Phys 20 (1971) 167-192; and Commun Math Phys 23 (1971) 343-4

[Sell1]   Sell G, Arch Rat Mech Anal 69 (1979) 199-

[Sell2]   Sell G, Resonance and bifurcation in Hopf-Landau dynamical systems, in Nonlinear Dynamics and Turbulence, eds Barenblatt GI, Iooss G, Joseph DD (Pitman, 1983) 305-313

[Sw]   Swinney HL, Observations of order and chaos in nonlinear systems, Physica D 7 (1983) 3-15

[WKPS]   Walden RW, Kolodner P, Passner A, Surko CM, Nonchaotic Rayleigh-Bénard convection with four and five incommensurate frequencies, Phys Rev Lett 53 (1984) 242-5

[Z]   Zeeman EC, Bifurcation, catastrophe, and turbulence, in New Directions in Applied Mathematics, eds Hilton PJ, Young GS (Springer, 1982) 109-153

SHEAR TYPE INSTABILITIES IN THE BENARD PROBLEM

Josep M. Massaguer

Departament de Física Aplicada
Univ. Politécnica de Catalunya
08034 Barcelona, Spain

1. INTRODUCTION

The aim of the present paper is to argue that if a generic mechanism for the transition to turbulence in parallel flows exists, the same mechanism should also be generic in low Prandtl number thermal convection. In order to stress similarities between both types of flows we shall assume they share the same geometry: A doubly periodic domain in the horizontal directions, bounded on top and bottom by two horizontal surfaces. It will be shown below that a mean horizontal flow – i.e.: a wind – is possible even in the absence of both a head pressure and external shear stresses. Buoyancy force itself may be enough to produce a mean flow. Therefore, a unified treatment is possible. We shall talk in the following about doubly periodic channel flows, and this will explicitly include thermal convection.

It is generally believed that a turbulent flow should be largely independent of its driving and of the geometry of the container. Only large scale dynamics is expected to show some signature of both. A common view is to describe the turbulent flow as a large scale velocity field driven by some external force – pressure head, shear stresses or buoyancy force – and feeding energy into the smaller scales through a cascade process, perhaps with some back-scattering. In such a view the cascade of energy uncouples the bulk of the turbulent flow from its driving. But a different view may be taken by assuming the existence of a primary, externally driven flow, which at a large Reynolds number becomes unstable, so giving rise to a secondary flow. Then the primary flow will act as a buffer and any signature of either the external driving or of the geometry of the container will fade away in the secondary flow. In previous papers on low Prandtl number thermal convection (Massaguer & Mercader 1988, Massaguer, Mercader & Net 1990) we have explored the latter view on the basis of a simple model of thermal convection, and the main goal of the present paper is to extend such a discussion so as to include parallel flows.

In order to understand how a flow can forget the shape of its boundaries or the size of the container, it is helpful to take the example of thermal convection and to think of the system as proceeding in two steps. By heating the fluid at rest it becomes unstable and a primary flow starts up. The velocity for this flow scales with thermal time, $v \simeq \kappa/\ell$, and the Reynolds number takes the value $Re \simeq \sigma^{-1}$, where $\sigma = \nu/\kappa$ is the Prandtl number, the kinematic viscosity is ν and the thermal diffusivity κ. At low σ values, $\sigma << 1$, the primary flow is shear unstable for almost any value of the

Rayleigh number, and even if the primary flow is two-dimensional, which is possible under controlled initial conditions, the secondary flow will be three-dimensional. If the results in the previously quoted papers can be trusted, the secondary flow is the result of a spontaneous break of symmetry induced by the growth of a vertical vorticity component. This symmetry breaking gives rise to a temporal chaos, and any trace of the primary flow becomes strongly damped during the process. This scenario is indeed not very different from what is expected in parallel flow turbulence, which is usually taken as a paradigm.

As discussed in Bayly, Orszag & Herbert (1988) for parallel flows, the primary and the secondary instabilities are of a very different nature. The latter brings the flow into a neighbourhood of the turbulent solution. Thus, it shares the generic properties of a turbulent flow and it is expected to be largely independent of the external driving, whereas the former still reflects a balance between this driving and the viscous damping. In order to emphasize the difference between both types of instabilities we have found helpful to associate the concepts of primary and secondary flows with those of poloidal and toroidal components of the velocity field. This is not normal practice, but is consistent with it, as it will be shown below. Instabilities maintaining the poloidal structure of the flow are thought to be of no relevance in the present context, because they lead to a change in the geometry so as to keep the main balances in the primary flow.

2. THE GEOMETRY OF THE FLOW

A very large fraction of turbulence theory concerning parallel flows and thermal convection has been developed for doubly periodic rectangular boxes. It is an extended view that unwanted end-boundary effects in large aspect ratio containers can be avoided in this way. Thus, it is likely that the velocity field in a doubly periodic rectangular domain is a good representation for the flow in the bulk of the domain either for parallel flows or for thermal convection. Two-dimensional doubly periodic domains are topologically non-trivial objects, as they are equivalent to a torus, and this is a surface on top of which there can exist two different types of closed curves. Those shrinking into a point by continuous deformation, and those winding around its surface. As they are seen on the rectangle, the former are closed curves whereas the latter are open lines crossing the borders of the domain. Either class of curves defines a family of solutions of the Navier-Stokes equation in a way that will be specified below. We shall call them, for obvious reasons, closed and open flows.

2.1 Geometry open flows

For a large class of geometries any solenoidal velocity field can be split into a sum of a poloidal plus a toroidal velocity field, with either one being associated with a scalar function called a potential. In a doubly periodic channel such a statement is still true, provided that no additional requirement is imposed on the potentials. But if both potentials are forced to be periodic, the statement is wrong. To be precise, any velocity field periodic in two horizontal directions and with zero horizontal average velocity in the vertical direction can be written as

$$v(x, y, z, t) = U(z, t) + \nabla \times \nabla \times (\phi e_z) + \nabla \times (\psi e_z) \qquad (2.1)$$

where $U(z, t) = U_x(z, t)e_x + U_y(z, t)e_y$ is the horizontal average flow - often called a wind -, e_x, e_y and e_z are the unit vectors along the x, y and z axis, the z coordinate is in the vertical direction, and the scalar fields $\phi(x, y, z, t)$ and $\psi(x, y, z, t)$ have zero horizontal average. The second and third terms are the poloidal $v_P = \nabla \times \nabla \times (\phi e_z)$ and the toroidal $v_T = \nabla \times (\psi e_z)$ velocity components respectively. The former is associated with the vertical component of the velocity, $v_z = -\nabla_1^2\phi$, and the latter with the vertical component of the vorticity, $\omega_z = -\nabla_1^2\psi$. The operator ∇_1^2 is the horizontal laplacian. For simplicity, we shall also define the velocity fluctuation $\boldsymbol{v}' = \boldsymbol{v}_P + \boldsymbol{v}_T$.

A proof that (2.1) is the most general form for the velocity field in the aforementioned conditions can be easily obtained from the homology group theory used in Marqués (1990), but a plausibility argument will be given below. Difficulties with (2.1) come more from physical intuition than from mathematics, as can be realized from the analysis of the expression used by Zhan *et al.* (1974) to describe Poiseuille flow in a doubly periodic channel. A head pressure in the direction $\boldsymbol{e}_x$ was assumed by the authors to force a mean flow, $U(z,t)\boldsymbol{e}_x$, along this direction. A mean flow in the direction orthogonal to the head pressure, $\boldsymbol{e}_y$, was precluded. The assumption may be plausible under some circumstances – it is so in the quoted paper –, but indeed it shows a prejudice. And what is worse, it can give rise to inconsistencies. With such a prejudice in mind it is not surprising that most theory on thermal convection is derived on the assumption that no mean flow $U(z,t)$ can be present in (2.1) – see, for instance, Clever & Busse (1974) – and this is a hypothesis which is hard to justify from current numerical work on low Prandtl number convection. The reader's attention is called to figure 9 in Meneguzzi *et al.* (1988) where a flow and its counterflow along the roll axis can be realized, though the roll is bent and the presence of a mean flow $U(z,t)$ cannot be firmly established from reported data. Numerical work in progress by J.P. Zahn (private comunication) also supports a mean flow driven only by buoyancy.

The possibility of an axial flow along a bent roll was conjectured in Massaguer, Mercader (1988). The conjecture originated from numerical results on a truncated modal approach – see §4. Although at that time we were unclear about the limits of validity for our approach, the consistency of our numerical results with some laboratory experiments forced the view that we were modeling an axisymmetric flow (Massaguer, Mercader & Blázquez 1987), and there we called this instability the instability of swirl. The analogy with the axial flow in bent filaments was also stressed there, though a mean horizontal flow term was not included in the equations.

In order to show the plausibility of (2.1) for the case of a doubly periodic flow, let us write the velocity field in cartesian components

$$
\begin{aligned}
v_x &= U_x(z,t) + \partial_{xz}^2\phi + \partial_y\psi \\
v_y &= U_y(z,t) + \partial_{yz}^2\phi - \partial_x\psi \\
v_z &= -(\partial_{xx}^2 + \partial_{yy}^2)\phi
\end{aligned}
\tag{2.2}
$$

If $U_x(z,t) = U_y(z,t) = 0$, periodic velocity components may exist that cannot be recovered by taking periodic expansions for the potentials ϕ and ψ because the ∂_x and ∂_y operators kill the mean flow term -i.e.: the term independent of x and y. Whether this term can be neglected or not depends on the dynamic equations, and it will be shown that taking $U = 0$ often leads to inconsistencies. This is certainly not a problem in two-dimensional flows if the velocity fields are written in terms of a periodic streamfunction

$$
\boldsymbol{v}(x,z,t) = (\partial_z\chi,\, 0,\, -\partial_x\chi)
\tag{2.3}
$$

because in the two-dimensional case $v_y = \partial_y = 0$, the function χ can be isolated from (2.2) and (2.3)

$$\chi(x, z, t) = \int U_x(z, t)dz + \partial_x \phi(x, z, t)$$

showing that a streamfunction representation contains the mean horizontal flow, whereas a potential representation requires this term explicitly included.

To further confuse the situation it is worth mentioning here that in the case of a doubly periodic channel with stress-free top and bottom boundaries the components v_x and v_y can be taken, and have been taken – see Meneguzzi *et al.* (1987) and McLaughlin & Martin (1975) –, as the two required dynamic variables instead of the potentials ϕ and ψ. By taking advantatge of the Fourier series expansion, it is straightforward in such a formulation to obtain a system of equations with the pressure term canceled. This can be done by using the formal analogy $\nabla \sim i\boldsymbol{k}$, where $\boldsymbol{k}$ is the wavevector, and subsequent projection onto the space of solenoidal velocity fields by using the well known operator $P_{ij} = \delta_{ij} - k_i k_j / k^2$. The resulting formulation is almost analogous to that for potentials, but it is not equivalent. The potentials ϕ and ψ can be obtained from v_z and ω_z which can in turn be computed from v_x and v_y by using the continuity equation $\boldsymbol{k} \cdot \boldsymbol{v} = 0$ and the definition of vorticity $\boldsymbol{\omega} = \boldsymbol{k} \times \boldsymbol{v}$. But any term independent of the horizontal coordinates x and y – i.e.: $k_x = k_y = 0$ – present in the expansion of v_x and v_y will be killed in v_z and ω_z. Taking v_x and v_y as dynamic variables is a safe procedure, and a mean horizontal flow is not precluded.

2.2 The dynamic equations

In order to solve a dynamic problem we require one equation for every variable. The mean velocity $U(z, t)$ can be computed from the averaged Navier-Stokes equation, and the functions ϕ and ψ from the z-components of the curl and double curl of the Navier-Stokes equation. The equivalence between such a formulation and the Navier-Stokes equation has been proved for some non-trivial geometries by Marqués (1990). Moreover, it has also been shown that for some geometries the equivalence requires additional boundary conditions – see Marqués *et al.* (1990) for the case of thermal convection in cylinders –, but this is not the case here.

The full set of equations to be solved can be written as

$$(\partial_t - \nu \partial_{zz}^2)U_x + \partial_z \overline{v_x' v_z'} = \Delta p / \rho \ell \tag{2.4a}$$

$$(\partial_t - \nu \partial_{zz}^2)U_y + \partial_z \overline{v_y' v_z'} = 0 \tag{2.4b}$$

$$(\partial_t - \nu \nabla^2)\nabla^2 v_z' + (\nabla \times \nabla \times (\boldsymbol{v}' \cdot \nabla \boldsymbol{v}'))_z = \boldsymbol{U} \cdot \nabla(\nabla^2 v_z') - \partial_{zz}^2 \boldsymbol{U} \cdot \nabla v_z' + F_z \tag{2.4c}$$

$$(\partial_t - \nu \nabla^2)\omega_z' + \boldsymbol{v}' \cdot \nabla \omega_z' - \boldsymbol{\omega}' \cdot \nabla v_z' = -\boldsymbol{U} \cdot \nabla \omega_z' + (\partial_z \boldsymbol{U} \times \nabla v_z')_z + G_z \tag{2.4d}$$

with $\boldsymbol{F} = \nabla \times \nabla \times \boldsymbol{f}$ and $\boldsymbol{G} = \nabla \times \boldsymbol{f}$, where $\boldsymbol{f}$ is the bulk force – for instance buoyancy force in thermal convection. The velocity fluctuation, $\boldsymbol{v}' = \boldsymbol{v}_P + \boldsymbol{v}_T$, has been defined before, and $\boldsymbol{\omega}' = \nabla \times \boldsymbol{v}'$ is the vorticity fluctuation. With $\Delta p / \rho \ell$ we denote the mean head pressure, and with an overbar the horizontal average.

System (2.4) complemented with appropriate boundary conditions describes parallel flows such as Couette or Poiseuille flows but it also describes thermal convection. It is important to realize that even in the absence of a head pressure the Reynolds stresses $\overline{v_x' v_z'}$ and $\overline{v_y' v_z'}$ can drive a mean flow, thus showing that thermal convection or Couette flow between cylinders may display two very different regimes, depending on they being open or closed flows. In addition, in parallel flows suchs as plane Couette or Poiseuille, the mean flow U need not be parallel to the external forcing, no matter whether the forcing is a head pressure or some shear boundary stresses. The so-called barber-pole instability in Couette flow between counter-rotating cylinders (Coles 1965) can be called forth because it produces a spiral flow, whereas forcing is in the azimuthal direction.

An important implication of periodicity is that the circulation along the borders of the rectangular domain is zero, $\oint \boldsymbol{v} \cdot d\boldsymbol{l} = 0$, and by the Stokes theorem the mean value over the rectangle of the vertical component of the vorticity ω_z should be zero. A second implication concerns the conservation of angular momentum. Because of the absence of lateral boundaries, the moment of the external forces on the rectangle will have a zero vertical component. Therefore, in a periodic box with stress free top and bottom boundaries angular momentum is conserved, in contrast with the non-slippery case, where conservation does not follow from first principles. To be precise, the equation for the mean vertical component of the angular momentum $l_z = \boldsymbol{e}_z \cdot (\boldsymbol{r} \times \boldsymbol{v})$ can be obtained from the Navier-Stokes equation as

$$(\partial_t - \nu \partial_{zz}^2)\bar{l}_z + \partial_z \overline{v_z' l_z} = 0$$

and $\bar{l}_z$ can be shown to decay only for stress free top and bottom boundaries or, also, for flows showing symmetries such that $\overline{v_z' l_z} = 0$ – see §3.3.

2.3 Geometrical constraints for closed flows

We say a flow is closed if $U = 0$ in (2.1). From the integration of (2.4a) and by using the boundary condtion $v_z'(\pm d/2, t) = 0$, it can be concluded for these solutions $\Delta p = 0$. Thus, for a flow to be closed the mean Reynolds stresses are to be zero

$$\overline{v_x' v_z'} = \overline{v_y' v_z'} = 0 \tag{2.5}$$

and (2.4a,b) can be dropped. Conditions (2.5) may be thought as being dynamic conditions or as being geometric restrictions on the flow. In the following we shall adopt the latter view.

2.3.1 *Shear breakup in two-dimensional convection.* In the two-dimensional case $\partial_y = v_y = 0$ the velocity field can be given by (2.3) and the mean Reynolds stress can be written

$$\overline{v_x' v_z'} = \int_{-1/2}^{+1/2} \partial_x \chi \, \partial_z \chi \, dx \tag{2.6}$$

The mean stress $\overline{v_x' v_z'}$ is zero, a least in those cases where $\chi(x, z, t)$ is an even or an odd function of x. By breaking the integral in (2.6) into two pieces $\int_{-1/2}^{+1/2} = \int_{-1/2}^{0} + \int_{0}^{+1/2}$ and using the definition of symmetry $\chi(x, z, t) = \pm\chi(-x, z, t)$ it can be shown that $\overline{v_x' v_z'} = 0$. Therefore, *breaking the symmetry is a necessary condition for the existence of an open structure.*

Near the threshold of instability for thermal convection the streamfunction χ is symmetric, as can be shown by solving the corresponding linear eigenvalue problem. Thus, the real question is whether such a break of symmetry will ever take place. To the best of our knowledge the question is still open, but Krishnamurti & Howard (1981) reported on some laboratory experiments on thermal convection where slanted, two-dimensional convection patches indicating strong horizontal shearing could clearly be identified. Therefore, in our doubly periodic channel analog we may also expect, a symmetry break that triggers a mean horizontal flow. At this point the reader must notice from (2.4a,b) that global conservation of the linear momentum $\int \rho U \, dz$ needs not be the rule. Conservation can only be proved for flows with stress-free boundaries and $\Delta p = 0$. It is worth noticing that in a closed container, say in a two-dimensional one, the mean flow is zero, $U_x = 0$, but because of the lack of periodicity the head pressure across the container width need not be zero. The mean momentum equation can be written

$$\partial_z \overline{v'_x v'_z} = \Delta p / \rho \ell$$

where the boundary condition $v_x = 0$ has been used. Thus, the price paid to stop the mean flow at the lateral walls is building a head pressure.

3. GEOMETRY AND DYNAMICS OF 3-D CLOSED FLOWS

By definition a flow is closed if $\overline{v} = 0$. Closure does not imply in any sense the existence of a well defined material volume with permanent boundaries. Nevertheless, it is easier to think about closed flows in terms of cells with well defined geometries, say rectangles or hexagons, perhaps a little shaky, paving the channel. On the topologically equivalent torus surface these cells are bounded by closed lines shrinking into a point.

3.1 Kinematics of a toroidal field

In the following we shall assume no mean flow, $U = 0$, and no forcing for the vertical vorticity, $G_z = 0$, as in thermal convection. In terms of potentials, the equation (2.4d) for the toroidal component v_T can be written

$$(\partial_t - \nu \nabla^2) \nabla_1^2 \psi + \Lambda(\phi, \psi) + \frac{\partial(\nabla_1^2 \psi, \psi)}{\partial(x, y)} = \frac{\partial(\nabla^2 \phi, \nabla_1^2 \phi)}{\partial(x, y)} \tag{3.1}$$

where Λ is a bilinear operator – see Massaguer & Mercader (1988) for details. By using the appropriate boundary conditions it can be shown that only two terms in (3.1) can play the role of source terms for ψ. The Λ term, and the jacobian at the right hand side of (3.1) – for a term Q to be called a source term we shall require that $< \psi Q >$, the volume integral of Q times ψ, is neither negatively defined nor identically zero.

Equation (3.1) can be integrated once the scalar function ϕ is given. This is indeed an oversimplification for it neglects the feedback of ψ on ϕ, but gives some useful results concerning the stability of the poloidal velocity field. Let us notice, first of all, that if ϕ is a two-dimensional field – i.e.: is a function of z and one horizontal coordinate, not necesarily cartesian – the jacobian on the right hand side is zero and $\psi = 0$ is a solution. Moreover, if $\partial_y \phi = 0$ it can be shown that the Λ term does not provide a source and the solution decays towards $\psi = 0$. If ϕ is two-dimensional but not invariant along a straight line, say there exists an horizontal line $\beta = \beta(x, y)$ such that $\partial_\beta \phi = 0$, the jacobian at the right hand side is still zero but Λ can now be a source term.

A purely poloidal flow, $v_T = 0$, is two dimensional in the sense that it is a flow invariant along a β-line – see (3.1). By choosing an orthogonal coordinate system $\{\alpha, \beta, z\}$, with α and β being some specified functions of the x and y coordinate, it can be shown that $\partial_\beta \phi = \psi = 0$ implies $\boldsymbol{\omega} = \partial_\alpha (\nabla^2 \phi) e_\beta$. Therefore, any two-dimensional poloidal flow is made of horizontal vorticity tubes. However appealing such a physical picture may be, it should be handled carefully, for there are very few non-trivial *orthogonal* coordinates systems supporting one *global* invariance of this type. As can be realized from a straightforward application of the chain rule derivative, $\nabla_1^2 \phi = \partial_\alpha \phi \, \nabla_1^2 \alpha + \partial_{\alpha\alpha} \phi \mid \nabla_1 \alpha \mid^2$ which, for consistency, should be invariant against ∂_β. Therefore, only those systems of coordinates where $\nabla_1^2 \alpha$ and $\mid \nabla_1 \alpha \mid^2$ are functions of α are admissible. The polar coordinates r and θ are obvious examples of α and β.

3.2 Linear stability of a poloidal field

Let us now take for the unperturbed state the two-dimensional field $\{\phi_0, \psi = 0\}$, with $\partial_\beta \phi_0 = 0$, and let us call the perturbation field $\{\phi', \psi'\}$. The linearized version of (3.1) is

$$(\partial_t - \nu \nabla^2) \nabla_1^2 \psi' + \Lambda(\phi_0, \psi') = \frac{\partial(\nabla^2 \phi_0, \nabla_1^2 \phi')}{\partial(x, y)} + \frac{\partial(\nabla^2 \phi', \nabla_1^2 \phi_0)}{\partial(x, y)} \qquad (3.2)$$

For any perturbation of the poloidal field ϕ' preserving the geometry of the flow, $\partial_\beta \phi' = 0$, both jacobians on the right hand side of the equation are identically zero, (3.2) becomes uncoupled from the ϕ equation, and the only possible source of instability is the Λ term. But, as mentioned before, Λ is a source term only if the poloidal velocity field is not invariant along a straight line, meaning that for this type of instability to be possible the vorticity tubes should be bent. To be precise, the source $< \psi' \Lambda(\phi_0 \psi') >$ is proportional to the curvature of the β-line. Solutions of (3.2) are to be sought as $\psi' = \Psi e^{i\mu\beta}$, with $\partial_\beta \Psi = 0$. In particular, solutions with $\mu = 0$ can exist. In this case the toroidal velocity field is $v_T = -\partial_\alpha \psi e_\beta$, and the flow is invariant along the vorticity tube. Elsewhere we called this type of instability *swirl instability*.

As example of perturbation not preserving the geometry, let us take for the unperturbed flow a straight vortex tube, $\partial_y \phi_0 = 0$. The only source terms are now the jacobians on the right hand side of (3.2), and the perturbation field is $\{\phi', \psi'\} = \{\Phi, \Psi\} e^{i(\mu y - \Omega t)}$, with $\partial_y \Phi = \partial_y \Psi = 0$. In order to solve (3.2) the linearized equation for ϕ' must specified. This type of instabilities is very well known in thermal convection, and there is a large list of them – see Busse (1978) for a review. In most cases the non-linear system evolves towards a new poloidal flow, $\{\phi_1 \neq \phi_0, \psi = 0\}$, and neglecting the toroidal component does not substantially change the dynamics, as corresponds to the toroidal field being a slave mode. An important exception is the so called *oscillatory instability*, where neglecting ψ kills the instability. In this case Ω has a non-zero real part and the perturbation is periodic in x with the periodicity of the basic flow. Thus the perturbation propagates along the straight roll like a wave on a rope. Elsewhere we called all these instabilities *shape instabilities* for their existence strongly depends on the particular geometry of the flow.

Current knowledge on parallel flows suggests the existence of a generic mechanism to drive the secondary instability – see the review by Bayly, Orzsag & Herbert (1988). There are many instabilities that may act on the primary flow, so they could be called secondary instabilities, but the discussion in §3.1 puts forward the view that only instabilities triggering a jump from a poloidal to a toroidal structure deserve that name. Although the two instabilities that generate a toroidal flow, oscillatory and swirl, are very different in nature, they originate in the non-linear advection term and we can therefore expect them to be triggered in the same range of Reynolds numbers. In real flows without controlled initial conditions both instabilities can be triggered at once. In order to decide which one will dominate, a detailed computation is required, and the possibility of the final result being dependent on the prescribed geometry, the initial conditions and the external stresses cannot be ruled out.

However, besides thermal convection, the oscillatory instability has also been found in two-dimensional vortices with elliptic core by Pierrehumbert (1986). He showed that in the limit of small wavelength perturbation, $\mu \to \infty$, the eigenmodes must be independent of μ, and the growth rates increase with the eccentricity of the elliptical core. He conjectured this instability to be a generic mechanism for the transition to turbulence in parallel flows. With respect to thermal convection, both theory and numerical computations (Busse & Clever 1981, Clever & Busse 1981) show that in the limit of small Prandtl number, and for stress free boundaries, the cross section of roll vortices becomes circular. But Pierrehumbert (see also Bayly 1986) showed that the growth rates go to zero with the eccentricity going to zero. Therefore the oscillatory instability will not be acting in the low Prandtl limit, in agreement with the result of Busse & Bolton (1984), Bolton & Busse (1985). Although a mean flow was not included in Busse & Bolton's model, the symmetry of circular streamlines precludes a wind, so encouraging the view that at large Reynolds numbers it is the instability of swirl that becoming dominant.

3.3 Symmetry breaking instabilities

In the previous section we have only considered local instabilities, but global instabilities may also provide genuine mechanisms for transition to turbulence, and they are good candidates for the so-called catastrophic or explosive transitions to turbulence. A system invariant against a finite symmetry group is a candidate to show such global instabilities. If a system presents a symmetry group invariance, every solution can be replicated as many times as elements belong to the group. If all these replicas are identical up to a translation in time, the system is said to be symmetric. Thus, we shall speak of symmetry breaking if by increasing the forcing the system jumps from a symmetric solution towards an asymmetric solution. As a consequence of the group properties, symmetry breaking gives rise to a full set of asymmetric solutions. Working backwards, the lack of unicity provides a frame for the merging of solutions, and the outcome of merging is a symmetric solution if the group is simple or, in any other case, a solution displaying a higher order symmetry that the initial one.

There are two well known groups of symmetry in doubly periodic channels. One is the skew-symmetry of the velocity field against a plane at mid height

$$\begin{aligned}
\{x, y, z, t\} &\to \{x + x_0, y + y_0, -z, t + t_0\} \\
\{v_x, v_y, v_z\} &\to \{v_x, v_y, -v_z\}
\end{aligned} \tag{3.3}$$

where the coordinate system is taken at the center of the layer, $\{x_0, y_0\}$ is a constant space shift and t_0 is a constant time lag. And the second group of symmetries is a reversal in the vertical vorticity, defined by

$$\begin{aligned}
\{x, y, z, t\} &\to \{-x - x_0, y + y_0, z, t + t_0\} \\
\{v_x, v_y, v_z\} &\to \{-v_x, v_y, v_z\}
\end{aligned} \tag{3.4a}$$

and

$$\begin{aligned}
\{x, y, z, t\} &\to \{x + x_0, -y - y_0, z, t + t_0\} \\
\{v_x, v_y, v_z\} &\to \{v_x, -v_y, v_z\}
\end{aligned} \tag{3.4b}$$

Symmetries (3.3) and (3.4) are such that their square is the identity

$$\{x, y, z, t\} \rightarrow \{x + 2x_0, y + 2y_0, z, t + 2t_0\}$$
$$\{v_x, v_y, v_z\} \rightarrow \{v_x, v_y, v_z\}$$

Therefore, for a symmetric solution the constants $2x_0$, $2y_0$ and $2t_0$ are integer multiples of the fundamental period. In the following we shall take for x_0 and y_0 either half a period or zero without loss of generality. We should also notice that every one of these symmetries changes the sign of the helicity $\boldsymbol{h} = \boldsymbol{\omega} \cdot \boldsymbol{v}$, as they are symmetries acting on a pseudoscalar. Thus, a symmetry breaking with $x_0 = y_0 = 0$ of a solution with $\boldsymbol{h} = 0$ is physically meaningful because for the two new born solutions $\boldsymbol{h} \neq 0$, so they being physically different.

Symmetry (3.4a) with $x_0 = y_0 = 0$ implies $U_x = 0$ and $\overline{v_x' v_z'} = 0$, and indeed it cannot be fulfilled for a parallel flow. Symmetry (3.4b) implies $U_y = 0$ and $\overline{v_y' v_z'} = 0$. Breaking either symmetry produces two possible solutions with mean flows in opposite directions. Each one of these solutions is equally like, and the reconnection of their orbits in, for example, a pulsating flow is also possible. The breakup described in §2.3.1 is in this class of symmetries. On the other part, the breaking of the symmetry (3.3) has been found to play a fundamental role in several transitions to chaos. It has been described in connection with the Lorenz model (Sparrow 1982, Guckenheimer & Holmes 1983), thermohaline convection (Knobloch et al. 1986) and low Prandtl number convection Massaguer, Mercader & Net (1990), and this is a symmetry breaking independent of whether or not the mean flow is zero, which makes it a good candidate to provide a generic mechanism for the secondary instability.

4. CHAOS FROM SYMMETRY BREAKING

In Massaguer, Mercader & Net (1990), Mercader, Massaguer & Net (1990) we have examined the transition to chaos of a very simple dynamic model. It was obtained as a Galerkin truncation for thermal convection, but it was devised so as to include the swirl instability described in §3.2, but not the oscillatory instability. In this simplified model no provision has been made for a mean flow, which has been taken consistently as zero. Here we assumed there for the velocity field the expression (2.1) with

$$U = 0$$
$$\phi = f(x, y) W(z, t)$$
$$\psi = f(x, y) \xi(z, t) \tag{4.1}$$

where $f(x, y)$ is a periodic eigenfunction of the laplacian operator, $\nabla_1^2 f = -a^2 f$, $\overline{f^2} = 1$ and $\overline{f^3} \neq 0$. The corresponding truncated equations, in non-dimensional form, can be written

$$(\partial_t - \nabla^2)\nabla^2 W = -Ra^2\theta - (W\partial_z\nabla^2 W + 2\partial_z W\nabla^2 W + 3\xi\partial_z\xi) \tag{4.2a}$$

$$(\partial_t - \nabla^2)\xi = -(W\partial_z\xi - \xi\partial_z W) \tag{4.2b}$$

$$\nabla^2\theta = W \tag{4.2c}$$

where R is the Rayleigh number. In order to make the present discussion simpler, we have written here the heat equation in the small Péclet number approximation so as to give some hints about the bulk force. From (4.2) it can be realized that a pure poloidal flow, $\xi = 0$, is always possible.

System (4.2) was numerically explored in the previously quoted papers for several boundary conditions, and several scenarios for transition to chaos were found, but only those appearing just after the onset of a toroidal component were thoroughly explored. A sequence of reconnections of asymmetrical solutions induced by (3.3) and subsequent symmetry breakings dominated the scenarios. And, more relevant in our opinion, coherence in time between the poloidal and the toroidal component was found to be very small, as can be seen, for instance, from their Fourier spectra. This reduced coherence was ascribed there to the the fact that the fundamental period for ϕ was found to be double that for ψ – this is so because of the symmetry (3.3) –, so giving a different sequence of events in their transition to chaos. This mechanism for transition to chaos is thought to be generic in the sense that it depends only on the symmetries displayed in §3.3, which are quite general. In addition, and because of the different way that ϕ and ψ enter into (2.1), a lack of coherence between them both results in a *spatial intermittence* – i.e.: the values of W and ξ for fixed z change in a random way, so do ϕ and ψ and, therefore, the geometry of the flow changes in an unpredictable way. Thus, we may conclude that, in spite of the simplicity of (4.1), it includes most of the basic ingredients of a turbulent flow.

Acknowledgements. It is pleasure to acknowledge fruitful discussions with Dr. F. Marqués while writing this paper. The present work was supported by the Dirección General de Investigación Científica y Técnica (DGICYT) under grant PS-87-0107.

References

Bayly, B.J. 1986 *Phys. Rev. Lett.* **57**, 2160

Bayly, B.J., Orszag, S.A. & Herbert, T. 1988 *Ann. Rev. Fluid Mech.* **20**, 359

Bolton, E.W. & Busse, F.H. 1985 *J.Fluid Mech.* **150**, 487

Busse, F.H. 1972 *J.Fluid Mech.* **52**, 97

Busse, F.H. 1978 *Rep. Prog. Phys.* **41**, 1929

Busse, F.H. & Bolton, E.W. 1984 *J.Fluid Mech.* **146**, 115

Busse, F.H. & Clever, R.M. 1981 *J.Fluid Mech.* **102**, 75

Clever, R.M. & Busse, F.H. 1974 *J.Fluid Mech.* **65**, 625

Clever, R.M. & Busse, F.H. 1981 *J.Fluid Mech.* **102**, 61

Coles, D. 1965 *J.Fluid Mech.* **21**, 385

Guckenheimer, J. & Holmes, P. 1983 *Non-linear Oscillations, Dynamical Systems and Bifurcations of Vector Fields.* Springer

Knobloch, E., Moore, D.R., Toomre, J. & Weiss, N.O. 1986 *J.Fluid Mech.* **166**, 409

Krishnamurti, R. & Howard, L. 1981 *Proc. Natl. Acad. Sci. USA* **78**, 1981

McLaughlin, J.B. & Martin, P.C. 1975 *Phys. Rev. A* **12**, 186

Marqués, F 1990 *Phys. Fluids A* **2**, 729

Marqués, F., Mercader, I., Net, M. & Massaguer, J.M. 1990 "Thermal Convection in Vertical Cylinders I: A Method Based on Potentials of Velocity" (submitted *Phys. Fluids*)

Massaguer, J.M. & Mercader, I. 1988 *J.Fluid Mech.* **189**, 367

Massaguer, J.M., Mercader, I. & Net, M. 1990 *J.Fluid Mech.* **214**, 679

Massaguer, J.M., Mercader, I. & Blázquez, S. 1987 In "Advances in Turbulence" (Ed. Compte-Bellot & J.Mathieu). Springer

Meneguzzi, M., Sulem, C., Sulem, P.L., Thual, O. 1987 *J.Fluid Mech.* **182**, 169

Mercader, I., Massaguer, J.M., & Net, M. 1990 *Phys. Lett. A*, **149**, 195

Pierrehumbert, R.T. 1986 *Phys. Rev. Lett.* **57**, 2157

Sparrow, C.T. 1982 *The Lorentz Equations: Bifurcation, Chaos and Strange Attractors.* Springer

Zahn, J.P., Toomre, J., Spiegel, E.A. & Gough D.O. 1974 *J.Fluid Mech.* **64**, 319

INTERPRETATION OF INVARIANTS OF THE BETCHOV-DA RIOS EQUATIONS
AND OF THE EULER EQUATIONS

H. Keith Moffatt *and* Renzo L. Ricca

Department of Applied Mathematics and Theoretical Physics
Silver Street, Cambridge CB3 9EW, U.K.
and
Trinity College
Cambridge CB2 1TQ, U.K.

INTRODUCTION

Interest in the study of invariant quantities is generally motivated by the need to interpret and to understand their meaning and their fundamental role in the theory. The invariants we shall consider in this paper emerge in two contexts. In the context of the localized induction approximation (LIA) for the motion of an inextensible vortex filament in a perfect fluid flow, we shall deal with certain conserved quantities that emerge from the Betchov-Da Rios equations; the polynomial invariants for the related nonlinear Schrödinger equation (NLSE) are calculated employing the recurrence formula of Zakharov and Shabat (1974); these quantities are constants of the motion for the vortex as long as self-intersection does not occur and long-distance effects are neglected. Some of them are then interpreted in terms of kinetic energy, momentum, angular momentum, "pseudo-helicity," and some inequalities between physical and global geometrical properties are stated. In the context of the Euler equations, given that the topology of the vortex structure is conserved and the helicity invariant is the natural measure of the topological complexity of the field (Moffatt, 1969; Arnol'd, 1974), we comment further on this role of helicity; moreover, by associating an "energy spectrum" with any knot or link present in the fluid, the lowest ground-state enstrophy is interpreted as an even more powerful invariant for evaluating the knot or link complexity of the field structure.

INVARIANTS OF THE BETCHOV-DA RIOS EQUATIONS

We consider a vortex filament $\mathcal{F}$ in a perfect fluid flow, with the vorticity vector ω everywhere parallel to the tangent $\mathbf{t}$ on the curve. In the context of the local induction approximation (LIA) the motion of the vortex is governed by the simple equation

$$\mathbf{V} = c(s,t)\mathbf{b} \qquad (1)$$

where $\mathbf{V}$ is the Eulerian velocity of $\mathcal{F}$, $c(s,t)$ is the local curvature, a function of arclength s and time t, and $\mathbf{b}$ the binormal (Da Rios, 1906; Arms and Hama, 1965). In the LIA context, the mean diameter of the vortex core-size is assumed negligible compared with the local radius of curvature and self-interactions of different parts of the filament are believed not to occur. The equation (1) led Da Rios and Betchov (1965), independently, to derive the equations which govern the time derivative of the curvature c and the torsion ϑ of $\mathcal{F}$

$$
\left.\begin{aligned}
\dot{c} &= -(c\vartheta)' - c'\vartheta \\
\dot{\vartheta} &= \left(\frac{c'' - c\vartheta^2}{c}\right)' + c'c
\end{aligned}\right\}
\tag{2}
$$

where dots and primes stand for partial derivatives with respect to t and s, respectively. Hasimoto (1972) by the remarkable transformation

$$
\psi(s,t) = c(s,t)\exp\left[i\oint \vartheta(s,t)\,ds\right] \qquad \psi(s,t) \in \mathbb{C}
\tag{3}
$$

reduced the set (2) to the nonlinear Schrödinger equation (NLSE)

$$
\frac{1}{i}\frac{\partial\psi}{\partial t} = \frac{\partial^2\psi}{\partial s^2} + \frac{1}{2}|\psi|^2\psi
\tag{4}
$$

which in one dimension is completely integrable (Zakharov and Shabat, 1972). This means that we have an infinite set of quantities which are constants of the motion. Among these, there is a countable family of so-called 'polynomial conservation laws'; they have the form of an integral with respect to s of a polynomial expression in terms of the function $\psi(s,t)$ and its derivatives with respect to s. Using the Hasimoto transformation (3), we can employ the recurrence formula given by Zakharov and Shabat to calculate this family of invariants in terms of c and ϑ. The formula is

$$
f_{n+1} = q\frac{\partial}{\partial s}\left(\frac{1}{q}f_n\right) + \sum_{j+k=n} f_j f_k \quad ; \quad f_n = f_n(s,t) \in \mathbb{C} \quad n = 1, 2, \ldots
\tag{5}
$$

where

$$
q = \frac{1}{2}i\psi \qquad f_1 = \frac{1}{4}|\psi|^2 = \frac{1}{4}c^2
$$

and the associated invariants are then

$$
(2i)^n \mathcal{I}_n = \oint f_n(s,t)\,ds = \text{constant} \qquad n = 1, 2, \ldots
\tag{6}
$$

The first five conserved quantities are

$$
2i\mathcal{I}_1 = \frac{1}{4}\oint |\psi|^2\,ds = \frac{1}{4}\oint c^2\,ds
\tag{7a}
$$

$$(2i)^2 \mathcal{I}_2 = -\frac{1}{8} \oint (\bar{\psi}\psi' - \psi\bar{\psi}') \, ds = -\frac{1}{4} i \oint c^2 \vartheta \, ds \qquad (7b)$$

$$(2i)^3 \mathcal{I}_3 = \frac{1}{4} \oint \left(\frac{c^4}{4} - c'^2 - c^2 \vartheta^2 \right) ds \qquad (7c)$$

$$(2i)^4 \mathcal{I}_4 = \frac{1}{4} \oint \left[cc''' - i\left(\frac{3}{4}c^4 - c'^2 - c^2\vartheta^2 + 2cc'' \right)\vartheta \right] ds \qquad (7d)$$

$$(2i)^5 \mathcal{I}_5 = \frac{1}{4} \oint \left[c^2 \left(\frac{c^4}{8} - c'^2 \right) - \frac{3}{2}cc'^2 + c''^2 + c^2 \left(\vartheta^4 - \frac{3}{2}c\vartheta + \vartheta'^2 \right) - 2\vartheta^2 \left(2cc'' + c'^2 \right) \right] ds. \quad (7e)$$

where $\bar{\psi}$ is the complex conjugate of ψ. We recognise the first two integrals (7a, 7b) as invariants discovered by Betchov; note that the total torsion of the filament

$$\mathcal{I}_T = \oint \vartheta \, ds \qquad (8)$$

is also an invariant of equations (2) (Keener, 1990 — note added in proof).

INTERPRETATION OF SOME SCALAR AND VECTOR INVARIANTS OF THE BETCHOV-DA RIOS EQUATIONS

Suppose for simplicity that the vorticity is uniform across the core of the vortex, i.e. $\omega = \omega_0 \mathbf{t}$, and let

$$\Gamma = \omega_0 A_0 \quad , \quad d\mathcal{V} = A_0 ds = \frac{\Gamma}{\omega_0} ds$$

where A_0 is the mean cross-sectional area. Apart from numerical coefficients, the integral $\mathcal{I}_1$ can be related to the Eulerian kinetic energy T associated with the motion of the filament simply applying LIA (eq. 1):

$$T = \frac{1}{2} \int_{\mathcal{F}} \mathbf{V}^2 \, d\mathcal{V} = \frac{1}{2} \frac{\Gamma}{\omega_0} \oint c^2 \, ds = \text{constant.} \qquad (9)$$

The interpretation of $\mathcal{I}_2$ is more subtle: it is asymptotically related to the helicity of $\mathcal{F}$; in this context we can call it "pseudo-helicity" and write

$$\tilde{\mathcal{H}} = \int_{\mathcal{F}} \mathbf{V} \cdot \mathrm{curl}\mathbf{V} \, d\mathcal{V} = -2\frac{\Gamma}{\omega_0} \oint c^2 \vartheta \, ds = \text{constant} \qquad (10)$$

[hint: from the definition, apply LIA and consider the "abnormality" b·curlb of unit binormals to the filament; the surface swept out by the vortex filament minimises the Hamiltonian for the system (2) (Weatherburn, 1927, p. 197; Marris and Passman, 1969; Rasetti and Regge, 1975; Ricca, in preparation)]. The integrand of $\mathcal{I}_3$ can be recognised as the Lagrangian density for the "lossless" NLSE, via the Hasimoto transformation (3).

Furthermore, the LIA implies that the total length L and the enstrophy Ω of $\mathcal{F}$ are conserved. By the Schwarz integral inequality we can state the following constraints: the helicity $\mathcal{H}$ has an upper bound, given by

$$\mathcal{H}^2 \leq \Gamma^2 L \oint c^2 \, ds = \text{constant.} \tag{11}$$

For an unknotted filament, the total curvature is bounded as follows

$$2\pi \leq \oint c \, ds \leq \left(L \oint c^2 \, ds \right)^{1/2} = \text{constant} \tag{12}$$

[hint: apply Fenchel theorem (do Carmo, 1976, p. 399)] and for a knotted filament, we have

$$4\pi \leq \oint c \, ds \leq \left(L \oint c^2 \, ds \right)^{1/2} = \text{constant} \tag{13}$$

[hint: apply Fary-Milnor theorem (do Carmo, 1976, p. 402)].

It is also possible to prove the conservation of two vector quantities which can be interpreted as momentum invariants; the linear momentum

$$\mathbf{M} = \frac{1}{2} \int_{\mathcal{F}} \mathbf{X} \wedge \omega \, d\mathcal{V} = \frac{1}{2} \Gamma \oint \mathbf{X} \wedge \mathbf{X}' \, ds \tag{14}$$

and the angular momentum

$$\mathbf{P} = \frac{1}{3} \int_{\mathcal{F}} \mathbf{X} \wedge (\mathbf{X} \wedge \omega) \, d\mathcal{V} = \frac{1}{3} \Gamma \oint \mathbf{X} \wedge (\mathbf{X} \wedge \mathbf{X}') \, ds \tag{15}$$

are constants of the motion [hint: use LIA and the "expansion theorem" for the vector triple product]. The first of these last two results in particular (the conservation of the right-hand side of (14)) was obtained by Arms and Hama (1965) but the interpretation given here is believed to be new. These results imply that the localized induction approximation respects conservation of both momentum and angular momentum.

TOPOLOGICAL INVARIANTS OF THE EULER EQUATIONS

Since, under evolution governed by the Euler equations, the vorticity field is frozen in the fluid, the complete topology of the vorticity field is conserved. It is interesting to inquire what is the complete set of topological invariants that characterise this topology. One subset of invariants are the helicity invariants (Moreau, 1961; Moffatt, 1969), one such invariant being defined for every closed surface within the fluid on which the normal component of vorticity is zero. The helicity of linked vortex tubes is directly related to the Gauss linking number, the most basic topological invariant of two linked curves (see figure 1). However, there exist simple linkages (e.g. the Whitehead link of figure 1c) for which the linking number is zero; if an untwisted vortex tube of small cross-section is constructed around each of the curves of figure 1c, then the helicity of the resulting flow is zero, although the topology is obviously non-trivial. The question then arises as to whether there is any other topological invariant of the vorticity field which may provide a measure of the topological complexity of the structure. A possible approach to the problem, which is related to the fundamental problem of the topological classification of knots and links in $\mathbb{R}^3$ has been described by Moffatt (1990). Here, we simply indicate the essential features of this approach, in the context of the Euler equations.

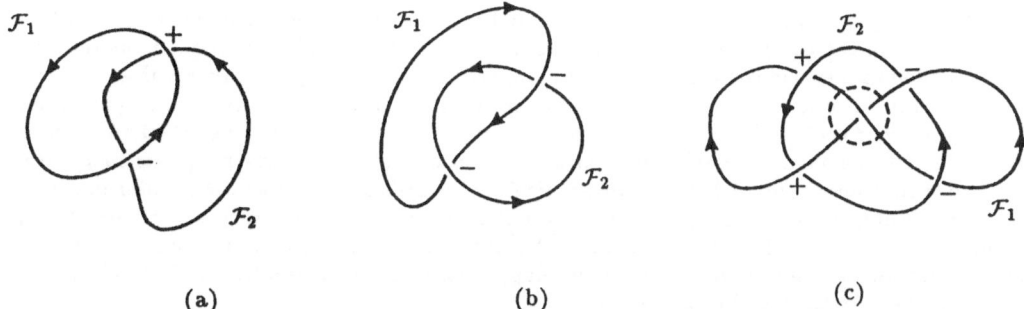

Fig. 1. Examples of (a) unlinked and (b), (c) linked filaments $\mathcal{F}_1$ and $\mathcal{F}_2$. The linking number $\mathcal{L}_{12}$ is respectively: (a) $\mathcal{L}_{12} = 0$; (b) $\mathcal{L}_{12} = -1$; (c) $\mathcal{L}_{12} = 0$ (Whitehead link) (see for example Maxwell, 1873, vol. II, p. 41).

Suppose first that we have an arbitrary closed knotted curve C, and let us construct around C a vortex tube of circulation κ, and small cross-section. We suppose that, within this tube, each vortex lines runs essentially parallel to the axis C. There is a certain arbitrariness here as regards the twist of the vorticity field around the axis of the tube: for we can imagine cutting the tube at any section, twisting it about the axis through an arbitrary angle, and reconnecting. As shown in Moffatt (1990), the effect of this is to change the helicity of the associated flow by an amount $h\kappa^2$ where $2\pi h$ is the angle of twist. Whatever the initial helicity may have been, there is always a twist which can be introduced in this way, and which makes the resulting helicity zero. Let us suppose that this condition is satisfied, although it is by no means the only possibility. This assumption pins down the topology of the vorticity field and relates it unambiguously to the topology of the knot C.

Under any volume-preserving diffeomorphism, the topology of the vorticity field is preserved, but the enstrophy of the associated flow is not conserved. It is obvious that there can be no theoretical *upper* bound on the enstrophy, because natural evolution governed by the Euler equations is generally such as to increase enstrophy, and this increase can proceed without limit when the effects of viscosity are ignored; this type of behaviour is well known, if not fully understood, for turbulent flow.

However, it is relevant to ask whether, given the topology of the vorticity field, there is any *lower* bound to the enstrophy, under the action of volume-preserving diffeomorphisms.

If we were to integrate the Euler equations backwards in time, it might be natural in general to expect a decrease of enstrophy, and the question is how far can this decrease proceed, or equivalently what is the minimum level of enstrophy from which flow with given vorticity topology may evolve?

It was shown by simple application of the Schwarz inequality (Moffatt, 1969, section 4) that when the helicity of a flow is non-zero, the enstrophy under Euler evolution (whether forwards or backwards in time) is bounded below, essentially because indefinite contraction of any vortex ring necessarily leads to ultimate extension of any vortex ring with which it is linked. This is a simple and illuminating example of the way that a topological barrier implies the existence of a lower bound. As shown by Freedman (1988) even when the helicity is zero (as for the Whitehead link topology) the enstrophy has a lower bound, under the action of volume-preserving diffeomorphisms, provided only that the topology of the vorticity field is non-trivial. This important result provides a crucial link between formal topological techniques and fluid mechanics.

A technique by which the lower bound may be attained was developed by Moffatt (1985), and is applied in the context of knots and links in Moffatt (1990). The technique is more easily comprehended in terms of the magnetic field in a perfectly conducting fluid rather than in terms of vorticity, but the end result is the same. In terms of the magnetic field, the volume-preserving diffeomorphism is seen as the effect of an incompressible flow which is driven by the Lorentz force associated with the magnetic field. This construction is such as to dissipate magnetic energy (the analogue of enstrophy) and to drive the magnetic field towards a magnetostatic equilibrium state, and in this state, the magnetic energy is minimal with respect to arbitrary volume-preserving frozen-field distortions. The flow that takes the initial magnetic field to this final equilibrium state defines also a limit volume-preserving mapping $\mathbf{x} \to \mathbf{X}$ which relates the final position of fluid particles to their initial position.

If we now think in terms of vorticity rather than magnetic field, the same mapping converts the vorticity field (via the usual Cauchy transformation $\omega_i(\mathbf{X}) = \omega_{j0}(\mathbf{x})\partial X_i/\partial x_j$) to a state of minimum enstrophy.

For the case in which the vorticity is confined to a single closed knotted vortex tube, as described above, the minimum enstrophy must, on dimensional grounds, have the form

$$\Omega = m\kappa^2 \mathcal{V}^{-\frac{1}{3}} \tag{16}$$

where $\mathcal{V}$ is the volume of the vortex tube and κ the circulation (both conserved under volume-preserving frozen-field distortions). The coefficient m depends only on the topology of the tube, i.e. on the topology of the knot C around which it is constructed. This dimensionless positive real number is evidently a topological invariant of the knot.

The situation becomes a little more complicated when we consider linked vortex tubes. If each tube has the same circulation κ and the same volume $\mathcal{V}$, then the minimum enstrophy state is characterised by just the two dimensionless parameters κ and $\mathcal{V}$, and by the topology of the linkage, and so the minimum enstrophy again has the form $m\kappa^2 \mathcal{V}^{-\frac{1}{3}}$ where the real number m is a topological invariant characteristic of the linkage. If however the circulations are different, then the conclusion is not so simple. For two tubes of circulation κ_1, κ_2 exhibiting an arbitrary linkage (e.g. the Whitehead linkage) each tube having volume $\mathcal{V}$, the enstrophy, on dimensional grounds, has the form

$$\Omega = \Sigma m_{ij}\kappa_i\kappa_j \mathcal{V}^{-\frac{1}{3}} \tag{17}$$

where m_{ij} is a real symmetric positive-definite matrix. Under frozen-field deformation, this matrix varies, and, for given κ_1, κ_2, there is a unique minimising configuration. However, m_{ij}

in the minimum enstrophy state is a function of the ratio κ_1/κ_2. This may be seen explicitly for the Whitehead link, for which m_{11} is non-zero if κ_2 is non-zero. However, if κ_2 is zero (so that the second vortex tube vanishes), then m_{11} in the minimum enstrophy state is zero, because the topology is trivial.

Many questions on the frontier between fluid mechanics and topology are raised by considerations of this kind, and these present challenging problems for the future.

CONCLUDING REMARKS

Some comments need to be made in conclusion. Either in the LIA context or in the context of the Euler equations, global geometrical considerations play a critical part. In the LIA context the topology is not conserved; but the invariants are constants of the motion for the isolated vortex filament as long as it does not interact with itself and the long-distance effects are negligible; under these circumstances these invariants are consistent with the assumed model. Some of these conserved quantities have been interpreted in terms of kinetic energy, momentum, angular momentum, "pseudo-helicity," and some inequalities between physical and global geometrical properties have been stated. It is remarkable that even under this approximation the momentum and the angular momentum are still conserved.

In the context of the Euler equations the topology is fully conserved. Helicity, being the natural and simplest measure of topological complexity of a convected vector field, plays a prominent role. Here, we can reasonably expect to have a wider family of invariants of such a topological nature. Adopting the method of magnetic relaxation, which conserves vorticity topology, a new type of invariant has been found; it has been interpreted as the minimum enstrophy state for knotted vorteces and its nonzero value is due to the topological barrier in the relaxed field.

Finally, we observe that this invariant should be computable, at least for reasonably simple knots. Given the equation of the knot in parametric form $\mathbf{x} = \mathbf{x}(s)$, it should be straightforward to construct a knot field $\mathbf{B}_C(\mathbf{x})$ and then to implement the relaxation procedure numerically. In the equilibrium end-state, delicate questions concerning tangential discontinuities (i.e. vortex sheets or vortex gradient sheets) may arise. This means that particular care has to be taken in the asymptotic stage of relaxation.

REFERENCES

Arms, R. J., and Hama, F. R., 1965, Localized-Induction Concept on a Curved Vortex and Motion of an Elliptic Vortex Ring, Phys. Fluids, 8(4):553.

Arnol'd, V. I., 1974, The asymptotic Hopf invariant and its applications, (in Russian), in: "Proc. Summer School in Differential Equations," Armenian S.S.R. Acad. Sci., Erevan. (English translation: Sel. Math. Sov., 5:327).

Betchov, R., 1965, On the curvature and torsion of an isolated vortex filament, J. Fluid Mech., 22:471.

Da Rios, L. S., 1906, On the motion of an unbounded fluid flow with an isolated vortex filament, (in Italian), Rend. Circ. Mat. Palermo, 22:117.

do Carmo, M. P., 1976, "Differential Geometry of Curves and Surfaces," Prentice-Hall, Englewood Cliffs.

Freedman, M. H., 1988, A note on topology and magnetic energy in incompressible perfectly conducting fluids, J. Fluid Mech., 194:549.

Hasimoto, H., 1972, A soliton on a vortex filament, J. Fluid Mech, 51:477.

Keener, J. P., 1990, Knotted vortex filaments in an ideal fluid, J. Fluid Mech., 211:629.

Marris, A. W., and Passman, S. L., 1969, Vector Fields and Flows on Developable Surfaces, Arch. Rat. Mech. Anal., 32:29.

Maxwell, J. C., 1873, "A Treatise on Electricity and Magnetism," MacMillan & Co., Oxford.

Moffatt, H. K., 1969, The degree of knottedness of tangled vortex lines, J. Fluid Mech., 35:117.

Moffatt, H. K., 1985, Magnetostatic equilibria and analogous Euler flows of arbitrarily complex topology. Part 1. Fundamentals, J. Fluid Mech., 159:359.

Moffatt, H. K., 1990, The energy spectrum of knots and links, Nature, (to appear).

Moreau, J. J., 1961, Costantes d'un îlot tourbillonaire en fluid parfait barotrope, C. R. Acad. Sci. Paris, 252:2810.

Rasetti, M., and Regge, T., 1975, Vortices in He II, Current Algebras And Quantum Knots, Physica, 80(A):217.

Weatherburn, C. E., 1927, "Differential Geometry of Three Dimensions," C.U.P., Cambridge.

Zakharov, V. E., and Shabat, A. B., 1972, Exact theory of two-dimensional self-focusing and one-dimensional self-modulation of waves in nonlinear media, Sov. Phys. JETP, 34(1):62.

STRETCHING OF VORTEX LINES

Alain Pumir[1] and Eric D. Siggia[2]

[1] L. P. S., E.N.S. 24, rue Lhomond, 75231 Paris Cedex, France
[2] LASSP, Cornell University, Ithaca, N.Y. 14853, USA

Decades of active research in turbulence have allowed to identify several important effects responsible for the very rich and original behavior observed in high Reynolds number flows. The stretching of vortex lines is one of the physical mechanisms that are commonly evoked in this context[1]. Indeed, available experimental data show that the stretching is active in 3-dimensional turbulent flows, and contributes to vorticity production[2]. Because of the complexity of real flows, it is hard to obtain more precise information on this phenomenon. Computation is not easy either. If one restricts oneself to the incompressible case, the origin of the stretching is obvious if one writes the Euler (Navier-Stokes) equations in terms of the vorticity :

$$\frac{D\omega_i}{Dt} = \partial_t\omega_i + (\mathbf{u}.\nabla)\omega|_i = e_{ij}\,\omega_j\,(+\,\nu\nabla^2\omega_i) \qquad (1\text{--}1)$$

$$e_{ij} = 1/2\left(\partial_i u_j + \partial_j u_i\right) \qquad (1\text{-}2)$$

$$\omega = \nabla \times \mathbf{u} \qquad (1\text{-}3)$$

$$\nabla.\mathbf{u} = 0 \qquad (1\text{-}4)$$

where $\mathbf{u}$ is the velocity field, ω the vorticity, $D/Dt \equiv \partial_t + \mathbf{u}.\nabla$ the Lagrangian derivative and ν the viscosity. The rate of strain tensor, $\mathbf{e}$, is symmetric, hence its eigenvalues are real and its eigenvectors are orthogonal to one another. Furthermore, because of incompressibility, $\mathbf{e}$ is traceless, thus, the sum of the eigenvalues is zero. In the following, we denote these eigenvalues by α, β and γ, the associated eigenvectors being $\mathbf{e}_\alpha$, $\mathbf{e}_\beta$ and $\mathbf{e}_\gamma$. The eigenvalues are sorted in a decreasing order : $\alpha \geq \beta \geq \gamma$. Equation (1-1) then shows that the component of ω in the direction given by $\mathbf{e}_\alpha$ grows as the flow evolves. In the direction given by the smallest eigenvalue, the component of the vorticity decreases. Equation (1-1) for the vorticity vector is similar to the equation that governs the evolution of two infinitely close particles in the flow. The effect of stretching refers to the modification of the length of the vorticity vector, as described by the right hand side of equation (1-1). It has to be noted that in 2 dimensions, the stretching term identically vanishes.

In order to study quantitatively equation (1-1,2,3,4), one has to solve a nonlinear partial integro differential equation. A very crude simplification of equation (1-1), compatible with dimensional analysis yields the following equation for the norm of Ω of the vorticity vector :

$$\partial_t\Omega = \Omega^2 \qquad (2)$$

This equation takes into account the nonlinear character of the stretching. However, it misses completely the nonlocality of the rate of strain. It is obvious that equation (2) gives rise to finite time singularities. This naturally raises the question of the existence of finite time singularities in the fluid equations. More generally, the question one can ask is : given an initial condition with a given distribution of vorticity, how much stretching develops in time ? How does the stretching depend on viscosity ?

From an experimental point of view, the intermittent occurence of very large 'signals' (typically a derivative of the velocity field) is a well established fact[3]. A configuration generating a great deal of stretching (possibly an infinite amount in a finite time) would be an ideal candidate to explain such experiments.

The Global Geometry of Turbulence
Edited by J. Jiménez, Plenum Press, New York, 1991

The possibility of a blow-up of a solution of the Euler equations has been studied several times in the past. Back in 1937, G.I. Taylor and A.E. Green considered the initial value problem, with an initial condition consisting of a few Fourier modes[4]. No conclusive evidence of the existence of a finite time singularity has been found, despite some recent efforts[5]. Mathematicians have not come up with a definite answer for the singularity problem. Nevertheless, they have shown the interesting property that a singularity in the fluid equations implies that the norm $\| \cdot \|_\infty$ of the vorticity (defined by $\| \omega \|_\infty = \sup_{x \in R} | \omega(x) |$) itself must diverge. For the Navier Stokes equations, the velocity also must diverge, according to :

$$\| u \|_\infty(t) \geq cst. \sqrt{\frac{\nu}{(t^* - t)}} \qquad (3)$$

as shown by Leray [6]. Vorticity must also diverge[7], like $1/(t^*-t)$. In the case of the Euler equations, the following inequality :

$$\left(\int (\partial_k u)^2 \right)(t) \leq \left(\int (\partial_k u)^2 \right)(0) \times \exp\left(cst. \int_0^t \|\omega\|_\infty(t') \, dt' \right) \qquad (4)$$

has been obtained by Beale, Kato and Majda [8]. The two first integrals in Eq.(4) are space integrals. This inequality shows that any singular behavior must have a signature on the vorticity. Vorticity is therefore the right quantity to investigate, both for the physicist and for the mathematician!

In this paper, we summarize some of our work, aimed at studying the stretching and the possibility of the existence of finite time singularity in the 3d Euler equations[9]. The method we have followed in this work consists in studying numerically the evolution of a relatively simple vorticity field, and trying to understand how the vorticity field stretches itself.

A direct numerical simulation of a solution of the Euler equations that involves a large stretching and hence the generation of very small scales requires an adequate resolution all the way down to very short distances. With a uniform mesh, and with the current supercomputers available, it is impossible to go to very small scales ! To circumvent this difficulty, we used adaptive mesh technics. More precisely, we mapped the interval $(-1,1) \rightarrow (-\infty, +\infty)$ by the following function :

$$X = \left\{ a - b \sin(\pi\xi/2) - c \cos(\pi\xi) \right\} \times \tan(\pi\xi/2) \qquad (5)$$

and then discretized $(-1,1)$. The change of variables, given by Eq.(5) allowed us to build a mesh of the entire space R^3, by choosing separately the x, y and z coordinate systems. The three free parameters: a, b and c can be varied, so as to change the resolution of the numerical grid. Whenever the resolution becomes inappropriate, we modified the parameters a, b and c in equation (5) so as to move grid points where they are necessary.

In order to keep the total number of points reasonable, we had to cut off our solution far away from the region where small scales were building up. This can be done by multiplying the velocity field by a Schwartz function. This procedure of course affects the energy and circulation conservations. However by systematically monitoring the errors, we showed that the stretching was affected by a few percent only, and the conclusion of our study were not qualitatively changed by the cutoff we used.

There is of course a huge phase space, so the choice of a small number of initial condition is very arbitrary. Our choice of an initial condition has been guided by our previous study of the Biot-Savart model for a vortex tube. This model describes very slender tubes of vorticity, with the very strong simplification that the vorticity distribution is frozen inside the tube. We found that a vortex tube spontaneously tended to fold, so as to pair two antiparallel pieces of filament[10]. A very violent stretching unavoidably followed, leading to a fnite time singularity[11]. More precisely, we found that the velocity diverged like : $\| u \|_\infty \sim 1/(t^*-t)^{1/2}$ and that the vorticity diverged like $\| \omega \|_\infty \sim 1/(t^*-t)$ (up to logarithmic corrections). Of course, this increase in vorticity and velocity was associated with the generation of very small scales of motion. The study of this model raised, among others, two obvious questions :

(i) what is the role of viscosity in the process ?

(ii) what is the effect of the internal degrees of freedom of the core ?

The first of these questions has been attacked by many authors[12], and is related to the reconnection issue. The second question motivated the study summarised below.

In what follows, we will report the results obtained for two antiparallel vortex tubes in some detail, and comment briefly on some of the results obtained with other initial conditions.

The initial condition we chose was symmetric with respect to the x = 0 and z=0 planes.

Each tube lies in a half space x>0 or x<0. They interact strongly close to the origin, where their mutual distance is minimal. Close to this region, the vorticity is mostly parallel to the z direction, pointing towards positive (negative) z when x is positive (negative). We have been able to follow the growth of vorticity by a factor 40 approximately, and a contraction in the x direction by a factor of 3000.

The growth of $\| \omega \|_\infty$ as a function of time is initially faster than an exponential as revealed by figure 1. In fact, $1/\| \omega \|_\infty$ seems to decay linearly in time, in the early stage of the collapse. As time goes on, the rate of growth of $\| \omega \|_\infty$ saturates, and an exponential regime is observed; rather than the algebraic growth observed previously. This shows in figure 1, where the logarithm of $\| \omega \|_\infty$ is plotted versus time. Also, the maximum of the ratio: $\| \omega.e.\omega \|_\infty / \| \omega \|_\infty$ saturates to a finite value, indicating that the rate of growth of $\| \omega \|_\infty$ no longer grows.

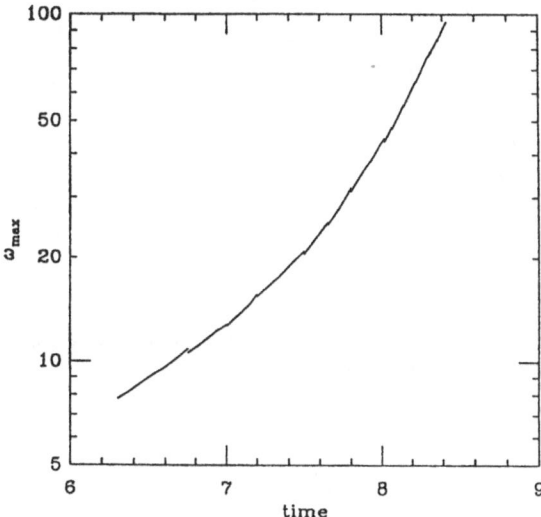

Figure 1 . The evolution of the maximum vorticity on a logarithmic scale as a function of time.

In order to try to understand qualitatively this property, it is interesting to study the solution in physical space. One of the very remarkable aspects of our vortex tubes is the formation of very thin sheets. This can easily be seen, for instance on figure 2, where a cut of the vortices through the mid-plane z=0 was made at time 7.2. Only the x>0 part of the solution is shown in this figure. The solution moves towards the y<0 region. The very elongated aspect of the solution is obvious since the x-scale has been blown up 5 times more than the y-scale. As times goes on, the sheets of vorticity near the x=0 plane remain very stable, and become thinner and thinner. Also more and more vorticity is ejected on the side, thus building up tongues of vorticity, that continue the sheets. A 3 dimensional view of the solution is shown on figure 3. The sheet like aspect of the solution is obvious. Also, it is clear from figure 3 that the solution involves a very small scale in the x-direction, and another, much larger length scale in the y and z-direction.

Another aspect of the evolution of our solutions is the failure to concentrate the circulation in regions where the vorticity is large (i.e., in the sheet region). This can be seen by systematically plotting the isovalue lines of the function :

$$f(x,y) = \int_0^x \omega_z(x',y,z=0)dx' \qquad (6)$$

This function represents the amount of circulation, computed in the middle plane z=0, per unit of length along the y-axis, in between 0 and x. A typical plot of this function is shown in figure 4. Figure 4 corresponds to the same solution as shown on figure 2, and with the same scales. A comparison between these two figures shows that the region of high vorticity contains a rather weak amount of the total circulation (about 15%). As times goes on, this trend becomes even more pronounced. This failure to confine the circulation where the vorticity is large can be related, at least qualitatively, to the saturation of the rate of growth of the vorticity. The regions that generate the stretching, and the region where the vorticity is intense are not the same. Furthermore, the dynamics is much slower in the former than in the latter regions. Hence, on the fast time scale, the stretching experienced by the sheets of vorticity is almost constant, and thus the growth of vorticity becomes exponential.

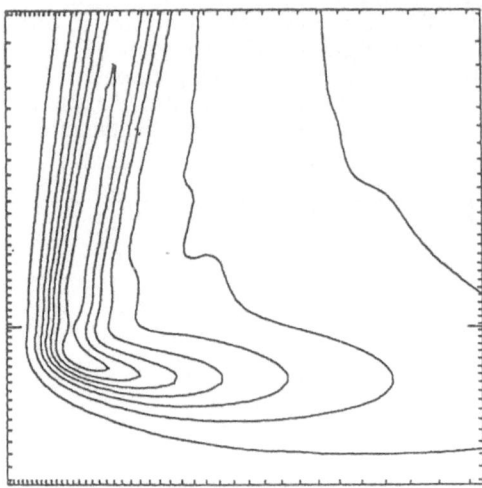

Figure 2 .Contour plot of ω_z at $T=7.2$ and $z=0$. The coordinate ranges are $0 \leq x \leq 0.12$ and $-0.2 \leq y \leq 0.4$. The contour intervals are 2.0

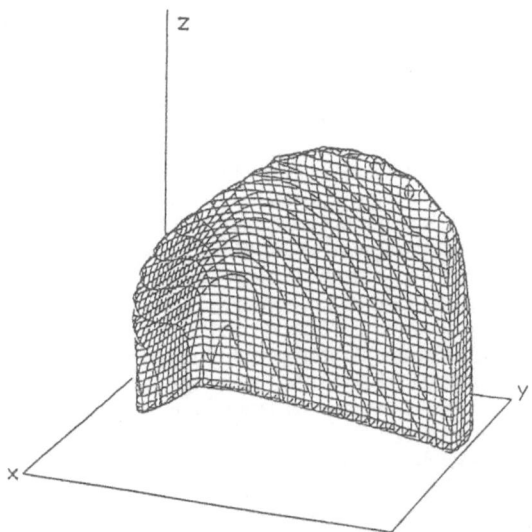

Figure 3 . Isovorticity surface at $T=7.2$. The region shown is drfined by $0 \leq x \leq 0.15$; - $0.15 \leq y \leq 0.90$ and $0 \leq z \leq 0.45$. The threshold on ω is 40% of the maximum.

Another interesting property of our solutions concerns the rate of strain tensor, and its correlations with the vorticity field. The eigenvalues α, β and γ happen to be preferentially such that $\alpha >> \beta > 0 >> \gamma \, (= -\alpha - \beta)$. During the evolution in time, the ratio between β and α becomes smaller and smaller (in a statistical sense). Furthermore, the vorticity field tends to be preferentially aligned with the second eigenvector. The saturation of the rate of growth of the vorticity thus corresponds to the fact that β saturates, and the decay of the ratio between β and α corresponds to the fact that α grows roughly like $\| \omega \|_\infty$. It is interesting to compare these results with those of direct numerical simulations of the Navier-Stokes equations [13], where it was found that (α,β,γ) were in the ratio $(3,1,-4)$ and that ω was preferentially aligned with the intermediate eigenvector. Of course, the Navier-Stokes simulations correspond to relatively viscous runs, because of computer limitations. It is nevertheless encouraging that the results of the two calculations seem to point in the same directions.

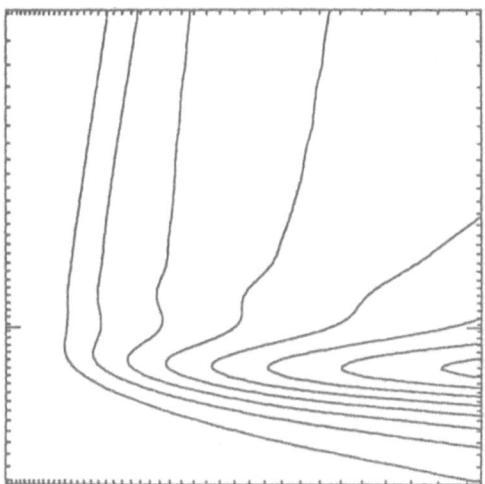

Figure 4 . Contour plot of the circulation per length $\int_0^x \omega_z(x',y,0)dx'$ at $T = 7.2$. The coordinate range is the same as figure 2. Contour interval is 0.1.

Let us stress that our other attempts to look for a singularity with other initial conditions also failed, for rather similar reasons. Namely, we never succeeded in maintaining a region of high vorticity to self stretch for a long enough time. Of course, our exploration of the phase space is far from complete. We obviously cannot rule out the existence of a singularity from the study of so limited number of initial conditions. Our calculations however show that contrary to what is commonly believed, solutions of the Euler equations can lead to an important stretching of the vorticity lines, and not to a blow-up in a finite time. Possibly more interesting is the fact that the properties that prevented the singularity from happening could be more general, and would result from the very structure of the equations. Obviously, much more needs to be done to study these questions. Also, the relation of these findings with the problem of flows with a very small viscosity (very high Reynolds numbers) is far from obvious and needs further investigations.

Acknowlegements .It is a pleasure to thank the organisers of the meeting 'The Global Geometry of turbulence', in particular J. Jimenez, for a very stimulating and enjoyable meeting. The computations described in this work have been done, thanks to generous grants of computer time from the 'Cornell Supercomputer Center', from the 'Service de Physique Théorique' (Saclay, France) and from the 'Centre de Calcul Vectoriel pour la Recherche', (Palaiseau, France).

References

1) H. Tennekes and J.L. Lumley, 1972, A first course in turbulence, MIT Press
2) R. Betchov, 1956, J. Fluid Mech. 1, 497
3) A.A. Townsend, 1951, Proc. R. Soc. Lond. A208, 534
4) G.I. Taylor and A.E. Green, 1937, Proc. R. Soc. Lond. A158, 499
5) M. Brachet, D. Meiron, S. Orszag, B. Nickel, R. Morf and U. Frisch, 1983, J. Fluid. Mech. 130, 411
6) J. Leray, 1934, Acta Math. 63, 193
7) L. Caffarelli, R. Kohn, L. Nirenberg, 1982, Comm. Pure Appl. Math. 35, 771
8) J. T. Beale, T. Kato and A. Majda, 1984, Commun. Math. Phys. 94, 61

9) A. Pumir and E.D. Siggia, 1990, Phys. Fluids A2, 220

10) E.D. Siggia, 1985, Phys. Fluids **28**, 794

11) A. Pumir and E.D. Siggia, 1987, Phys. Fluids **30**, 1606

12) see, e.g., the contribution of P.G. Saffman and N. Zabusky in these proceedings.

13) R. Kerr, 1987, Phys. Rev. Lett. **59**, 783 and W. Ashurst, A. Kerstein, R. Kerr and C. Gibson, 1987, Phys. Fluids **30**, 2343

THEORETICAL MODELS:

AMPLITUDE AND MODEL EQUATIONS

BIFURCATIONS AND SUBCRITICAL INSTABILITIES

Yves Pomeau

Laboratoire de Physique Statistique, E.N.S.
24 rue Lhomond, 75231 Paris Cedex 05 - France

Abstract

Any theoretical approach to fluid mechanics of real flows faces the sad truth that the Navier-Stokes equations have no solution general enough to account for turbulence, not only fully developped turbulence, but also the "mild" or "wimpy" (following Alan Newell) turbulence. This is particularly true in 3D flows, where the numerical approach is rather restricted, both from the point of view of the Reynolds number, and for the time duration of the simulations. On the other hand, there is (almost) no doubt that the computer may reproduce the characteristics of real flows, but certainly cannot generate, at least directly more global pictures and ideas that we need to develop a theory or at least an understanding. Here I claim that one can understand a lot of phenomena occuring in real flows by a reduced amplitude approach, which is a short-cut to the full solution of the full equations.

Those amplitude equations aim at describing the dynamics of weak fluctuations and are defined usually under restrictive assumptions: large aspect ratio (many typical wavelength), supercritical or weakly subcritical instabilities, control parameter near threshold. Dropping anyone of those assumptions leads to major difficulties linked to the summation of poorly controlled perturbation series. It seems better then to try to study simple amplitude models, with the most general structure one can think of, and to compare their predictions with real life experiments.

This is the line of thought followed here. To be a little more concrete, I shall consider first the connection between the subcritical character of the bifurcation in parallel flows and the occurence of localised patches of turbulence therein. This can be explained with a simple model of the reaction diffusion (RD) type. However this dynamical model is quite special in the sense that it has a variational structure. This entails some nongeneric qualitative properties, that are no more valid for more general system, as one expects the turbulent flows are. I shall consider some of those nongeneric properties, and refer briefly to other works on other ones. Then I will single out a remarkable theoretical discovery, that is the possibility of stable localised structures in a nonvariational system, although in gradient/variational systems those localised structures are always linearly unstable.

1.Finite amplitude instabilities in open flows: generic features

Fluid mechanics is a fascinating domain of science: the basic equation are known now for more than a century and they remain still many unsolved problems, the least one being not fully developed turbulence. One may get an idea of the difficulty of the field by thinking how much ingenuity and nontrivial mathematics it took to understand (via strange attractors and chaotic dynamics) the transition to chaos in "small boxes".

Below I do not intend to consider such a grand problem as fully developed turbulence, but instead report some significant progress made in the understanding of transition to turbulence in parallel flows. This is a very much studied field, but very frustrating too, because the best analytical theory one can do, by using formally the basic Navier-Stokes equations only, is the linear and weakly linear stability analysis, that does say very little or nothing at all on real world experiments. This comes from a very fundamental character of the relevant instabilities: they are subcritical (or hysteretic-the pair of words subcritical/supercritical are sometimes used with a different meaning, referring to the absolute/convective character of instabilities in the linear theory). As already noticed by Osborne Reynolds, circular pipe flows are stable against small amplitude perturbations but unstable against large amplitude, for the same Reynolds number, if large enough. This seemingly innocuous remark about the subcritical character of the instability (in modern words) can be put at the basis of a more formal theory explaining many experimental facts at the price of a moderate analytical investment. This requires an act of faith, because one cannot derive everything from the Navier-Stokes equations all the way down to finite amplitude equations. In some sense, the situation is like in dynamical system theory applied (successfully) to "small boxes": one assumes there is nothing special to the Navier-Stokes equations and then try to get "universal" results, applicable to fluid flows from model systems.

To make more concrete what I have in mind, let me consider a classical phenomenon in Fluid Mechanics, that is the formation of localized patches of turbulence in an otherwise laminar flow. As I argued[1] some time ago this may be interpreted by analogy with the behavior of solutions of reaction-diffusion equations. Let us assume that the local state of the fluid can be represented via an 'order parameter', a scalar quantity that changes continuously from a definite value in the turbulent state to another one in the laminar state. The most simple example of such a quantity would be the amplitude of the turbulent fluctuations. The dynamics of this amplitude may be derived through a systematic small amplitude expansion "à la Landau", although in a more rigorous theory(still to come) this amplitude would be an intrinsic parameter of the statistical ensemble describing the turbulent fluctuations. There is likely a long way to go before we can do that starting from the mere Navier-Stokes equations.

This concept is already nontrivial, in the sense that it implies that, in an experiment where one observes a localised patch of turbulence in an otherwise laminar fluid, the turbulent state *inside the patch* is homogeneous, up to boundary effect indeed. As far as I know this has never bee checked in real fluid experiments (as for instance in the turbulent flash of Reynolds). This homomegeneity would imply that (time) statistical averages inside the patch are independent on the position-again up to end effects. In the case of a pipe flow experiment, I have in mind the statistical homogeneity in the flow direction, not across the pipe.

Suppose now that both the turbulent state and the laminar one are stable against small amplitude fluctuations, thus it is natural to represent the dynamics of the order parameter as a gradient flow in a two well potential, each well corresponding to a locally stable state of the global system. Then it is still natural to represent the tendency of the system to become spatially uniform, say because of the molecular diffusion effects, by adding a linear diffusion term to the equation of motion for the order parameter. This yields a reaction-diffusion equation of a kind that has been much studied[2]. In good cases it is even possible to derive this sort of equation from the first principles in a convenient limit [3]. The main property that interests us here is that those reaction-diffusion(like) equations have, as an asymptotic solution to a rather large class of initial datas, a front separating the two possible uniform stable states of the system and moving at a constant speed. But this is not the full story, mainly for two reasons:

i) the reaction-diffusion equations (or the amplitude equations) are local in space, whereas the turbulent fluctuations couple to large scale flow, via the Reynolds stress [9]. This may be included into the picture by writing a coupled system of reaction-diffusion equation plus an elliptic equation for the large scale flow with the Reynolds stress as source term. This induces a feedback between the growth of the localised patches of turbulence and their control parameter (the Reynolds number, for instance) and this feedback may even stop the growth of the turbulent spot at a given size. This is what is (partly) responsible of the well definite azimuthal size of the turbulent spirals in counterarotating Taylor-Couette flows [4].

ii) the reaction-diffusion equation with a single unknown, as well the amplitude equation with real coefficients have an important non-generic property of being variational. This implies in particular very specific qualitative properties. One of those properties is that the localised structures are always linearly unstable against a growth/shrinking mode. On the contrary, it has been observed in computer experiments[5] and explained analytically [6] that *non*variational systems may have linearly stable localised structures. This mathematical observation is relevant to the transition in pipe flows, where they are either continuously expanding "puffs" or localised globally stable structures, called "slugs".

As explained in [1] one can understand the dynamics of the localized patches in turbulence through the solutions of reaction-diffusion (RD) equations with the general structure:

$$\frac{\partial A}{\partial t} = D_{ij} \frac{\partial^2 A}{\partial r_i \partial r_j} - \frac{dV}{dA} \qquad (1)$$

, where D_{ij} is a symmetric real and positive "diffusion tensor" and where $\frac{\partial^2 A}{\partial r_i \partial r_j}$ is the second derivative of the amplitude A with respect to the Cartesian coordinates of index i and j. In equation (1) one has applied the Einstein convention for the index summation and V(A) is the two well potential. The equation (1) has a variational structure in the sense that it may be written as:

$$\frac{\partial A}{\partial t} = - \frac{\delta G}{\delta A},$$

where $\frac{\delta G}{\delta A}$ is the Fréchet derivative of the real functinal G[.] defined as:

$$G[A] = \int d\mathbf{r} \, [D_{ij} \frac{\partial A}{\partial r_i} \frac{\partial A}{\partial r_j} + V(A)]$$

The evolution described by equation (1) tends to lower as much as possible the functional G[.], so that the system tends to be as uniform as possible in space, because the gradient term $D_{ij} \frac{\partial A}{\partial r_i} \frac{\partial A}{\partial r_j}$ is positive and it tends too to put A at the lowest possible value of V(.), as far as the boundary conditions allow it.

As already said, this equation (1) has solutions attracting a large set of initial datas and representing moving fronts separating two linearly stable regions. Those fronts are sometimes called the ZFK[7] fronts to help to distinguish them from the KPP[8] fronts separating stable from linearly unstable regions. Fronts moving at speed u are represented by solutions of (1) with a dependence on one space coordinate, say x, and on time through the combination z=(x-ut), u constant, and with the condition that A(z) reaches each of the two stable equilibria of V(.) at z equal plus and minus infinity. There are many important differences between this picture and the real front separating two different flow regimes as for instance observed in pipes. Before to come to this let us emphasize however some striking similarities: first of all, in the RD picture as for real flows those fronts are rather independent on the details of the initial conditions because of the metastability of the system, a consequence itself of the subcritical character of the bifurcation yielding one of the two states from the other one. Then the front velocity is a smooth function of any parameter controlling deformations of the potential V(.). Now there are also some qualitative differences between real fronts as observed in pipes for instance and what follows from the RD picture. Below I shall consider them one after the other.

(i) Indeed one expects that in most parallel flows, but for plane Couette, such fronts are convected at some constant speed by the mean flow. Those advection terms appear through first order derivative in space added to equation (1). They have been already introduced[9] in the amplitude theory for inclined Taylor vortices by Tabeling. This is not enough, as even with those advection terms the trailing and leading edges are deduced from each other through a reflection, although all experimental datas point to the absence of such a symmetry[10]. This symmetry may be broken, for instance by adding third order derivatives in space to the amplitude equations, as allowed from the basic symmetries of parallel flows, since no Galilean frame can get rid of all possible effects of the advection.

(ii) The amplitude equations are actually written for complex amplitudes and, when some of their coefficients are complex they do not have a Lyapunov functional, as equation (1). Complex amplitudes represents linearly unstable dispersive waves with an amplitude dependent dispersion relation. However even in those nonvariational cases the notion of linearly stable state persists and one may reasonnably assume that the moving front solutions still attract a large class of initial conditions with the two different states at + and - infinity. As one cannot compare the energies of the two possible states to determine which one is metastable or stable this is decided now (i.e. for those nonvariational systems without any energy as the functional G) through the sign of the velocity of the front.

(iii) Another problem which has not yet been looked at is the extension of this theory to the development of "turbulent" patches of turbulence in anisotropic 2D (as plane Poiseuille or Couette) or 3D flows (as a Blasius boundary layer). Indeed a theory in the form of equation (1) can model at least in part the anisotropy of such flows through a nonisotropic tensor D. However this is not enough. As said before, in this gradient picture the sign of the front velocity is completely determined by the potential difference between the two equilibria. This implies in particular that this velocity vanishes in all possible directions for the same value of the control parameter. This is presumably a nongeneric situation for nonpotential systems [see ref. 12 for more details on this].

(iv) Comparing this to usual thermodynamics, one gets the impression that there is an important difference: in thermodynamics, there are thermal fluctuations and so there is a possibility for a system to jump spontaneously from a metastable to a stable state. On the contrary, in a pure gradient flow system, as (1), there is no such fluctuations and no homogeneous nucleation either. However, it seems reasonnable to expect that turbulence generates fluctuations more or less similar to the thermal fluctuations in thermodynamics. This leads to a rather interesting lack of symmetry between the two coexisting states in a system with a local patch of turbulence: there is no fluctuation in the laminar state, although they are present in the turbulent part of the fluid. This is at thhe origin of the so-called transition through spatiotemporal intermittency that has been predicted [1] and observed in a numerical model [11] and in a physical system too[12].

(v) In the next section 2, I shall consider in more details the theory of stable localised states (called s-waves) in a complexified version of equation (1).

2. Stable localised structures in nonvariational systems

As shown below, variational system described by amplitude equations may present localised and stationary structures, but those structures are always unstable. On the contrary, both in transition flows [10 and references therein] as well as in various theoretical models, one has observed such localised structures that are thus certainly linearly stable. Theoretically this occurence of solitary (s-) patterns or waves has been for some time more like a set of isolated facts found when studying model equations, as Benney-Lin for the Kapitza instability[14] or Ginzburg-Landau (G.L. is a generic name for amplitude equations with complex coefficients) for nonlinear waves in shear flows [5]. Below I present a rather detailed analytical discussion of those s-waves in the framework of this G.L. equations. This approach by perturbation near the two extreme cases: first the neighbourhood of the fully conservative case, where one can show the possibility of stable s-waves, then the vicinity of a gradient flow system (or fully dissipative). Those two limits have to be dealt with rather different mathematical techniques that may have their own interest.

Our purpose below will be to study a class of solutions of the G.L. equation for a complex amplitude q (this is the same as the amplitude A before, but the letter "q" is more used for that purpose in the soliton literature):

$$\frac{\partial q}{\partial t} = i\frac{\partial^2 q}{\partial x^2} + 2i|q|^2 q + \varepsilon\left(-\frac{\partial U}{\partial q^*} + \beta\frac{\partial^2 q}{\partial x^2}\right) \qquad (2).$$

In this equation, t is the time, x the position in a frame of reference moving with the group velocity of the unstable waves, $U(|q|)$ is a real function of $|q|$ and β a real positive number. The equation (2) becomes at $\varepsilon = 0$ the nonlinear Schrödinger (NLS) equation:

$$\frac{\partial q}{\partial t} = i\frac{\partial^2 q}{\partial x^2} + 2i|q|^2 q \qquad (3)$$

At $\varepsilon \longrightarrow +\infty$ (ε is always real), equation (2) becomes the gradient flow equation with scaled time $T = \varepsilon t$:

$$\frac{\partial q}{\partial T} = -\frac{\partial U}{\partial q^*} + \beta\frac{\partial^2 q}{\partial x^2} \qquad (4).$$

As shown by Thual and Fauve [5] an equation similar to (2) has solutions representing linearly stable s-waves of zero speed. These s-waves appear when the "potential" $U(|q|)$ has two sets of stable minima, one at $|q| = 0$, and a continuum at $|q_0|e^{i\phi}$, ϕ real arbitrary and $|q_0|$ real $\neq 0$. Let us consider now some relevant solutions of (2) and (3).

278

i) Solitons of NLS equation

Although the NLS-equation (3) is known to be integrable, we shall not really use this property here, and we could as well consider a nonintegrable equation instead of NLS, but with a unitary (and not catastrophic) evolution. This would amount to replace the last term on the r.h.s. of the equation (3) by an arbitrary function $2iF(|q|^2)q$, F real, positive, vanishing at zero argument and increasing not too fast at infinity, to avoid catastrophic self-focussing. NLS has a manifold of 1-soliton solutions in the general form

$$q(x,t) = Q_0(x;\Omega)\, e^{i\Omega t}$$

where Q_0, as a function of x, is a solution of

$$(\Omega - \frac{d^2}{dx^2})Q_0 = 2|Q_0|^2 Q_0 \qquad (5)$$

together with the boundary condition (b.c.) $Q_0 \longrightarrow 0$, $x \longrightarrow \pm\infty$. This yields

$$Q_0(x;\Omega) = \Omega^{1/2} \mathrm{sech}(\Omega^{1/2}x)\, e^{i\phi} \qquad (6),$$

ϕ is an arbitrary phase, and the frequency Ω is any positive number and parametrizes the 1D manifold of solitons.

ii) s- waves and moving-front solutions of (4)

In a sense, the long-term dynamics of solutions of (4) is trivial, because this equation has a Lyapunov functional that has to decay. But this does not exclude non trivial unstable and metastable solutions either.

Here we shall be interested in 2 kinds of solutions of (4). First, we introduce the moving front solution. This tends to one minimum of $U(|q|)$, say $U(q_0)$ at $x = -\infty$ and to $q= 0$ at $x = +\infty$. For large times, any initial condition satisfying these b.c. will become a front moving at a constant speed u, the sign of u is such that the optimal "state" -i.e., the one with the lowest $U(.)$ replaces at a constant rate the metastable state. This moving front solution is a solution of (4) of the form $q(x,T) = R(\xi = x - uT)$, R real such that

$$-u\frac{dR}{d\xi} = -\frac{dU}{dR} + \beta \frac{d^2R}{d\xi^2} \qquad (7),$$

together with the b.c. $R \longrightarrow q_0(/0)$ at $\xi \longrightarrow -\infty$ (/ $+\infty$). The velocity u in (7) is then uniquely determined as a nonlinear eigenvalue.

An s-wave solution would be a homoclinic trajectory in x-space of solutions of (7) with u = 0 and R → 0 as x →±∞. Non trivial (i.e., not identically zero) solutions of (4) satisfying those conditions exist if $U(q_0)<U(0)$ only, as can be seen from the conservation of "energy" in the Newton-like equation

$$\beta\frac{d^2H}{dx^2} - \frac{dU}{dH} = 0 \quad (8),$$

when the even function U(H) has three minima, one H = 0, and two symmetric ones at $H = \pm q_0$. When it exists, the relevant solution of (8) may be found by quadratures. However it can be shown that this s-wave solution is always linearly unstable. Consider a solution of (4) in the form $H(x) + \tilde{q}(x)e^{\sigma t}$, $\tilde{q}$ small. This perturbation is the solution of

$$\sigma\tilde{q} = -\frac{d^2U}{dq^2}\bigg|_{q=H}\tilde{q} + \beta\tilde{q}_{xx} \quad (9),$$

with $\tilde{q} \to 0$ as x→±∞.

The eigenvalue problem (9) has $\tilde{q} = H_x$, $\sigma = 0$ as solution. From Sturm-Liouville this is not be ground state because H_x has at least one zero, due to the b.c. H → 0 at x → ±∞. This means that the ground state of (9) (the solution with the largest σ with our signs) has a positive σ, which shows that H(x) is a linearly *unstable* solution of (4).

Now I shall interpolate in between the properties of the limit forms of equation (2). by considering two "generic" parameters: ε and another one, not made explicit yet and representing continuous changes of U(.). This last parameter could be, for instance the level difference ΔU of the the two stable equilibria of U(.): $\Delta U=U(q_0)-U(0)$.

By perturbation near $\varepsilon=0$, I show first that the manifold of NLS solitons collapses into finitely many s-waves . Then, in the opposite limit ($1/\varepsilon \to 0$) an unstable and a stable s-wave merge at $\Delta U=U(q_0)-U(0) =0 =1/\varepsilon$ at the same point in parameter space as the steady front.

(2.1) s-waves near ε=0

Below, I show by perturbation that, as ε is turned on from 0, the continuum of NLS solitons collapse on a finite number of s-waves. This can be shown in a number of ways, in particular [15] by relying upon the integrability of NLS. A more elementary approach is sufficient here. The NLS part on the right-hand side of (2) conserves the L^2-norm $\int |q|^2 dx$ in the course of time. Multiplying (2) by q* and adding the complex conjugates, one gets an evolution equation for this norm depending formally on the nonunitary piece :

$$\frac{d}{dt}\int|q|^2\, dx = -\varepsilon \int_{-\infty}^{+\infty}dx\ [\beta\ |\frac{dq}{dx}|^2 + q*\frac{dU}{dq*} + q\frac{dU}{dq}] \quad (10).$$

280

We are looking now for s-wave solutions of (2). By analogy with the NLS case, those solutions have to have the form

$$q(x,t) = e^{i\Omega't} Q'(x,\Omega')$$

with

$$Q' \to 0, \quad x \to \pm\infty$$

If such a solution of (2) exists, it has to cancel the right-hand side of (10). This is indeed insufficient to find it in general (that is for ε finite), but if ε is close to zero, one expects $Q'(x, \Omega')$ to be very close to the NLS-soliton solution $Q_0(x, \Omega)$ as given by equation (6). Thus, in this limit, the dominant contribution to the integral on the right-hand side of (10) is obtained by plugging in a NLS-soliton. The condition that the result is equal to zero yields an equation for Ω. To be more definite, let us take a potential in the form (all coefficients are real):

(11)
$$U(|q|) = \mu |q|^2 + \frac{\alpha}{2} |q|^4 + \frac{\Upsilon}{3} |q|^6$$

They are two stable equilibria at $q = 0$ and $q = q_0$ whenever, $\mu < 0$, $\alpha^2 > 4\Upsilon\mu$, $\Upsilon > 0$ and $\alpha < 0$. Moreover the condition that the r.h.s. of (10) vanishes when computed with a NLS soliton reads :

$$2 \frac{\Omega^{1/2}}{5} \left(20\mu + \Omega(3\beta + 4\alpha) + \frac{16\Upsilon\Omega^2}{21} \right) = 0 \qquad (12),$$

which has two real roots, as sought, if

$$(3\beta + 4\alpha)^2 > \frac{1280}{21} \Upsilon\mu$$

A more refined calculation based upon a systematic perturbation expansion in ε, and done following the method of ref. [15] shows that a finite number of solitons only survive the ε-perturbation, and their amplitude is given by equation (12). Dropping the term of degree 6 in $U(.)$, one gets only one (instead of 2) s-waves, and this one is always unstable. On the contrary, when a pair of s-waves exist with the 6th degree term, one is linearly unstable and the other is linearly stable. This branch of linearly stable s-wave is the one found by Thual and Fauve [5].

<u>(2.2) s-waves and fronts at ε large</u>

We know already that, if ΔU is negative, there is an unstable s-wave at 1/ε=0, as given by the relevant solution of equation (8). When ΔU→0., the width of this unstable s-wave diverges and we shall show that, in the parameter space, 1/ε=ΔU=O is at the merging of two branches: one branch of steady front solutions and another one where stable and unstable s-waves become identical and metastable.

Under the heading (2.2.i) below, I shall study first the steady front solutions near 1/ε=0 and then in (2.2.ii) the s-wave solutions to validate the above picture.

2.2.i : Front wave solutions near 1/ε=0.

For ε →∞, the equation describing steady fronts in the same as eq. (8) but with a possible nonuniform phase of Q :

$$- \frac{dU}{dQ^*} + \beta \frac{d^2Q}{dx^2} = 0 \qquad (13),$$

together with the b. c. Q→0 at x→-∞ and Q→q$_0$ at x→+∞. Putting q=re$^{\iota\phi}$ into (13), r and φ real functions of x:

$$- \frac{dU}{dr} + \beta \left(\frac{d^2r}{dx^2} - r \frac{d\phi^2}{dx} \right) = 0 \quad (14.a),$$

$$2 \frac{d\phi}{dx} \frac{dr}{dx} + r \frac{d^2\phi}{dx^2} = 0 \quad (14.b).$$

From (14.b), $r^2 \phi_x$ has to be independent of x and is zero from the b.c. r→0 at x→-∞. This is possible if $\phi_x = 0$ for a non trivial (=not r = 0 everywhere) solution. Thus, with those b.c., equations (14) becomes equivalent to (8) with r = H. Indeed this has a steady front solution iff U(q$_0$) = U(0) only, that is if the potential U(.) has two minima of equal depth at q = 0 and q = q$_0$. The extension of (14) to a non zero 1/ε reads :

$$- \frac{dU}{dr} + \beta \left(\frac{d^2r}{dx^2} - r \frac{d\phi^2}{dx} \right) = \frac{1}{\varepsilon} \left(2 \frac{dr}{dx} \frac{d\phi}{dx} + r \frac{d\phi^2}{dx} \right) \qquad (15.a)$$

$$\beta r^2 \frac{d\phi}{dx} = \frac{1}{\varepsilon} \int_{-\infty}^{x} dx'\, r \, (\Omega r - 2r^3 - r_{xx} + r\phi_x^2) \qquad (15.b),$$

with the b.c. r→0 at x→-∞ (subcripts are for derivatives: $\phi_x = \frac{d\phi}{dx}$).

We shall analyse now the steady front solution of equation (15) by expansion near 1/ε = 0. Let r(x) be expanded in Taylor series in 1/ε as :

$$r(x) = r^{(0)}\, (x) + \frac{1}{\varepsilon}\, r^{(1)}\, (x) + ...,$$

Similar expressions hold for ϕ_x as well as for other quantities. We have shown before that $\frac{d\phi^{(0)}}{dx} = 0$ and that $r^{(0)}$ is the solution of

$$- U_r^{(0)} + \beta \ r_{xx}^{(0)} = 0 \qquad (16)$$

with the b.c. $r^{(0)} \to 0$ ($/q_0$) at $x \to +$ ($/-$)∞. Thus we are looking for a solution of (15) such that $r(x) \to 0$ ($/Q_0$) at $x \to +$ ($/-$)∞, where Q_0 is the asymptotic value of r, not necessarily q_0 if $\varepsilon \neq \infty$ and that $\phi_x \to$ constant at $x \to +\infty$. With those b.c. and to make the integral on the r.h.s. of (15.b) convergent at large x positive, Ω has to be equal to

$$\Omega = 2Q_0^2 - (\frac{d\phi}{dx}|_{as})^2 \qquad (17) \ ,$$

where $\frac{d\phi}{dx}|_{as}$ is the limit value of $\frac{d\phi}{dx}$ as x tends to infinity.

Keeping now the dominant terms on the r.h.s. of (15.b) one gets:

$$\frac{d\phi^{(1)}}{dx}|_{as} = \frac{1}{\beta \ q_0^2} \int_{-\infty}^{+\infty} dx' \{2(r^{(0)})^2 \ [q_0^2 - (r^{(0)})^2] + [\frac{dr^{(0)}}{dx'}]^2 \} \qquad (18).$$

From (15.a), the "equilibrium" at $x = +\infty$ has to be shifted from $q = q_0$ to $Q_0 = q_0 + q^{(1)} + q^{(2)} + ...$ where $q^{(1)} = 0$, although $q^{(2)}$ is given by

$$- U_q|_{Q_0} \ q^{(2)} = \beta \ q_0 \ [\frac{d\phi^{(1)}}{dx}|_{as}]^2 \qquad (19).$$

But this "equilibrium" condition is not the only one to be imposed to $r(x)$ at $x \to +\infty$. One can deduce from (15.a) a relation extending to non zero $\frac{1}{\varepsilon}$ the previous condition $U(0) = U(q_0)$. By multiplying (15.a) by r_x , one gets :

$$U(0) - U(Q_0) = \int_{-\infty}^{+\infty} [\frac{1}{\varepsilon} \frac{dr}{dx} (2\frac{d\phi}{dx} \frac{dr}{dx} + r \ \frac{d^2\phi}{dx^2}) + \beta r \frac{dr}{dx} (\frac{d\phi}{dx})^2] \qquad (20).$$

This condition is independent on (19), as can be checked on particular examples by expansion in $1/\varepsilon$ and by noticing that the dominant order for the r.h.s. of (20) can be expressed in terms of the zeroth order solution. In other terms, the potential $U(.)$ has to satisfy a non trivial condition to show front solutions, and this condition becomes $U(0) = U(q_0)$ in the gradient limit ($\varepsilon = \infty$). We shall assume that this still holds at finite values of ε, so that a codimension 1 manifold in the $(\varepsilon, \Delta U)$ space defines the locus where steady fronts exist, this manifold ending at $U(q_0) = U(0)$ at $\varepsilon = \infty$. In this framework, the branch of stable s-wave (of codimension zero !) meets this manifold at a finite ε, say ε_c as ε changes at constant ΔU.

2.2.ii : s-waves near $1/\varepsilon = 0$

Near $\Delta U = 1/\varepsilon = 0$, s-waves are made of widely separated shelfs bounding a plateau where Q is very close to q_0. The analysis of this situation proceeds in two steps: first one computes ϕ_x inside the s-wave by solving (15.b) in the appropriate limit and then plugs the result into (15.a) to take into account the perturbation brought to the modulus $r(x)$ by this nonconstant phase.

The phase equation (15.b) defines first the boundary value of ϕ_x at the two ends of the plateau. At the dominant order in $1/\varepsilon$ the variation of ϕ_x across the shelf is:

$$[\phi_x] = \frac{1}{\beta \varepsilon q_0^2} \int_{-\infty}^{+\infty} dx \, [\Omega(r^{(0)})^2 - r^{(0)} r^{(0)}_{xx} - 2\,(r^{(0)})^4] \qquad (21),$$

where $r^{(0)}(x)$ is the modulus of the front solution of (4) at $\Delta U = 0$. Note that the contribution proportional to Ω on the r.h.s. of (21) will turn to be always subdominant with the estimates found below for ε large. As $\phi_x = 0$ at $x = +$ or $-\infty$, (21) defines the b.c. for ϕ_x at the two ends of the plateau, as announced. On this plateau, $r \approx q_0$ and, from (15.b):

$$\varepsilon \beta \phi_{xx} \approx \Omega + \phi_x^2 \qquad (22).$$

From the previous remarks the b.c. for (22) are $\pm[\phi_x]$ at the two ends of the plateau, that is at $x = \pm L/2$, L being the width of the s-wave, unknown for the moment. By solving (22) with the appropriate b.c. one relates Ω and L. From (21) ϕ_x is of order ε^{-1} and we shall assume that Ω is the dominant term on the r.h.s. of (22)., that is $\Omega >> \varepsilon^{-2}$, which implies $L << \varepsilon^2$, as we shall assume. This is consistent with the final estimate $L \sim \ln(\varepsilon)$ at ε large. With those assumptions the solution of (22) to be retained is $\phi_x = \frac{\Omega}{\varepsilon \beta}(x + x_0)$ (23).

The b.c. impose $x_0 = 0$ and $\Omega L = 2\varepsilon\beta\,[\phi_x]$. Now we shall put into the equation for $r(x)$ the above expression for ϕ_x. At large ε the dominant perturbation with respect to the pure gradient flow comes from the $-\beta r \phi_x^2$ term on the l.h.s. of (15.a), and so we shall try to find a s-wave solution of:

$$-U_{rr} + \beta r_{xx} = \beta r \phi_x^2 \qquad (24),$$

where ϕ_x is given by (23) in the plateau region. There $r(x)$ is close to q_0 and by putting $r \approx q_0 + \delta(x)$, we may linearize (24) near q_0 and assume that $\delta(.)$ is of the same order of magnitude as the r.h.s.. This yields:

$$-\alpha\delta + \Delta + \beta\delta_{xx} = \beta q_0 \phi_x^2 \qquad (25).$$

In this equation α is for $2U_{rr}$ at $r=q_0$ although the constant Δ accounts for the possibility that U(.) is a priori not exactly minimum at q_0, as we are exploring the vicinity of $\Delta U=0$ in the parameter space. The integration of the linear equation (25) is straightforward and the free parameter left (Ω or L) is found by matching the solution with the asymptotics of the front solution limiting the plateau.

A possible form for this solution, consistent with the parity $\delta(x)=\delta(-x)$, $x=0$ being the center of the s-wave, is:

$$\delta(x)=\frac{\Delta}{\alpha}+a(e^{kx}+e^{-kx})+\beta q_0 e^{kx}\int_0^x dx'\ e^{-2kx'}\int_0^{x'}dx''e^{kx''}\ \phi_{x''} \quad (26),$$

where $k=(\beta/\alpha)^{1/2}$. The parameter α is determined by the condition that this solution has to fit the exponential approach toward q_0 of the front solution near $x=\pm L/2$. This imposes $a=a'e^{kL/2}$, where a' is negative and of order 1 when measured with the quantities entering into U and with β. This coefficient a' can be computed by considering the extension of the equations for U in the complex x-plane. The same extension to the complex x-plane shows that any other contribution to $\delta(x)$ has to cancel in this matching region. They are three such contributions : the constant Δ/α, a term of order $a'e^{-kL}$ coming from the second term on the r.h.s. of (26) and another one, with a positive sign coming from the asymptotics of the double integral and of order $\frac{\beta a''}{\varepsilon^2}$, a" being a positive constant of order 1 (that is independent of L and ε). Those three contributions cancel if:

$$\Delta+\frac{\beta a''}{\varepsilon^2}e^{-kL/2}-a'e^{kL}=0 \quad (27).$$

A more precise theory would also make appear a change of Δ of order $1/\varepsilon$, that we shall not detail here. As a' and a" in (27) are both positive, this equation has two roots for L when Δ is negative, and small.

Coming back to the meaning of this last result, two s-waves exist and one is stable (the wider one) and one unstable. They merge as announced for Δ of order $1/\varepsilon^2$. Furthermore when Δ tends to zero one of the s-wave becomes infinitely wide and thus disappear for the value of the parameters where a steady front exists.

3.Concluding remarks

I have explained how amplitude theory may be used to describe the occurence of localised patches of turbulence in flows at moderate to large Reynolds number. There is no doubt that this may be extended to other situations. I have in mind in particular the so-called vortex breakdown problem [16], so important for practical applications and that does not seem to have been yet convincingly explained. I suggest that the two possible states of the vortex: "laminar" and expanded are like the two bottom well of a potential. This is conforted by the fact that the velocity distribution on the laminar side is linearly stable (Maurice Rossi, private communication). Then the well definite location of the front separating physically the two states of the vortex in the streamwise direction could be simply at the point where the front between the two states stops moving, a bit like the interface between water and ice would stop on the 0°C isotherm in an imposed nonhomogeneous temperature field. Such global explanations clearly require a better understanding of the statistical turbulent ensembles that we have now. This is certainly a field where a lot can be done and where we shall see more progress in the years to come.

References

[1] Y.Pomeau; Physica **23D**,3(1986).

[2] J.Smoller "Shock Waves and reaction-diffusion equations", Springer , Berlin(1983).

[3] A.C.Newell, J.Whitehead, J. of Fluid Mech.**38**, 79(1969); L.A.Segel, ibid; p.203.

[4] J. Hegseth, C.D. Andereck, F. Hayot and Y. Pomeau, Phys. Rev. Letters **62**, 257 (1989).

[5] O. Thual, S. Fauve, J. de Physique (Paris), **49**, 1829 (1988).

[6] V. Hakim, P. Jakobsen and Y. Pomeau, Europhys. letters **11**, 19 (1990).

[7] Ya.B.Zel'dovich, "Theory of combustion and detonation of gases", Moscow (1944).

[8] A.N.Kolmogorov, A.N.Petrovskii and N.S. Piskunov, Bulletin de l'Universite d'Etat a Moscou, Sec.A, Vol.**1**, Math. et Mec. ,1(1937).

[9] P.Tabeling, J. de Phys. Lettres **44**, 665(1983); P.Hall, Phys.Rev. **A29**, 2921(1984).

[10] I.J.Wygnansky and F.H. Champagne, J.of Fluid Mech. **59**,281 (1973).

[11] H.Chaté and P.Manneville, Phys.Rev. Lett. **58**,112(1987).

[12] Y.Pomeau, in "Instabilities and nonequilibrium structures II", E. Tirapegui and D. Villaroel eds., Kluwer Academic press publishers (Boston, 1989).

[13] F. Daviaud, M. Dubois and P. Bergé, Europhys. Lett. **9**, 441 (1989)

[14] A.Pumir, P. Manneville and Y. Pomeau, J. of Fluid Mechanics, **135**, 27 (1983).

[15] A.C. Newell, "Solitons in Mathematics and Physics", SIAM pub. Philadelphia (1985).

[16] M.G. Hall, Annual review Fluid Mech. **4**, 195 (1972); S. Leibovich, ibid. **10**, 221 (1978).

TRANSITION TO CHAOS IN NONLINEAR OSCILLATORS

J. C. Antoranz and M. A. Rubio

Departamento de Física Fundamental
Universidad Nacional de Educación a Distancia
Apartado Correos 60141, E-28080 Madrid, Spain

INTRODUCTION

During the first half of last decade, the development of nonlinear dynamical systems theory and experiments /1,2/ produced great expectations in the scientific community. For some time it appeared that understanding the transition to turbulence in terms of transitions to chaos as they appeared in systems with small number of degrees of freedom. Today these expectations have weakened significantly; however, still some hydrodynamic flows have been shown to display chaotic transitions prior to entering into turbulent states /3/. Therefore studying chaotic temporal dynamics in initially oscillatory systems may still have some relevance for the understanding of the complex behavior of turbulent hydrodynamic systems.

Several nonlinear oscillatory systems have been extensively studied. The Duffing oscillator /4/ and the driven damped pendulum or its Josephson-junction analog /5,6/ are the most popular among them. Some of these systems have recently received added interest due to its appearance as normal forms in codimension-two bifurcation theory, e.g., the Duffing-Van der Pol oscillator /7/.

One of these systems is the so-called Helmholtz oscillator /8/ which is a driven damped oscillator with a quadratic nonlinearity in the elastic term. It was proposed by Helmholtz to explain the combinational tones generated in the ear's drum, and its analytical expression represents a canonical form in catastrophe theory /9,10/.

Systems like the Helmholtz or Duffing oscillators can show only one fundamental frequency in their evolution and therefore the universal scenarios of transition to chaos to be expected are the well known period-doubling cascade /11/ and the type I and III Pomeau-Manneville intermittencies /12/.

In this report, we illustrate some codimension-one bifurcations leading to deterministic chaos in the Helmholtz driven damped nonlinear oscillator. Using both analog and numerical simulations we have studied the bifurcation leading to subcritical period-doubling, and types I, and III intermittencies. We also illustrate the process of identification of the intermittency type, and examine the relative efficiency of the different tools at hand. Finally we introduce a onedimensional map which displays the whole fenomenology of the nonlinear oscillator.

THE HELMHOLTZ OSCILLATOR

The Helmholtz oscillator is a driven damped oscillator with a single well asymmetric cubic potential, which allows the system to escape from the well under certain conditions. The nonlinearity was introduced by Helmholtz to explain the combinational tones in the ear's drum. The equation that rules the dynamics of the system, with time normalized respect to the natural period of the linear oscillations, is

$$\frac{d^2x}{dt^2} + g\frac{dx}{dt} + x + x^2 = A\cos\omega t \qquad (1)$$

where g, ω and A stand respectively for the damping coefficient, the frequency and the amplitude of the forcing term.

A multiple time scale perturbation calculation of the nonlinear resonance curve /13/ shows that it is a *soft* oscillator. This means that, independently of the sign of the nonlinear term, the resonance curve always bends towards frequencies lower than the natural one (here $\omega = 1$). This implies that for frequencies higher than 1 there is a *single* stable limit cycle of period $T = 2\pi/\omega$. In what follows, we have worked in the region corresponding to $\omega > 1$ to avoid hysteretic phenomena due to the existence of two T-period stable limit cycles,

With the advent of very fast desktop minicomputers, direct numerical simulation is becoming the most appropriate way to locate the interesting regions in parameter space. However analog electronic simulators are still an alternative in this process. For our studies we built up an analog simulator based on two Miller integrators and using as nonlinear element a four quadrant analog multiplier (AD533) /14/. The periodic forcing was provided by a signal generator Interstate F30. We studied the evolution in phase space by feeding the signals corresponding to variables x and dx/dt into an oscilloscope. The power spectrum was obtained with a Spectral Dynamics SD375 Fourier analyzer. Pictures of the time evolution and averaged power spectra of the simulator output were directly obtained through a digital plotter HP7470. With this analog simulator, the most interesting region was located around $g = 0.52$ and $\omega = 1.3$. For this values an intermittent signal with power-law low-frequency behavior was found. The experimental results are depicted in Figures 1 and 2.

Figure 1 corresponds to the time evolution of variable x for $g = 0.52$, $\omega = 1.3$ and $A = 0.5$. Its corresponding power spectrum (Fig. 2), shows a definite $1/f^\alpha$ trend over two decades with an exponent α around -1.2 ± 0.1. We point out that the low frequency noise in the periodic regimes was three orders of magnitude lower and, therefore, this low frequency divergence in the power spaectrum had an unambiguous dynamical origin.

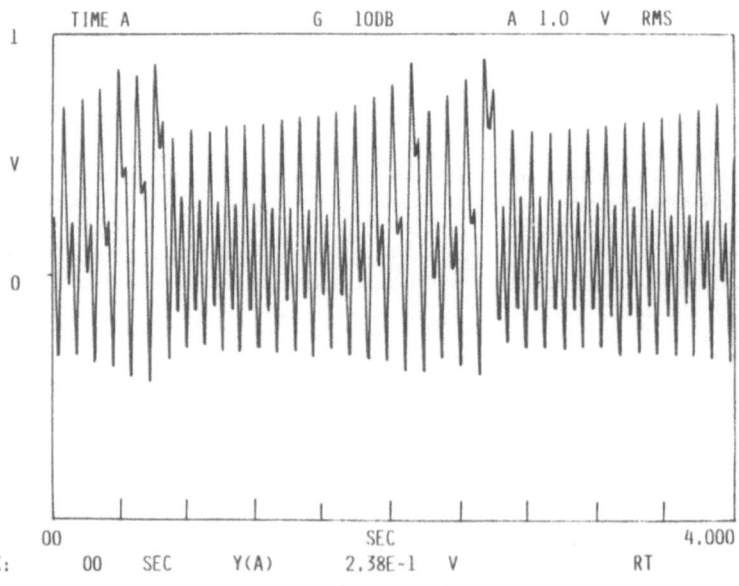

Figure 1. Time evolution of the analog simulator signal for $g = 0.52$, $\omega = 1.3$, and $A = 0.5$.

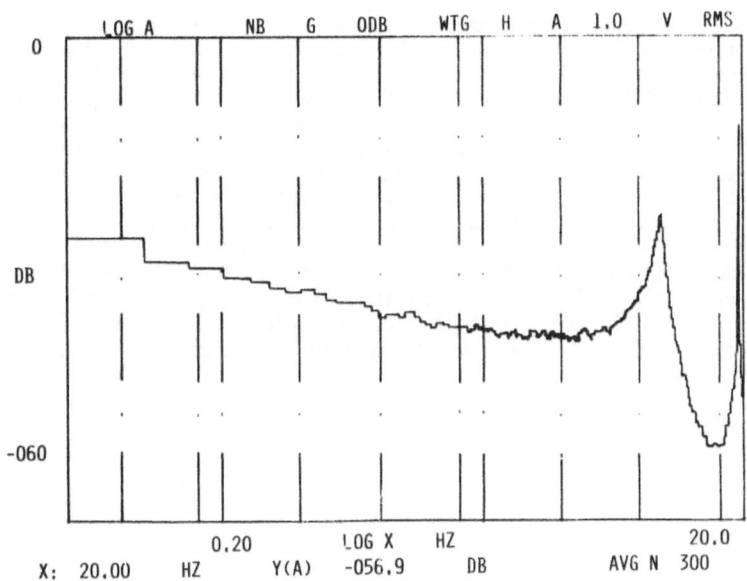

Figure 2. Averaged power spectrum of the analog simulator signal for $g = 0.52$, $\omega = 1.3$, and $A = 0.5$. The spectrum has been averaged over 300 times and it yields $\alpha = -1.2 \pm 0.1$.

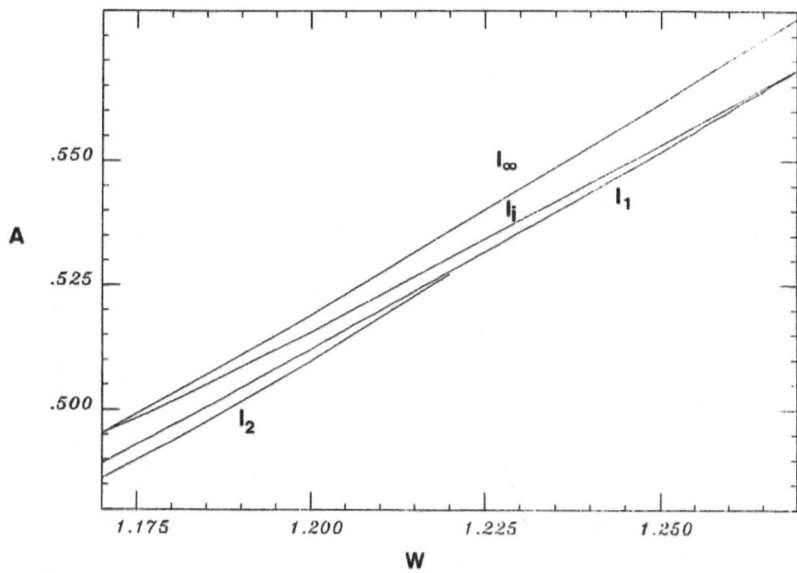

Figure 3. Regions of periodic and chaotic solutions in parameter space for $g = 0.52$. l_∞ - boundary for evolution inside the potential well. l_1 - supercritical period doubling. l_i - bifurcation from period $2T$ to chaos. l_2 - relaxation of the chaotic attractor of the subcritical branch to the T-periodic regime.

INTERMITTENCY IN THE HELMHOLTZ OSCILLATOR

With the hints provided by the analog simulation, we started a direct numerical simulation /15/. All computations have been performed with a fourth order Runge-Kutta method with double precision arithmetics. In Figure 3 we plot the critical curves corresponding to the main bifurcations in the plane $\omega - A$ for $g = 0.52$.

We have labeled these curves as follows. l_∞ is the limit of bounded evolution inside the single well potential. At l_1 a supercritical period doubling bifurcation separates the T-periodic solution from a stable $2T$-periodic solution. l_i is the stability boundary for this $2T$-periodic solution; beyond this curve no periodic attractor exists. This bifurcation shows no hysteretic behavior. Finally, l_2 represents the boundary of a chaotic attractor arising from another $2T$-periodic solution that undergoes a complete subharmonic bifurcation cascade. More explicitly, below l_2 the chaotic attractor is stable while above this line, it relaxes to the stable T-periodic orbit.

Two typical bifurcation diagrams, obtained by plotting the minima of $x(t)$, are shown in Figures 4a and b. Figure 4a corresponds to $\omega = 1.2$ and $g = 0.52$, while Figure 4b has been computed for $\omega = 1.3$ and $g = 0.48$. The two bifurcation diagrams show some similarities. For certain initial conditions a subcritical $2T$-periodic solution appears, and upon increasing A this orbit undergoes a complete period doubling sequence leading into chaos. Further increasing A this chaotic attractor finally decays into the T-periodic solution.

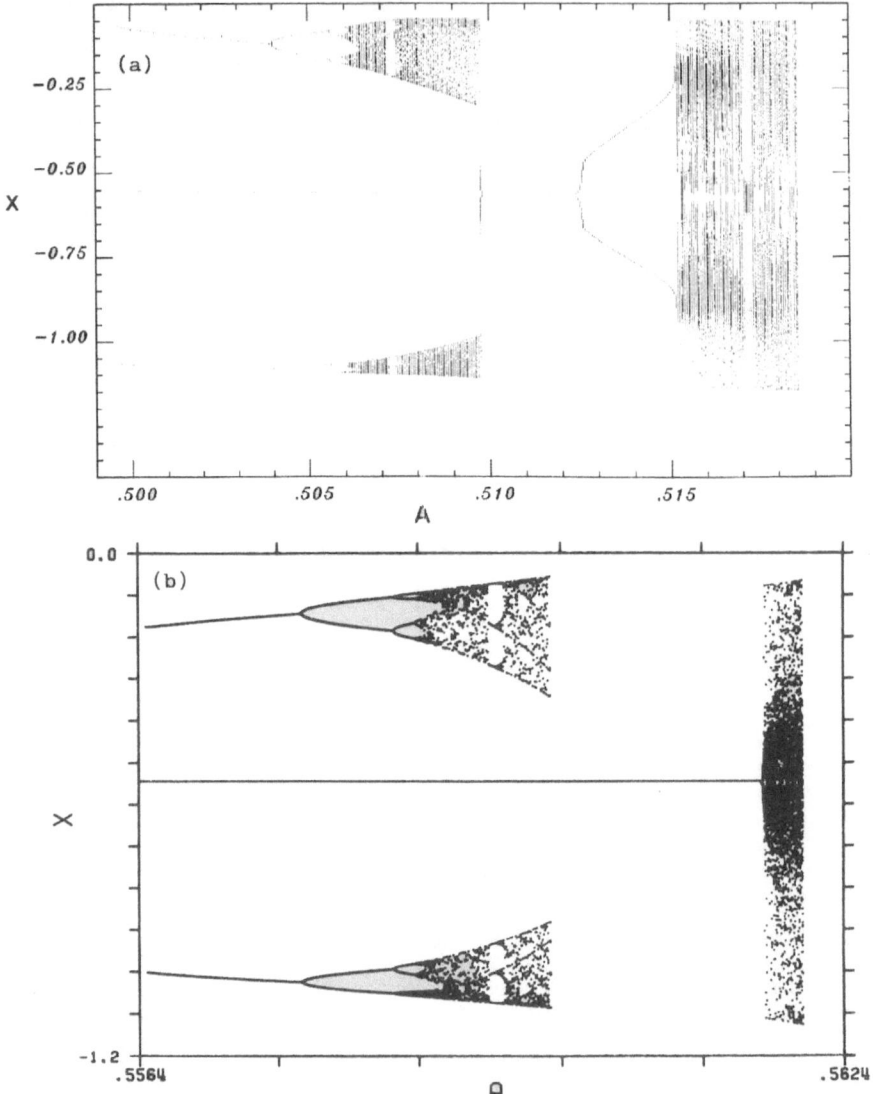

Figure 4. Bifurcation maps obtained for the minima of x. (a) $g = 0.52$, $\omega = 1.2$; the intermittent transition occurs at $A_I = 0.51515$. (b) $g = 0.48$, $\omega = 1.3$; the intermittent transition occurs at $A_{III} = 0.5610$.

290

However the differences in the Figs. 4a and b are by far more interesting. In Fig. 4a there is a supercritical period doubling bifurcation at $A = 0.5122$ and the emerging $2T$-periodic solution losses its stability at $A_I = 0.51515$ where an intermittent regime appears. Upon decreasing A these two bifurcations show no hysteresis. On the other hand, in Fig. 4b there is a direct transition from the T-periodic solution into an intermittent state at $A_{III} = 0.5610$. Then the question naturally arises: are the two intermittent regimes of the same type?. For this identification magnitudes corresponding to evolution in phase space such as the time evolution, phase space diagrams, or Poincaré return maps, provide very little information.

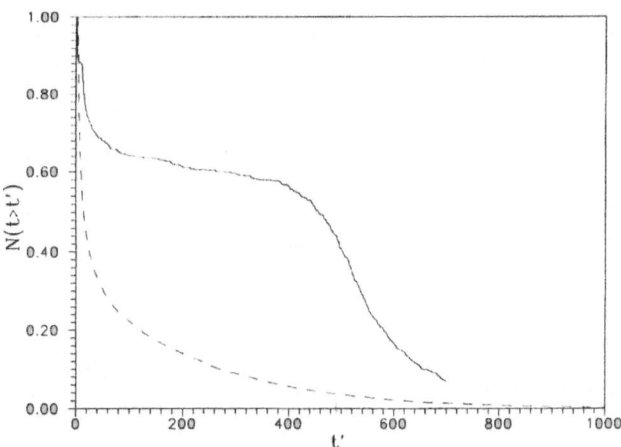

Figure 5. Statistics of the intermittent regimes, $N(t \geq t')$. (a) Type-I regime at $A = 0.5152$ (solid line), and $A = 0.518$ (dotted line). (b) Type-III regime at $A = 0.5612$.

The different types of intermittent behavior are classified according to the value of the Floquet multiplier at the bifurcation point. If a real Floquet multiplier crosses the unit circle at $+1$, corresponding to an inverse saddle-node bifurcation, then the intermittency is labeled as type-I. Conversely, if it crosses at -1, corresponding to a subcritical period doubling, it is labeled as type-III. Therefore, a definite identification of an intermittent regime can be done only by means of the Floquet multipliers and some statistical properties directly related to them; for instance, the distribution of laminar period lengths in the vicinity of the bifurcation point /16/.

For non-autonomous systems, the Floquet multipliers can be computed numerically /17/. In this case we obtained crossings at $+1$, for the bifurcation in Fig. 4a, and at -1, for the bifurcation in Fig. 4b, Therefore the intermittent regimes correspond to type I and III respectively.

Turning to the statistical properties, it is usual to study the probability distribution for the length of the laminar periods. However, for limited amounts of data, it is more convenient to speak in terms of an integral of this function, namely the number of laminar periods whose length is equal or higher than a certain time t' as a function of $t'(N(t \geq t'))$. This function shows a different shape for both types of intermittency. For type-III it is an exponentially decaying function, whereas for type-I it has a certain sigmoid shape, i.e., it decays for short lengths, after it shows a certain plateau, and finally it has another decaying part for the maximal laminar period lengths.

In Figure 5a we have plotted the function $N(t \geq t')$ correspondign to the bifurcation in Fig. 4a. The solid lines corresponds to $A = 0.5152$, while the dotted line corresponds to $A = 0.518$. The shape of the solid line corresponds to the one that should be expected for type-I intermittency. However, the dotted line is much closer to the shape of a type-III distribution. On the other hand, for the bifurcation in Fig. 4b, the statistics coincide with the expected type-III shape (Fig. 5b).

The answer is in the second iterate return map. Let us have a close look to the separation rate of the return map respect to the bisector. In the case of the type-III intermittent transition, the separation rate from the unstable T-periodic fixed point, is a monotonically increasing function. However, in the type-I regime, the situation is different. For A values below the critical one, the return map presents three fixed points. One of them corresponds to the unstable T-periodic orbit. The other two belong to the couple of stable and unstable $2T$-periodic orbits which will undergo the inverse saddle-node bifurcation. Therefore the return map crosses the bisector at three places. Two with slope higher than one, corresponding to the unstable fixed points, and one with slope smaller than one corresponding to the stable $2T$-periodic solution.

When the system undergoes the intermittent transition, the two $2T$-periodic fixed points collide and disappear, leaving a hump in the return map, which looks somewhat like the one sketched in Fig. 6. Therefore the map presents only one unstable fixed point corresponding to the T-periodic orbit, but there is a reattachment of the map to the bisector corresponding to the previous $2T$-periodic fixed points. In our system the global structure of the attractor allows for reinjection near the unstable T-periodic fixed point. Hence, the maximal length of the laminar regions will be ruled by the relative importance of the two processes involved: escape from the neighbourhood of the unstable fixed point and crossing the channel created by the intermittent transition.

Both processes may be analyzed in terms of the relative distance to the critical point, labeled by e. Escaping from the unstable fixed point is equivalent to a type-III intermittent regime and therefore, the average time required goes as e^{-1}. Instead, for the type-I channel crossing, the maximal length goes as $e^{-1/2}$.

In our system, when the type-I intermittent transition takes place, the return map has a certain slope near the unstable T-periodic fixed point, but the width of the channel can be made arbitrarily small, therefore giving a behavior with the whole type-I features. As A increases the hump goes away from the bisector faster than the slope near the unstable fixed point increases. Therefore the behavior of the system approaches that corresponding to type-III intermittency.

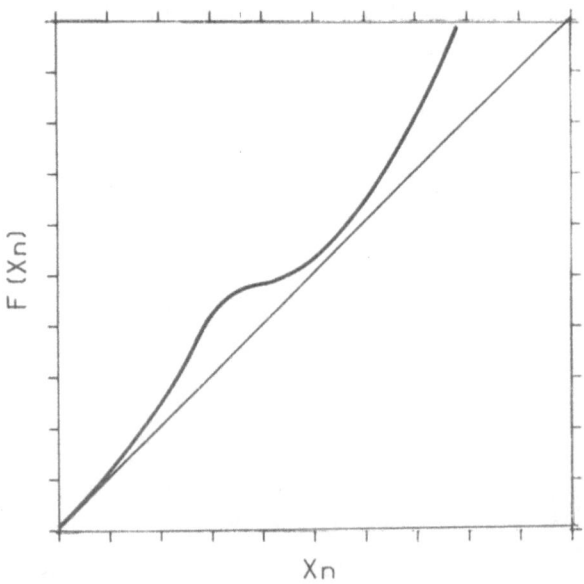

Figure 6. Qualitative sketch of the return map yielding the behavior explained in the text.

EQUIVALENT ITERATIVE MAP

All the above described behavior may be extracted from a one dimensional iterative map designed to mimmic the actual return maps found in the oscillator. The map has inversion symmetry (even terms are not involved) and a fixed point at $x = 0$. This symmetry guarantees the existence of a period doubling bifurcation as soon as the slope at $x = 0$ becomes higher than minus one. The map also allows for the existence of other fixed points, reinjection, and to escape to infinity for high values of the control parameter. To comply with these requirements, the coefficient of the seventh and fifth-order terms had to be negative and positive, respectively. The second iterate of the map can be written in the following form:

$$x_{n+2} = (1 + e)x_n + ax_n^3 + bx_n^5 + cx_n^7 \qquad (2)$$

The general shape of this map can be observed in Figure 7, in which the following set of parameter values was used: $a = 0.03$, $b = 4.4$, $c = -6.0$, and $e = 0.01$.

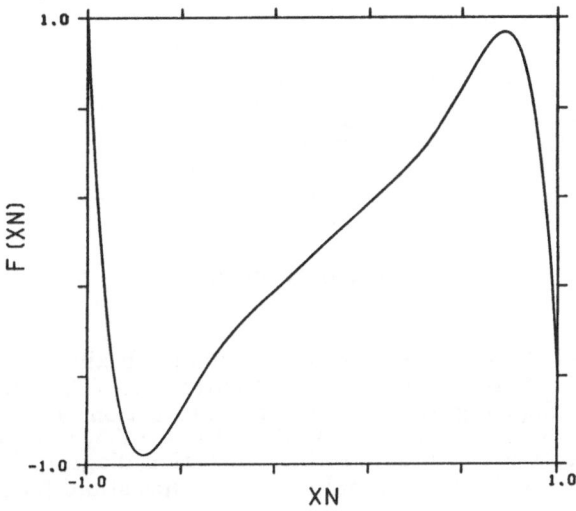

Figure 7. Iterative map for $a = 0.03$, $b = 4.4$, $c = -6.0$ and $e = 0.01$.

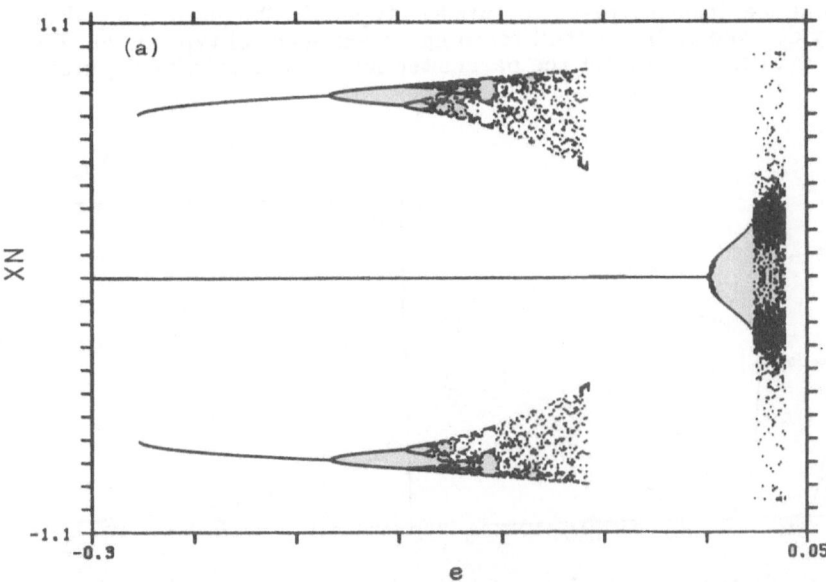

Figure 8. Bifurcation diagrams for (a) type-I intermittency and (b) type-III intermittency (sets $S1$ and $S3$, respectively).

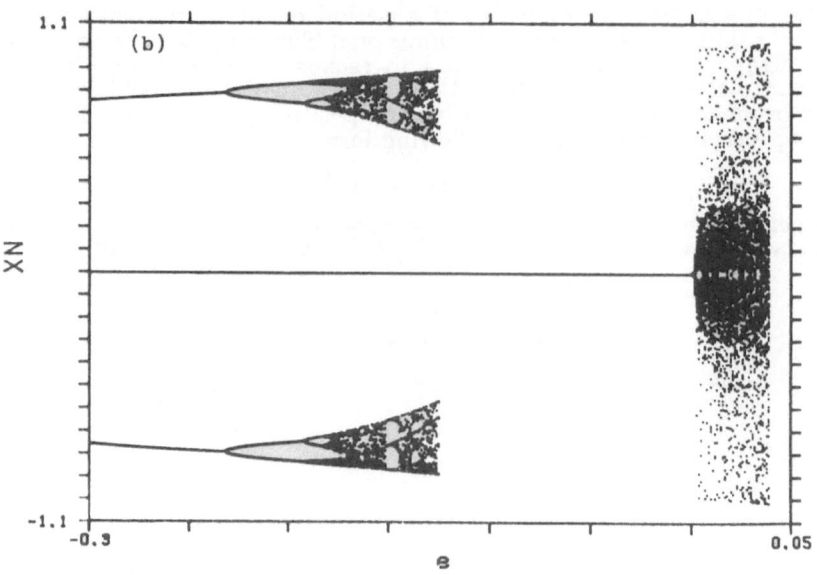

Figure 8. (Continued)

Figures 8(a) and (b) represent bifurcation maps obtained by fixing parameters values a, b and c, and taking e as the control parameter. The similarities with Figs. 4(a) and (b) are striking. Figure 8(a) shows the bifurcation diagram for the following set of parameter values $a = -0.746$, $b = 6.5$ and $c = -7.8$, that we will label as $S1$. On the other hand, Figure 8(b) shows the bifurcation diagram for a different set of parameter values ($a = 0.03$, $b = 4.4$ and $c = -6.0$, that stand for set $S3$).

The nature of both transitions to chaos has been checked by computing again $N(t \geq t')$. The results are shown in Figs. 9(a) and (b). In Fig 9(a) we show the statistics corresponding to the set $S1$ for three representative values of the control parameter e. The typical aspect for the type-I intermittent regime appears only very close to the bifurcation point ($e = 0.0235$). Upon increasing e the global shape of the statistics changes from a type-I structure to a type-III. The plateau shrinks ($e = 0.025$) and finally disappears ($e = 0.052$) reaching an exponential type shape. On the other hand Fig. 9(b) shows that for the parameter set $S3$, the bahvior is purely type-III.

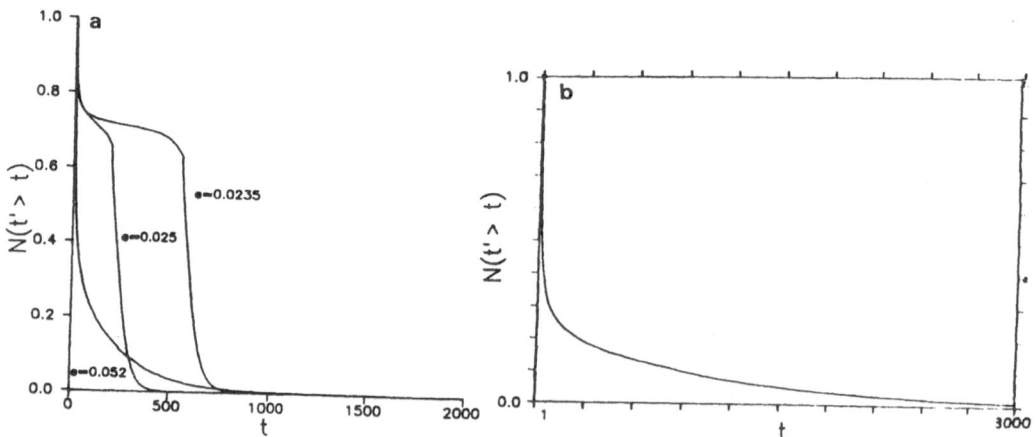

Figure 9. Statistics of the intermittent regimes, $N(t \geq t')$. (a) Type-I regime for set $S1$; the curves correspond to $e = 0.0235$, $e = 0.025$ and $e = 0.052$. (b) Type-III regime for set $S3$, at $e = 0.001$.

CONCLUSIONS

With the help of a nonlinear oscillator we have illustrated some codimension one bifurcations leading to deterministic chaotic behavior. Particularly we have focused on the intermittent transitions of type I and III. We have clasified the transitions using the Floquet multipliers and the distribution of the laminar periods. We have shown that the two bifurcations are different in nature; one is an inverse saddle-node, while the other is a subcritical period doubling. Moreover, due to the competition between the channel and the unstable fixed point, the statistical properties above the type-I transition approach those belonging to the type-III. All of these scenarios have been reproduced in a suited one-dimensional iterative map.

ACKNOWLEDGEMENTS

This research has been partially supported by DGICYT under contract PM88-0071.

REFERENCES

1. P. Bergé, Y. Pomeau and Ch. Vidal, *L'Ordre dans le Chaos*, Hermann, Paris, (1984).

2. H. L. Swinney, N. B. Abraham and J. P. Gollub, Physica 11D (1984), 252.

3. K. R. Sreenivasan and R. Ramshankar, Physica 23D (1986), 246.

4. F. T. Arecchi, F. Lisi, Phys. Rev. Lett., 49 (1982), 94; F. T. Arecchi, A. Califano, Phys. Lett. 101A (1984), 443.

5. R. F. Miracky, M. H. Devoret, J. Clarke, Phys. Rev., A 31 (1985), 2509.

6. E. G. Gwinn, R. M. Westervelt, Phys. Rev. Lett., 54 (1985), 1613.

7. J. Guckenheimer and P. Holmes, *Nonlinear Oscillations, Dynamical Systems, and Bifurcations of Vector Fields*, Springer-Verlag, Berlin, (1983).

8. H. Helmholtz, *Die Lehre von den Tonempfindungen als Physiologische Grundlage fur die Theorie der Musik*, Friedrich Viewveg, Braunschweig, (1870).

9. T. Poston and I. Stewart, *Catastrophe Theory and its Applications*. Pitman, London (1978).

10. J. M. Thompson, S. R. Bishop and L. M. Leung, Phys. Lett. 121A (1987), 116.

11. R. M. May, Nature 261 (1976), 459; M. J. Feigenbaum, J. Stat. Phys. 21 (1979), 669.

12. P. Manneville and Y. Pomeau, Physica 1D (1980), 219; Y. Pomeau, P. Manneville, Comm. Math. Phys., 74 (1980) 189; P. Manneville in *Symmetries and Broken Symmetries*, N. Boccara Ed., Paris (1981).

13. A. H. Nayfeh and D. T. Mook, *Non-linear Oscillations*. John Wiley, New York (1979).

14. M. A. Rubio, M. de la Torre, J. C. Antoranz and M. G. Velarde in *Computational Systems - Natural and Ar'ificial*, H. Haken Ed., Springer-Verlag, New York (1987).

15. M. A. Rubio, M. de la Torre and J. C. Antoranz, Physica D, 36 (1989), 92; J. C. Antoranz, M. A. Rubio and M. de la Torre, Prog. Theor. Physics, 81 (1989), 544.

16. M. Dubois, M. A. Rubio and P. Bergé, Phys. Rev. Lett., 51 (1983), 1446.

17. H. Bunz, H. Ohno and H. Haken, Z. Phys., B 56 (1984), 345.

TURBULENT COUPLED MAP LATTICES

Tomas Bohr

The Niels Bohr Institute
Blegdamsvej 17
DK-2100 Copenhagen

INTRODUCTION

The study of coupled map lattices has become a very fruitful way of building up intuition and knowledge about extended dynamical systems in chaotic states -like hydrodynamical turbulence or inhomogenuos chemical reactions- and testing and generalizing the methods used to describe low-dimensional chaotic systems in a context where spatial degrees of freedom are important. When the word *turbulence* is used in such systems it does not necessarily imply that we are modelling hydro-dynamical systems, nor even that a velocity field is defined. We use the term simply to describe an extended system in a chaotic state, which is spatially disordered as well. As in low-dimensional systems we define a chaotic state as one having a positive Lyapunov exponent, and we then define a turbulent state as one having a number of positive Lyapunov exponents, which grows with the system size, such that this number becomes arbitrarily large for systems large enough.

It is well known that the direct numerical simulation of hydrodynamical systems in the regime of fully developed turbulence is at present impossible, and thus we must try any possible shortcut into this interesting physics. Of course, if we put chaos into the maps that we couple, we are not going to explain the origin of chaos; but maybe we'll be able to say something about the important length- and time scales in the systems, about how the power is shared among the different modes and how chaotic invariants like the Lyapunov exponents depend on the system parameters (including the size) and one the properties of the single maps.

In this brief review I shall not be able to do justice to the large body of work, which has appeared in this field, and I shall restrict myself to reviewing a few topics in which I've been involved in the last years. The pioneering work in the study of chaotic coupled map lattices was done by K. Kaneko [1] and many of the models used in the following were introduced by him.

FROM COHERENT TO INCOHERENT CHAOS

Consider a two-dimensional, regular lattice indexed by (i,j), $i,j = 1,2,...,L$. On each site we have a variable $u_n(i,j)$ where n is the "discrete time" and we assume periodic boundary conditions in i and j. The dynamics is given by

$$u_{n+1}(i,j)=(1-\varepsilon)f(u_n(i,j)) + \frac{\varepsilon}{4} \sum_{n.n.\ i'j'} f(u_n(i',j')) \tag{1.1}$$

where (i',j') are nearest neighbors of (i,j). The function $f(x)$ is taken to be some non-linear map that can sustain chaotic motion, e.g. the logistic map $f(x) = Rx(1-x)$ with a period doubling cascade with accumulation point at $R_c =3.5699456...$ For $R_c <R<4$ the attractor is either chaotic or periodic.

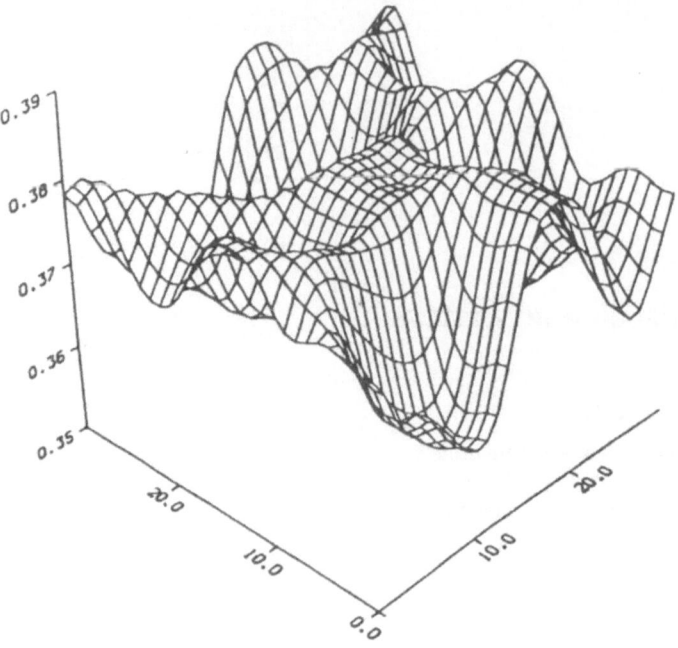

Fig.1a. Typical state of the coupled logistic map (1.1). $R = 3.5732$ and $\varepsilon = 0.4$.

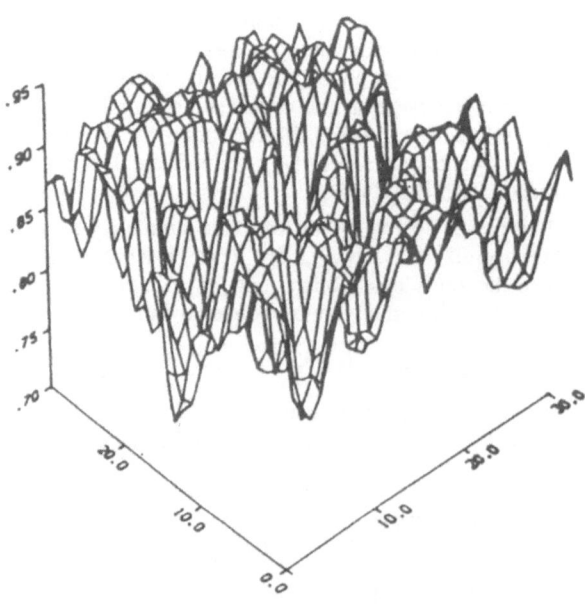

Fig.1.b. As in fig. 1a, but $R = 3.6786$ and $\varepsilon = 0.4$.

Typical states, differing in R, obtained by iterating (1.1) with $L=30$ and noisy initial conditions is shown in fig.1a-b. The figures show snapshots of the system at a given time. The next timestep looks different - the system is both spatially and temporally disordered - but certain statistical features are preserved. Small R ($> R_c$) and large ε (<1) gives smooth spatial variations whereas large R (<4) and small ε (>0) makes the picture rugged with rapid spatial variations. The variations with R is clearly illustrated by the figures and it is plausible that a "correlation length" of fig.1a is larger than fig.1b.

If we choose parameters corresponding to a periodic window for each of the logistic maps in the chaotic regime we see something surprizing: although (1.1) certainly has a stable uniform cycle the system never finds it. The state again looks spatially disordered like fig.1. To understand this one must realize that the periodic states of the logistic map are very different above and below R_c. Above R_c the stable cycle coexists with a *chaotic repellor* There are points in the unit interval that never settle down on the cycle, but they form a Cantor set of dimension less than 1 so they don't fill anything. If a starting point is not exactly on the Cantor set it will be repelled and finally end up on the cycle. For the coupled system this is different. When the size of the system grows the typical transient time to get into the cycle becomes larger: since the repellor has a positive Lyapunov exponent, the motion on it will be characterized by a finite correlation length, as argued in general in [2], and thus the transient times become very large as the system size is increased. A lot of work has been done on coupled map systems in one dimesion i.e. coupled chains. The reason for using two-dimensional lattices here is that the equilibration properties are much better. In one dimension one often gets kink-like structures, which, although only "metastable", take forever to decay. In two dimensions, analogosly to the equilibrium case, "surface tension" will speed up the equilibration process.

To calculate Lyapunov exponents one looks at differentials

$$du_{n+1}(i,j) = (1-\varepsilon)f'(u_n(i,j))du_n(i,j) + \frac{\varepsilon}{4}\sum_{n.n.}f'(u_n(i',j'))du_n(i',j') \qquad (1.2)$$

For a lattice of size L one can, in principle, define L^2 Lyapunov exponents, such that the sum of the first i characterize the exponential growth or decay of an i-dimensional hypervolume in phase space. For now we shall confine our attention to the *largest* one which describes the growth of distances in phase space, and we ask how this quantity depends on lattice size, whether it shows the characteristic lengthscales of the system. This question was addressed in [3] which we shall follow below.

A specific example is shown in fig.2, where the largest Lyapunov exponent of the map (1) with $R=3.5732..$ (i.e. $\log((R-R_c)/R_c) = -7$) and $\varepsilon=0.4$ is plotted against the size L of a quadratic lattice, varying from 1 to 50. For each lattice, noisy initial conditions close to uniform were used. The first 5000 iterates were discarded and the next $N=10000$ iterates were used to evaluate λ. The uncertainity was estimated by comparing to the value of λ obtained half-way (i.e. with $N=5000$) - it is of the order of a few times the dot size in the figure. For small lattices ($L\leq8$ in the specific example) the Lyapunov exponent is close (within 5%) to the single map value 0.059. Between $L=8$ and $L=9$ it drops to a value close to zero where it remains up to $L=13$. Between $L=13$ and 15 it rises to its "large system" value -all the way up to $L=50$ it remains close (again within 5%) to $\lambda_\infty=0.030$. A closer look at the motion reveals that the lattice becomes absolutely flat for $L\leq8$ i.e. the chaotic motion of each map is completely in phase (the differences in λ are numerical inaccuracy due to the finite waiting times). Between $L=9$ and $L=12$ the state is modulated in one direction and flat in the other. It is easy to see that this transition is simply determined by linear stability of the uniform chaotic state $u_n(\vec{x})=M_n$ where $M_{n+1}=f(M_n)$. In the chaotic state very long waves are unstable. A state which is almost uniform, i.e. of the form $u_n(\vec{x}) = M_n+\delta_n(\vec{x})$ (where $\vec{x}=(i,j)$) will, to linear order in δ, change to $M_{n+1}+\delta_{n+1}(\vec{x})$ where

$$\delta_{n+1}(\vec{x}) = f'(M_n)\left[(1-\varepsilon)\delta_n(\vec{x}) + \frac{\varepsilon}{4}\sum_{\vec{x}'}\delta_n(\vec{x}')\right] \qquad (1.3)$$

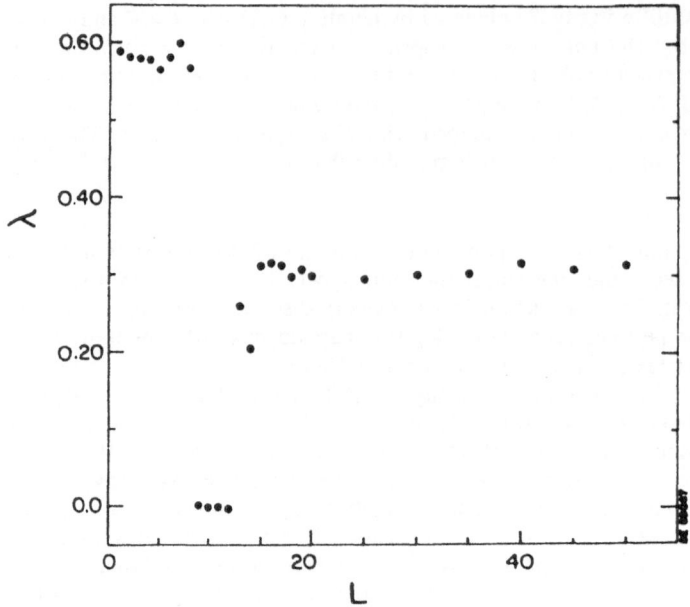

Fig.2. The largest Lyapunov exponent as function of system size. R=3.5732,
ε=0.4.

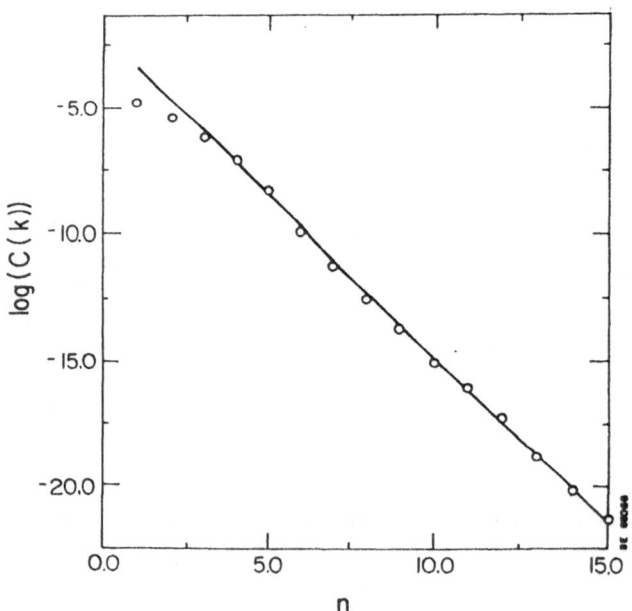

Fig.3. Spatial power spectrum for (1.1). R=3.5709, ε=0.4.

and where the sum is over nearest neighbors $\vec{x}'$ of $\vec{x}$. By Fourier transformation, i.e. expanding δ as

$$\delta_n(\vec{x}) = \sum_{\vec{q}} \delta_n(\vec{q}) e^{i\vec{q}\cdot\vec{x}} \tag{1.4}$$

where $\vec{q} = (q_1, q_2)$ we get

$$\delta_{n+1}(\vec{q}) = f'(M_n)(1-\varepsilon+\frac{\varepsilon}{2}(\cos q_1+\cos q_2))\delta_n(\vec{q})) \tag{1.5}$$

A perturbation $\delta_0 e^{i\vec{q}\cdot\vec{x}}$ of the flat state will therefore grow as $e^{\lambda_0 + \delta\lambda_\varepsilon(\vec{q})n}$, where λ_0 is the Lyapunov exponent of the single map f (i.e. $\lim\limits_{N\to\infty} \sum\limits_{n=1}^{N} \log f'(M_n)$) and

$$\delta\lambda_\varepsilon(\vec{q}) = \log|(1-\varepsilon(1-\frac{1}{2}(\cos q_1+\cos q_2)))| \tag{1.6}$$

which, for small q, becomes [4]

$$\delta\lambda_\varepsilon(\vec{q}) \approx -\frac{\varepsilon}{4}q^2 \tag{1.7}$$

The allowed wavevectors are of the form $\vec{q}=\frac{2\pi}{L}(n_1,n_2)$, where n_1 and n_2 are integers. Thus the first wavevector $\vec{q}=\frac{2\pi}{L}(1,0)$ or $\vec{q}=\frac{2\pi}{L}(0,1)$ becomes unstable at system size L_1 determined by $\lambda_\varepsilon(\vec{q}=(2\pi/L_1,0))=0$ which for large L_1 means $L_1 \approx \pi\sqrt{\varepsilon/\lambda_0}$.

For the parameters corresponding to fig.2 we get $L_1=8.02$ which means that $L=9$ is the first unstable system. Strangely enough at first sight the system responds to this instability by becoming *stable* - the vanishing Lyapunov exponent indicates (quasi)periodic behaviour. On further reflection this is perhaps not so strange. Above L_1 the system can suddenly use the spatial degrees of freedom (at least in one direction) and thus it has much more phase space to search for stable motion and can remain coherent by lowering the Lyapunov exponent. At $L=13$ modulations in both directions appear. This happens roughly at $L=\sqrt{2}L_1$ corresponding to $\lambda_\varepsilon(\vec{q}=\frac{2\pi}{L}(1,1))=0$ i.e. the instability of the second mode in the uniform state. Now the system begins to have problems keeping its different parts together and for increasing L the dynamics quickly approaches "large system" behaviour with a new Lyapunov exponent reflecting the fact that different parts are dephased. Already at $L=L_c=15$ the value of λ is indistinguishable from its value at $L=50$ indicating that the incoherent averaging responsible for its "large system" value is now effective. Fig.1a shows the system with $L=30$. The lowering of the Lyapunov exponent compared to the single map value is caused by the existence of both positive and negative slopes in $f(u)$. Continuity along the lattice (enforced by the coupling) implies that between regions with positive slopes and regions with negative ones there must be regions where the displacement is very small thus lowering the overall growth rate.

The number of positive Lyapunov exponents is a rough measure (lower bound if the measure is sufficiently smooth) of the dimension [5]. This number changes rapidly around L_c. For the above parameters it is zero between $L_1=9$ and $L=12$. From 3 at $L=13$ it changes to 6 at $L_c=15$ so indeed the system is quickly getting high-dimensional. This means that we can view L_c as a kind of *coherence length:* systems larger than L_c are effectively *large*. Naively one might have taken L_1 as a coherence length, but that would be misleading. At L_1 the *uniform* state is lost, but coherence is still maintained.

For other values of the parameter R one finds similar behaviour. There is a size L_1 at which the Lyapunov exponent jumps down to a value close to zero. At a later value L_2, λ starts moving up again and at L_c its value is indistinguishable from that of the infinite system (extrapolating from $L\approx50$).

Very roughly, L_c is a factor of (slightly less than) 2 larger than L_1 and they seem to diverge in the same way when R approaches R_c [3,6,7].

The spatial power spectra of the turbulent states are *exponetially* decreasing functions of k. This is brought out clearly by fig.3 which shows the angular average of the modulus of the Fourier transform of the state variable, u, at R= and L=. This means that there is a characterisctic lengthscale setting the exponential decay which scales like L_c very close to R_c. The fact that the power spectrum is exponential, rather than a power law as in fully developed hydrodynamical turbulence, is related to the boundedness of the variable: $u(i,j)$ is always between 0 and 1. We shall later see examples of coupled map lattices with power law spectra.

A SIMPLE MODEL FOR CONVECTIVE CHAOS

The coupled map lattice (1.1) describes a closed system with no preferred direction. Most hydrodynamical experiments are, however, performed under open flow conditions, where the flow comes in from one side and leaves through the other, as in pipe flows. One can break the symmetry between x and -x by putting gradient terms into (1.1) [8]. Forgetting the Laplace term for the moment the smplest model would be

$$u_{n+1}(x) = f(u_n(x)) - c(f(u_n(x)) - f(u_n(x-1)))$$ (2.1)

which includes coupling between sites x and x-1 but not between sites x and x+1. Here, again, u is a scalar field at site x (x=0,1,...,L) at discrete time n. The coefficient c is analogous to the mean flow velocity. Assume that the map f has an unstable fixed point, u^*, with eigenvalue $f'(u^*)=a>1$ and let us fix the one end (x=0) at u^*, i.e. $u_n(0)=u^*$ for all n. One should think of u^* as the laminar state and thus the incoming fluid is assumed to be laminar. Under these conditions linear stability theory gives, with $u_n(x)=u^*+\delta_n(x)$,

$$\delta_{n+1}(x) = (1-c)a\,\delta_n(x) + ca\,\delta_n(x-1)$$ (2.2)

for $x>1$, and

$$\delta_{n+1}(1) = (1-c)a\,\delta_n(1)$$ (2.3)

This can be written

$$\begin{bmatrix} \delta_{n+1}(L) \\ \cdot \\ \cdot \\ \cdot \\ \delta_{n+1}(1) \end{bmatrix} = \begin{bmatrix} (1-c)a & ca & 0 & \cdot & & \cdot \\ 0 & (1-c)a & ca & 0 & & \cdot \\ \cdot & 0 & (1-c)a & ca & 0 \\ \cdot & & 0 & (1-c)a & ca \\ \cdot & & \cdot & 0 & (1-c)a \end{bmatrix} \begin{bmatrix} \delta_n(L) \\ \cdot \\ \cdot \\ \delta_n(1) \end{bmatrix}$$ (2.4)

Since this matrix is tridiagonal the eigenvalues are in the diagonal and they all have the value $\lambda=(1-c)a$; thus, even if $a>1$, λ can be < 1 making the laminar state linearly stable.

The system is, however, *convectively* unstable. If we make a small disturbance $\delta_0(0)$ at time 0 (after which $\delta_n(0)=0$) we find

$$\delta_n(x) = \binom{n}{n-x}((1-c)a)^{n-x}(ca)^x$$ (2.5)

and if we look at $\delta_n(x)$ in a frame of reference moving down the chain with speed v and use Sterlings formula we get

$$\delta_n(vn) \sim e^{n\lambda(v)}$$ (2.6)

where

$$\lambda(v) = \log a - (1-v)\log\frac{(1-v)}{(1-c)} - v\log\frac{v}{c} \tag{2.7}$$

The maximal value of $\lambda(v)$ is $\log a > 0$ and this occurs at $v=c$. There exists an interval $[v_{min}, v_{max}]$ around $v=c$ in which $\lambda(v)$ is positive; outside of it λ is negative.

We conclude that the initial disturbance becomes a wave-packet with exponentially growing amplitude, whose front propagates downstream with leading velocity v_{max} and whose rear has a trailing velocity v_{min}. Consequently, the extent of the packet grows linearly in time.

If we want to consider more complicated problems, e.g. adding higher order spatial derivatives to (2.1) the combinatorial method presented above quickly becomes very cumbersome. We shall therefore give a more direct method based on the dispersion relation for the fields [9]. In fact the exponent $\lambda(v)$ turns out to be the Legendre transform of the analytical continuation of the dispersion relation.

Let us, for simplicity, return to (2.1) with $f(x)=ax$. Since we are looking for exponentially growing solutions we shall write $u_n(x)$ as

$$u_n(x) \sim \int d\alpha\, e^{nS(\alpha)+x\alpha} \tag{2.8}$$

and by using (2.1) we see that

$$S(\alpha) = \log[a(1-c)+ace^{-\alpha}] \tag{2.9}$$

In a reference frame moving with velocity v, we find

$$u_n(nv) \sim \int d\alpha\, e^{n(S(\alpha)+\alpha v)} \tag{2.10}$$

and by standard saddle point arguments the dominant exponent is found by choosing $\alpha=\alpha(v)$ where

$$S'(\alpha) = -v \tag{2.11}$$

and taking

$$\lambda(v) = S(\alpha(v))+\alpha(v)v \tag{2.12}$$

Further

$$\lambda'(v) = \alpha \tag{2.13}$$

so $S(\alpha)$ and $\lambda(v)$ are Legendre transform pairs and the maximal exponent is found, when v satisfies $\alpha(v)=0$. For the example at hand we get

$$v = -S'(\alpha) = \frac{ace^{-\alpha}}{a(1-c)+ace^{-\alpha}} \tag{2.14}$$

which gives

$$S(\alpha(v)) = \log(\frac{a(1-c)}{1-v}) \tag{2.15}$$

corresponding to (2.7) for $\lambda(v)$. This procedure can easily be used for more complicated coupled map systems or partial differential equations [9].

We shall now show a simple example of a coupled map lattice generating localized turbulent spots which move down stream [9]. Consider the two-dimensional lattice

$$u_{n+1}(x,y) = f(u_n(x,y))-c(f(u_n(x,y))-f(u_n(x-1,y)))+v\Delta f(u_n(x,y)) \tag{2.16}$$

where

$$\Delta f(u_n(x,y)) = f(u_n(x+1,y))+f(u_n(x-1,y))+f(u_n(x,y+1))+f(u_n(x,y-1))-4f(u_n(x,y)) \tag{2.17}$$

and where f is the piecewise linear map

$$f(x) = \begin{cases} 3x+2 & \text{if } -1 \leq x \leq -1/3 \\ ax & \text{if } -1/3 \leq x \leq 1/3 \\ 3x-2 & \text{if } 1/3 \leq x \leq 1 \end{cases} \qquad (2.18)$$

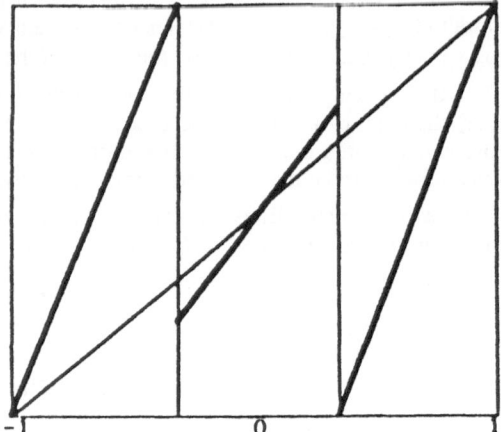

as shown in fig.4. The constants a and c are adjusted such that $a > 1$ whereas $(1-c)a < 1$ whereby the system is absolutely stable, but convectively unstable when the left edge is fixed at the unstable fixpoint: $u_n(0,y)=0$ for all y and n. Fig.5 shows the evolution for a=1.5, c=0.5 and v=0.1, when the system is perturbed slightly at n=x=y=0. An initial disturbance, say 10^{-6}, grows downstream and finally the field enters the region outside of the interval [-1/3,1/3], where it can undergo chaotic motion until it gets rein-jected into the "laminar" region [-1/3,1/3]. The velocities v_{min} and v_{max} fit nicely with the predictions from linear stability theory as sketched above.

Fig.4. Graph of the map given by (2.18).

POWER LAWS IN A COUPLED MAP LATTICE

The famous Kolmogorov theory of fully developed turbulence tells us that the velocity fluctua-tions in a given region grows with the size of the region to the power 1/3. Thus fluctuations at very large lengthscales become very large. This is reminiscent of the behaviour of noisy interphases between two media: when they are driven, they generically become *rough* which again means that the width grows with some power of the system size.

We shall now construct a coupled map lattice which is turbulent *and* at the same time rough, having power law behaviour of the spatial power spectrum [10]. Consider a one-dimensional chain indexed by $i = 1,2,...,L$. On each site we have a variable $u_n(i)$ where n is the "discrete time" and we assume periodic boundary conditions. The dynamics is given by

$$u_{n+1}(i) = u_n(i) + f(u_n(i) - [u_n(i)]) + \frac{\varepsilon}{2}(u_n(i+1) + u_n(i-1) - 2u_n(i)) \qquad (3.1)$$

The square brackets denote the integer part and the function f is again a nonlinear map like the logis-tic: $f(x) = Rx(1-x)$.

Fig.6 shows a chain of size L=1000, with R=10 and ε=0.4, after 250.000 iterates starting from a noisy but almost flat state. Since the maps mostly push upward the whole chain is moving up with some velocity, in the figure the mean displacement is subtracted, and it is seen that the chain indeed does become rough. The width grows with time but seems to saturate at a value, which is proportional to L^2, although the fluctuations, as seen in fig.7, remain of the same order of magnitude as the width itself. The spatial power spectrum of the final state is shown in fig.8. It is clearly a power law, with slope close to -2 which would result from the continuum limit of (3.1) with uncorrelated, white noise in stead of the chaotic map f, i.e. the Langevin equation

$$\dot{u} = \varepsilon \nabla^2 u + \xi \qquad (3.2)$$

where $<\xi(x,t)\xi(x',t')> = \Gamma\delta(x-x')\delta(t-t')$. The state has many positive Lyapunov exponents, theire number presumably growing as L. The largest one is 1.54 for the state shown. One should note that the local maps, f, again do not necessarily have chaotic attractors. They can be "mode-locked" and the resulting state can still look like fig.6 and have positive Lyapunov exponents.

Fig.5. Evolution of a turbulent spot. The initial disturbance had amplitude 10^{-6} and was applied to the center of the left-hand boundary. The plots at the left were done at times t = 191,193,197,201,205,219,229,248 and 260. The plots at the right, at t = 192,193,195,200,220,240 and 260. The parameters were a = 1.5, c = 0.5 and ν = 0.1.

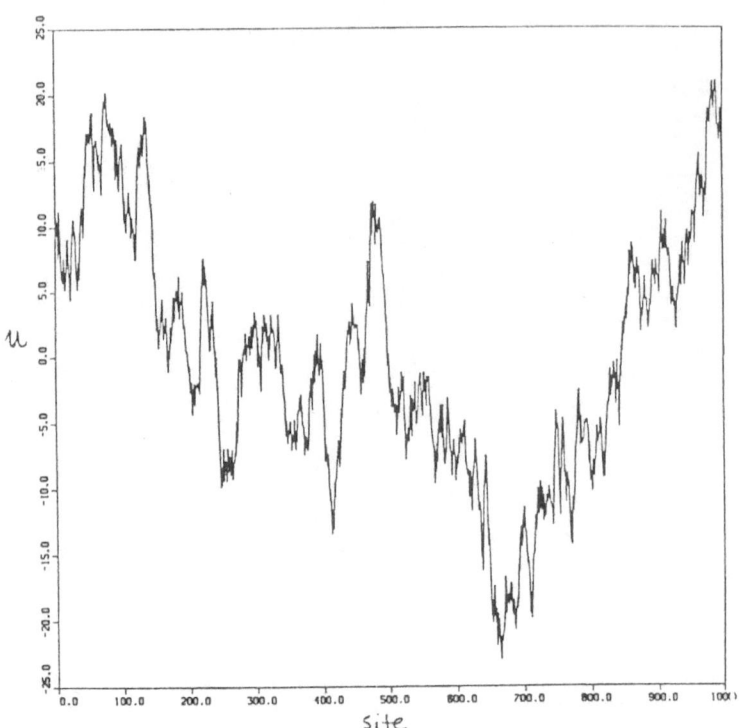

Fig.6. Turbulent "interface" state. R=10.0, ε=0.4. The state is shown after 250.000 iterates.

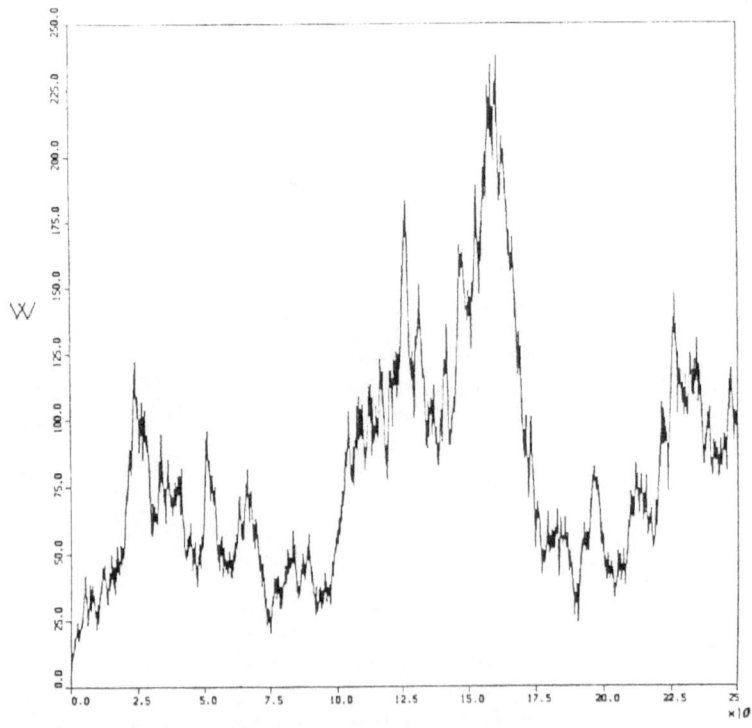

Fig.7. Width of the interface as function of time. The width is plotted for every hundred timesteps for 250.000 timesteps. Parameters as Fig.6.

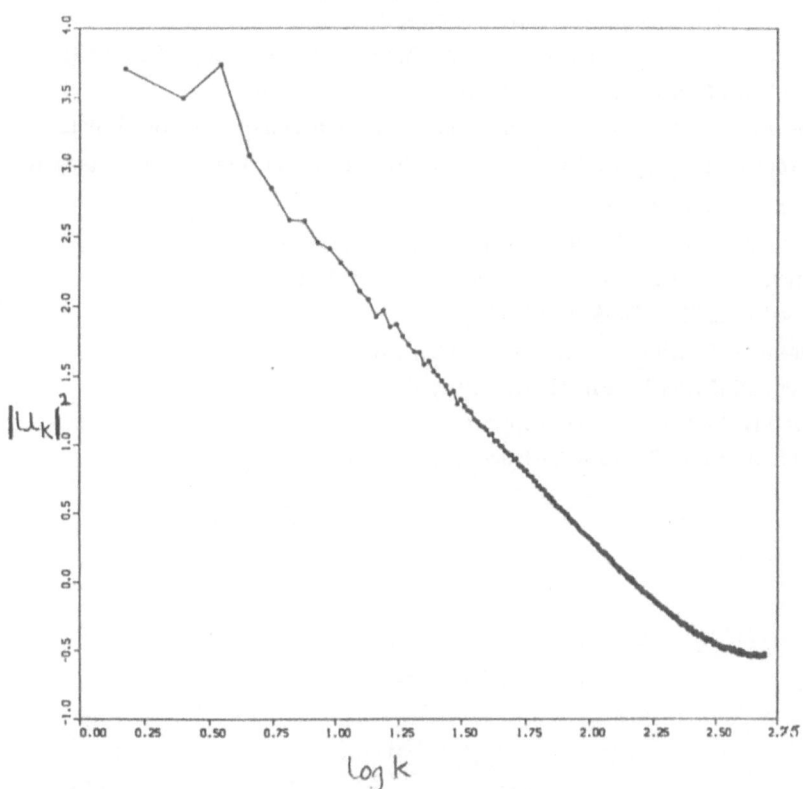

Fig.8. Spatial power spectrum for (3.1). The square of the Fourier component was averaged over the last 50.000 iterates. Parameters as in fig.6.

AKNOWLEDGEMENTS

It is a pleasure to aknowledge my debt to the collaborators in the work reviewed above. I thus thank Ole Bøssing Christensen, David Rand, Geoff Grinstein, C. Jayaprakash and Mogens Høgh Jensen for many happy hours of discussions. I would like to aknowledge support from Novo Nordisk through a Hallas-Møller scholarship.

REFERENCES

1. K.Kaneko, Prog.Theor.Phys. **74,** 1033 (1985).

   J.P.Crutchfield and K.Kaneko, "Phenomenology of Spatio-Temporal Chaos" in "Directions in Chaos" vol.1, ed. Hao Bai-Lin (World Scientific 1987).

2. T.Bohr, G.Grinstein, Yu He and C.Jayaprakash, Phys.Rev.Lett. **58,** 2155 (1987).

3. T.Bohr and O.B.Christensen, Phys.Rev.Lett. **63,** 2161 (1989).

4. Note that small wavelength instabilities can occur if ε becomes too large. It is easy to check that $|1-\varepsilon+\frac{\varepsilon}{2}(\cos q_1+\cos q_2)| \leq 1$ for all q_1, q_2 as long as $0\leq\varepsilon\leq1$. For $\varepsilon>1$ the "staggered" state with $q_1=q_2=\pi$ becomes unstable.

5. J.-P.Eckmann and D.Ruelle, Rev.Mod.Phys **67,** 617 (1985).

6. S.P.Kuznetsov and A.S.Pikovsky, Physica **19D,** 384 (1986).

   A.S.Pikovsky, Z. Phys. B **55,** 149 (1984).

7. F.Kaspar and H.G.Schuster, Phys.Lett. A **113,** 451 (1986).

8. R.Deissler and K.Kaneko, Phys.Lett. **119A,** 397 (1987)

9. T.Bohr and D.Rand, Physica **D** to appear.

10 T.Bohr, G.Grinstein, C.Jayaprakash and M.H.Jensen, unpublished.

WEAK AND STRONG TURBULENCE IN A CLASS

OF COMPLEX GINZBURG LANDAU EQUATIONS

J.D. Gibbon

Department of Mathematics
Imperial College, London SW7 2BZ UK

1 INTRODUCTION

Engineers tend to be interested mainly in results on real fluids and therefore have less interest in theoretical models other than the Navier Stokes equations. In contrast, physicists and applied mathematicians are more interested in the underlying mechanisms which cause and govern turbulence. This sometimes leads them to study idealised systems which are not wholly physical but which give a degree of insight into the mechanisms behind it. Only in 2 spatial dimensions are the incompressible Navier Stokes equations understood analytically in the sense that there is a rigorous proof of the existence of a finite dimensional global attractor. On a finite periodic domain, if G is the Grashof number, then it turns out[1,2,3] that the dimension of the global attractor for the 2D Navier Stokes equations is bounded above by $cG^{2/3}(1 + \log G)^{1/3}$ & below by $cG^{2/3}$. Computational methods[4] are generally good enough to resolve the smallest scale in a 2D flow and, for 2D homogeneous decaying turbulence, the vorticity obeys a maximum principle. No such maximum principle is known to exist in 3D & regularity remains to be proved in this case. Furthermore, numerical resolution of the smallest scale in a 3D flow is still a long way off.

To most fluid dynamicists, the only real turbulence is the fine scale 3-dimensional turbulence which occurs at high Reynolds numbers, with an energy cascade and an inertial subrange. Fully developed turbulence is a much more complicated affair than the low dimensional behaviour we have come to know in such phenomena as convection in a box[5]. In the former case the number of degrees of freedom in 3D turbulence is undoubtedly many orders of magnitude greater than in the latter where perhaps only a few spatial modes govern the dynamics. Both come under the heading of 'turbulence' but attempts to equate the two types of phenomena have led to much confusion. At the heart of the problem concerning the 3D Navier Stokes equations lies the question of finite time singularities in the Euler equations. These are the inviscid limit of the Navier Stokes equations and it is generally believed that, starting from smooth initial data, the 2D Euler equations do not blow up in finite time while the 3D equations do. However, there is no proof of either of these conjectures[6].

Because the 3D Navier Stokes equations are both analytically and computationally beyond us at present, it is a useful exercise to see whether it is possible to mimic some limited set of features of these with

The Global Geometry of Turbulence
Edited by J. Jiménez, Plenum Press, New York, 1991

an abstract PDE system which displays similar functional properties but in a lower spatial dimension. The motivation for this exercise is to get a better grip on the mechanisms involved although it must obviously be limited by the fact that models in lower spatial dimensions cannot display the vorticity properties possessed by the 3D Navier Stokes equations. Nevertheless, the lowering of the dimension must obviously make it easier to compute the dynamics.

Specifically we would like the PDE to have certain special properties. One ideal property would be to show the existence of a finite time singularity in the inviscid limit. Another would be to show that there exists a transition from spiky turbulence to a much weaker form of turbulence as dissipation increases and that where strong spiky turbulence occurs there would be a consequent cascade of energy to high k. It would also be nice if there was the possibility that these phenomena could occur in 2D or even 1D, thereby making numerical studies much easier. The existence of an inertial subrange in the strong region would also be essential.

What we have loosely called strong (hard) and weak (soft) turbulent behaviour needs some explanation. We define strong or hard turbulence to be a phenomenon where large fluctuations occur away from time and space averages. If this occurs, then solutions must be spatially and temporally narrow (i.e. spiky) with a consequent cascade of energy to high k. The spiking of the solutions would be caused by the singularity in the inviscid limit trying to force the solution to blow up, only for the dissipation to finally control the system & bring it back down again. In contrast, weak or soft turbulence would display no great fluctuations away from spatial and temporal averages. In such a system, one would naturally expect strong turbulent behaviour to occur near the inviscid limit where dissipation is small and the weak turbulent behaviour where dissipation is stronger.

One equation which we will show has the desired properties is a version of the complex Ginzburg Landau equation (CGL) in D spatial dimensions[7,8]

$$A_t = RA + (1+i\nu)\Delta A - (1+i\mu)A|A|^{2q} \tag{1.1}$$

where we use periodic boundary conditions on the domain [0,1]. R, μ and ν are real parameters with $R \geq 0$ and μ and ν of either sign. It is not our intention to treat it in its physical context where it occurs as a weakly nonlinear amplitude equation: indeed we shall divorce it from that entirely and treat it purely as an abstract model. In this sense, we do not pretend that it describes true fluid turbulence. Our intention in using it is to try and mimic certain features of the Navier Stokes equations with an equation over which we have slightly more control.

We note straight away that the first condition is fulfilled in the case of the CGL equation. The inviscid limit ($R \to \infty$, $|\nu|, |\mu| \to \infty$), produces the NLS equation in D dimensions

$$0 = iA_t + \nu\Delta A - \mu A|A|^{2q} \tag{1.2}$$

which is well known to blow up in finite time[9] provided $qD \geq 2$ and provided the energy,

$$E = \int [\,|\nabla A|^2 + \frac{\mu}{2\nu}|A|^{2(q+1)}] \tag{1.3}$$

is negative. Since E is a constant of the motion, this can be set negative by suitably chosen initial data, provided μ and ν are of opposite sign. It is particularly the so-called critical case $Dq = 2$ which interests us here because it allows us to to look at the $D = 2$ (with $q = 1$) and the $D = 1$ (with $q = 2$) cases together. When it is super-critical, for instance when $qD = 3$, then the $D = 1$, $q = 3$ case behaves in much the same way as the $D = 3$, $q = 1$ case. Consequently, we have a 1D system which will, to some limited degree, mimic the behaviour of a 3D system although it is also the criticality or supercriticality of the system which is as important as the dimension itself. This paper is a summary of the results in the papers by Bartuccelli, Constantin, Doering, Gibbon & Gisselfält[10] & Doering, Gibbon & Levermore[11].

2 A LATTICE THEOREM FOR THE CGL EQUATION

The CGL equation has a natural symmetry scaling: if A scales like λ^{-1} then ∇ scales like λ^{-q}. This motivates us to choose functionals whose terms are of equal weight[10,11]

$$F_{n,m} = \int [|\nabla^{n-1}A|^{2m} + \alpha_{nm} |A|^{2mn(q)}] \, d\underline{x} \tag{2.1}$$

where $\alpha_{nm} > 0$ for all m, $n \geq 1$ with $n(q) = q(n-1)+1$. The $F_{n,m}$ can be thought of as defining a 'lattice' where one steps up in n & along in m & is a generalization of the ladder idea described previously[10]. Here we state but do not prove[11] a series of theorems which will be useful in determining certain properties of the dynamics. Firstly, let us define

$$\beta_{m,\nu} = m - |1 + i\nu|(m - 1) \tag{2.2}$$

which appears in the following theorem:

THEOREM 1 :

For $m,n \geq 1$ and provided m and ν are chosen such that $\beta_{m,\nu} > 0$, then

$$\dot{F}_{n,m} \leq [2mn(q)R + c_{nm}\|A\|_\infty^{2q}]F_{n,m} - b_{nm}F_{n,m}^{1+1/m}/F_{n-1,m}^{1/m} \tag{2.3a}$$

where $c_{1,1} = 0$ and

$$b_{nm} = 2m\left[\tfrac{1}{2} \min \left\{ \frac{\beta_{m,\nu}}{(2m-1)^2} ; \frac{\alpha_{n-1,m}}{\alpha_{n,m}} \right\} \right]^{1/m} \qquad \blacksquare \tag{2.3b}$$

To close the hierarchy of the $F_{n,m}$ we must deal with the $\|A\|_\infty^2$ term.

THEOREM 2

$$\|A\|_\infty^2 \leq c \left\{ F_{n,m}^{\frac{2}{[2n(q)m-Dq]}} + F_{n,m}^{\frac{1}{n(q)m}} \right\} \tag{2.4}$$

provided $n > 1 + D/2m$. $\blacksquare$

Putting together Theorems 1 and 2, we obtain

THEOREM 3

$$\overline{\lim_{t \to \infty}} \; F_{n,m} \;\; \leq \; c \; \left[\; \overline{\lim_{t \to \infty}} \; F_{n-1,m} \; \right]^{\frac{[2mn(q)-Dq]}{[2mn(q)-Dq-2mq]}} \tag{2.5a}$$

provided

$$n > 1 + D/2m + (q-1)/q. \tag{2.5b}$$

We now return to (2.5b) which places a condition on the starting point of the lattice. Rewriting it as

$$n > 2 + (Dq - 2m)/2mq \tag{2.6}$$

we can see that we are able to choose $n = 2$ provided

$$Dq < 2m \tag{2.7}$$

With $n = 2$ the starting point of the lattice ($F_{n-1,m}$) is

$$F_{1,m} = \int |A|^{2m} \tag{2.8}$$

This motivates us to prove

THEOREM 4

$$\overline{\lim_{t \to \infty}} \int |A|^{2(1+\eta)} \leq R^{(1+\eta)/q} \qquad\qquad \eta \geq 0 \tag{2.9a}$$

where

$$|1 + i\nu| \leq 1 + 1/\eta \qquad\qquad \blacksquare \tag{2.9b}$$

We illustrate how (2.9) and (2.6) help with 3 special cases:

(a) $Dq = 1$ Choose $m = 1$ in (2.6) and the starting point is $F_{1,1}$. This is the L^2 norm which is bounded above for large times. Hence we have control over the whole ladder for $m = 1$. This case includes the standard 1D $q = 1$ CGL equation.

(b) $Dq = 2$ Take $m = 1 + \eta$ (for any $\eta > 0$) in (2.6) & the starting point is $F_{1,1+\eta}$ which is given by Theorem 4 with the choice of η depending on the value of ν.

(c) $Dq = 3$ The use of Theorem 4 is valid provided $m > 3/2$ (or $\eta > 1/2$). Equation (2.6) means that this is only valid in the region $|\nu| < \sqrt{8}$.

Case (c) simultaneously includes the $D = 3$, $q = 1$ case & the $D = 1$ $q = 3$ case. We therefore have control over the rungs of the ladder only in the region $|\nu| < \sqrt{8}$ for all μ. Outside of this region we have only the result of Theorem 4 which is not sufficient. In this region, control over the H^1-norm is necessary which we do not have *a priori*.

3 ESTIMATES AND TIME AVERAGES IN THE μ-ν PLANE

Let us begin by considering the quantity

$$F_{2,1} = \int [\, |\nabla A|^2 + \alpha_{21} |A|^{2(q+1)}] \tag{3.1}$$

We can easily find an upper bound on its time average by taking the CGL equation, multiplying by A*, taking the real part, integrating over space & then taking the time average. We get

$$<F_{2,1}> \leq cR <\|A\|_2^2> \leq c R^{1+1/q} \qquad (3.2a)$$

where the time average <.> is defined as

$$<f(t)> \leq \lim_{t\to\infty} \limsup_{f(0)} \frac{1}{t} \int_0^t f(s)ds \qquad (3.2b)$$

We note that the upper bound on $<F_{2,1}>$ in (3.2a) is true for all μ and ν. In the critical case $qD = 2$, computing $\overline{\lim}_{t\to\infty} F_{2,1}$ from (2.5a) & (2.5b) gives

$$\overline{\lim_{t\to\infty}} F_{2,1} \leq c R^{1/\eta + (q+1)/q} = c R^{|\nu|+(q+1)/q} \qquad (3.3a)$$

Furthermore, when $D = 2$ $q = 1$, for example, the pointwise bound on A is

$$\overline{\lim_{t\to\infty}} \|A\|_\infty^2 \leq c R^{1+1/\eta} = c R^{|\nu|+1} \qquad (3.3b)$$

Comparing the large time limsup of $F_{2,1}$ with its time average we see that the former can be *many* orders of magnitude greater than the latter if $|\nu| \gg 1$. This is precisely the behaviour for which we are looking in the inviscid limit. A comparison of the L^∞ norm with the L^2 norm produces the same conclusion.

These two comparisons show that for initial data which achieves, or nearly achieves, the suprema, then the large excursions away from temporal & spatial averages which can occur must be in the form of temporally and spatially narrow spikes. An obvious physical explanation for this is that the finite time singularity in the inviscid limit tries to make the solution blow up. The amplitude and gradients increase and the solution grows and narrows into a spike but finally the dissipative terms catch up with it and bring it back down. The spike must be temporally and spatially narrow to avoid violating the bounds on the averages.

The estimates found in (3.3) can be improved considerably in certain areas of the μ-ν plane: indeed, the ν-dependence of the exponent can be removed in these areas and the upper bounds brought down to be much closer to the temporal and spatial averages.

To achieve this, it is necessary to be much more precise with the quantity $F_{2,1}$ and not pull out the amplitude in L^∞ as in the lattice theorem. A careful and technical procedure[10] allows us to discard certain combinations of the nonlinear and the Laplacian terms which turn out to be negative definite. This, however, can only be done in a certain area of the μ-ν plane (away from the inviscid limit). We restrict ourselves to the case $q = 1$ for simplicity. The following theorem is stated and proved elsewhere[10]

THEOREM 5
        If μ and ν lie in the unshaded region of the μ-ν plane (see Figures 1 and 2); that is, if

313

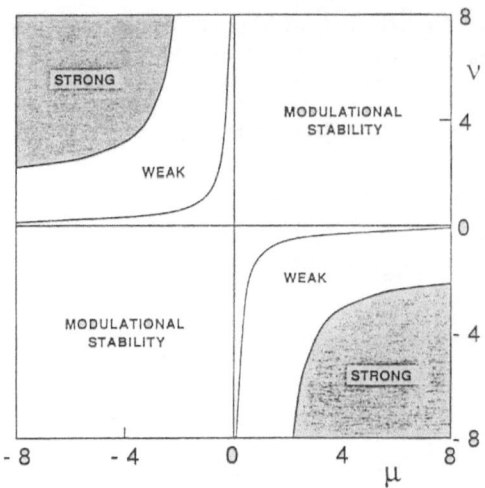

Figure 1. This summarises the behaviour of the q = 1, D = 2 CGL equation in the μ-ν plane. The label "modulational stability" inside the inner hyperbolae refers to the Lange and Newell criterion[13] $\varepsilon = 1 + \mu\nu$ where the spatially homogeneous solution is stable (unstable) when $\varepsilon > 0$ (< 0)[10,13]. The hyperbolae dividing the unshaded (weak) from the shaded (strong) regions are those given in Theorem 5 (see also reference 12). For D = 1 & q = 2, the results are essentially the same except the hyperbolae are shifted slightly. Estimates in the 'weak' region predict only small differences between time averaged & uniform norms indicating only weak departure from space & time averages. In the 'strong' region, large deviations away from averages are possible, indicating spatially & temporally localised intermittent spiking.

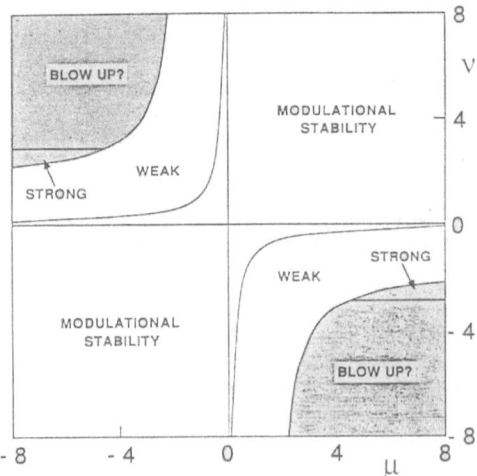

Figure 2. Behaviour of the CGL equation for D = 3 and q = 1. The designation of weak and strong areas is the same as in Figure 1 except that the bounds in the strong region are valid only for $\nu < |8|$ which goes out into the shaded region as far as the anvil shaped areas marked in the figure.

$$\nu > - \frac{\mu\sqrt{3}}{\mu - \sqrt{3}} \qquad\qquad \mu > \sqrt{3} \qquad\qquad\qquad (3.4a)$$

$$\nu < \frac{\mu\sqrt{3}}{\mu + \sqrt{3}} \qquad\qquad \mu < - \sqrt{3} \qquad\qquad\qquad (3.4b)$$

$or \qquad\qquad |\mu| < \sqrt{3} \qquad\qquad\qquad\qquad\qquad (3.4c)$

then coefficients $\alpha_{2,1}$ *and* b_2 *exist such that for every D*

$$\dot{F}_{2,1} \leq 4R\, F_{2,1} - b_2 F_{2,1}^2 /R \qquad\qquad \blacksquare \qquad (3.4d)$$

We conclude from (3.4) that in the unshaded region there is a ball

$$\overline{\lim_{t \to \infty}} F_{2,1} \leq c\, R^2 \qquad\qquad\qquad\qquad (3.5a)$$

$$\overline{\lim_{t \to \infty}} \|A\|_\infty^2 \leq c\, R^{4/(4-D)} \qquad\qquad\qquad (3.5b)$$

Upper bounds on all the other $F_{n,1}$ can be computed[10]. The upper bound on $\overline{\lim_{t \to \infty}} F_{2,1}$ compares favourably with the R^2 upper bound on $<F_{2,1}>$ as does the pointwise bound of A compared to the L^2 bound. Clearly, therefore, no large excursions away from averages can occur and spiky turbulence cannot happen. At most we would expect 'weak' turbulence where no cascade of energy or inertial range exists.

4   CONCLUSION

It is clear that the spiky region requires careful numerical handling because the length scales are much shorter than in the weak region. The occurrence of spikes is unusual in that it contrasts strongly with the dynamics of the usual 1D CGL equation (q = 1) on which most numerical studies have been made[14,15], although there have been limited integrations in the 2D cubic case[16]. Indeed, the lattice gives some lower bounds on the minimum scale in the flow. The CGL equation scales in such a way that we can construct a set of typical length scales $\lambda = \lambda(n,m)$

$$\lambda^{-2m} \equiv <F_{n,m}/F_{n-1,m}> \qquad\qquad\qquad (4.1)$$

From Theorem 1 we can find an upper bound on this directly from the lattice inequality

$$\lambda^{-2} \leq c_{n,m} <\|A\|_\infty^2> + 2mn(q)R \qquad\qquad (4.2)$$

In the weak or soft regime (unshaded region of Figures 1 & 2), Theorem 5 tells us that the constant in (4.2) is zero and so we have the minimum scale as $O(R^{-1/2})$. In contrast, in the strong or hard region we must keep the L^∞ term. The time average can be replaced by a limsup in the limit of large t but we then find that the exponent is $|\nu|$-dependent. Hence the minimum scale in terms of λ is *many* orders of magnitude smaller in the strong region than in the weak region. This is to be expected as most of the energy packs into very short scales on the order of the width of the narrowest spike.

Conventionally, the minimum scale of a system of volume L^D in D spatial dimensions is taken to be the dissipation length ℓ which is related to the number of degrees of freedom N by $N \approx (L/\ell)^D$. One would expect that ℓ would be approximately the same as the natural scale λ with N identified as the attractor dimension[7,8,19]. This appears to be the case in the weak region[10] as N is such that $\ell \geq cR^{-1/2}$. However, in the strong region this

315

is clearly not the case as $\ell \gg \lambda$. This discrepancy arises because the time average procedure in the calculation of the attractor dimension averages out the short scale motion. The idea that ℓ represents some smallest scale in the flow is therefore not true in the strong region.

In terms of the properties mentioned in §1, we have achieved most of those required. Indeed, the requirement that there exists a transition between the weak and strong regimes is precisely what is observed in gaseous helium convection experiments[20]. The property that we require that turbulent behaviour should exist in lower spatial dimensions is also fulfilled in the sense that the critical case $qD = 2$ contains the $D = 1$ $q = 2$ and the $D = 2$ and $q = 1$ cases simultaneously. Computing a 1D fifth order parabolic PDE is much easier than a 2D one. In the weak region and within the area $|\nu| < \sqrt{8}$ the $D = 1$ $q = 3$ and $D = 3$ $q = 1$ cases are equivalent. We have control in L^p only up to L^3 and this allows us to prove regularity for $|\nu| < \sqrt{8}$ but not beyond. Outside this region when $|\nu| \geq \sqrt{8}$, we need control over the H^1-norm which is not yet available. In fact, this condition is worse than that required for 3D Navier Stokes where control over L^3 of the velocity alone is sufficient to prove regularity [3,4]. However, it has been shown[11] that weak solutions exist for all ν when $D = 3$. One further condition we need is the existence of an inertial subrange. It turns out that preliminary work[21] indicates that an inertial range does exist for the $D = 1$ $q = 2$ case for R taken up to 1000 & $\nu = 30$, $\mu = -30$.

REFERENCES

1. A. Libchaber: Proc. Royal Soc. London 413A, 63, (1987) and "From chaos to turbulence in Benard convection", published in "Dynamical chaos", p63, Princeton University Press 1987.
2. P. Constantin, C. Foias & R. Temam; Physica 30D, 284, (1988).
3. P. Constantin, C. Foias; The Navier Stokes equations, Chicago University Press (1989).
4. R. Temam; Infinite dimensional dynamical systems in mechanics & physics. Springer Applied Math Series 68, (1988).
5. N. O. Weiss; Proc. Royal Soc. London 413A, 71, (1987) and "From chaos to turbulence in Benard convection", published in "Dynamical chaos", p71, Princeton University Press (1987).
6. A. Majda; Comm. Pure & Appl. Math. 39, 187-220, (1986).
7. C. Doering, J.D. Gibbon, D.D. Holm & B. Nicolaenko; Nonlinearity 1, 279-309, (1988).
8. C. R. Doering, J. D. Gibbon, D. D. Holm & B. Nicolaenko; Phys. Rev. Letts. 59, 2911-4, (1987).
9. M. J. Landman, G. C. Papanicalou, C. Sulem & P. L. Sulem; Phys. Rev. A38, 3837-43, (1988). K. Rypdal, I Rasmussen & K. Thomsen; Physica 16D, 339, (1985). I. Rasmussen & K. Rypdal; Phys. Scr. 33, 481, (1986). K. Rypdal & I. Rasmussen; Phys. Scr. 33, 498, (1986).
10. M. Bartuccelli, P. Constantin, C. R.Doering, J. D. Gibbon & M. Gisselfält; Physica D, 44, 421-444, (1990) and Phys. Letts. 142A, 349-56, (1989).
11. C. D. Doering, J. D. Gibbon & D. Levermore: "Weak and strong solutions of the CGL equation" preprint (1990).
12. C. D. Levermore; private communication.
13. C. Lange & A. C. Newell; SIAM J. Appl. Math. 27, 441-56, (1974).
14. L. Keefe; Stud. Appl. Math., 73, 91, (1985).
15. L. Sirovich & J. Rodriguez; Phys. Lett. A120, 211, (1988).
16. P. Coullet, L. Gil & J. Lega; Phys Rev Letts, 62, 1619-22, (1989).
17. L. Sirovich & P. K. Newton; Physica 21D, 115,25, (1986).
18. P. Constantin; "A construction of inertial manifolds"; Contemporary Mathematics 99, 27, (1989) (Proc.AMS Summer School, Boulder 1987).

19. J-M Ghidaglia & B. Heron; <u>28D</u>, 282, (1987).

20. F. Heslot, B. Castaing & A. Libchaber; Phys. Rev. <u>A36,</u> 5870-3, (1987).

21. C. D. Doering, J. D. Gibbon, D. D. Holm, J. M. Hyman, C. D. Levermore & G. Kovacic; in preparation 1990.

PATTERN FORMATION, EXTERNAL FIELDS AND FLUCTUATIONS

D. WALGRAEF†

Service de Chimie-Physique, Université Libre de Bruxelles
B-1050, Brussels, Belgium

ABSTRACT

Since the nucleation of spatio-temporal patterns in nonequilibrium systems is usually associated with continuous symmetry breakings, their selection and stability properties may be strongly affected by fluctuations or small external fields. These effects are discussed in the framework of amplitude equations of the Ginzburg-Landau type. In particular, it is shown that the presence of fluctuations modify qualitatively the bifurcation diagram and suppress any true long range order in low dimensional systems.

INTRODUCTION

The nucleation of spatio-temporal patterns in systems driven far from thermal equilibrium by external constraints remains the subject of intensive experimental and theoretical research [1-2]. It raises fundamental issues related to the spontaneous emergence of order in such diverse contexts as mechanics, hydrodynamics, chemistry, biophysics, nonlinear optics, geology, materials science and solid state physics. A general framework for the study of these phenomena has now emerged, based on bifurcation theory and normal forms, leading to amplitude equations for the spatio-temporal patterns which are analogous to the time dependent Ginzburg-Landau description of phase transitions [3-4].

Beyond many pattern forming instabilities, structures with different symmetries may be simultaneously stable (e.g. rolls and hexagons in non Boussinesq convection and ferro-fluids[5-7]; rolls and squares or bimodal structures in liquid crystal instabilities [8], cubic and planar defect microstructures in irradiated metals and alloys [9], etc.). The determination of parameter ranges for which these structures are stable and the study of the transitions and of the transition mechanisms between these structures are thus particularly interesting. Furthermore, since these instabilities are usually associated with continuous symmetry breakings, they are very sensitive to small external fields or fluctuations. As in equilibrium systems, Goldstone modes are present [10] and the absence of true long range order should also be observed in low dimensional systems [11]. Furthermore, as a result of the presence of fluctuations in

---

† Senior Research Associate, National Fund for Scientific Research (Belgium).

The Global Geometry of Turbulence
Edited by J. Jiménez, Plenum Press, New York, 1991

the system, supercritical bifurcations should in fact be slightly subcritical as shown below. Up to now, in contrasr with the theoretical analysis of these phenomena [11-13], their experimental observation has been scarce [14] and it was currently believed that fluctuation effects should be too small in non-equilibrium systems to practically affect pattern formation and stability. However, recent experimental observations of pre-transitional fluctuations in the electro-hydrodynamical instability of nematic liquid crystal may induce a renewed interest for these problems [15].

Hence, the aim of this paper is to show how small external fields may affect the selection mechanisms of nonequilibrium structuresbut also to discuss the role of fluctuations on pattern forming instabilities. Three specific examples will be discussed: the effect of imposed flows on pattern forming instabilities in convective or reaction-diffusion systems; the effect of spatial modulations on waves induced by Hopf bifurcations where strong resonances may occur between the unstable modes and fields which do not have the same symmetry, and finally, the effect of fluctuations on instabilities leading to the formation of steady spatial patterns.

FLOW FIELD EFFECTS ON PATTERN FORMING INSTABILITIES

Let us briefly discuss in this section, the effect of imposed flow fields associated to materials derivatives on pattern forming transitions. These effects have already been shown to affect the structures arising in convective or chemically active media [16-17] and some of their theoretical aspects have been discussed in the framework of amplitude equations and phase dynamics [18-19].

When an imposed velocity field $\vec{v}$ is preestablished in a medium where the bifurcation parameter λ reaches its critical value, the amplitude of the order parameter-like variable σ is sufficiently small to have no relevant feedback effect on the velocity field. Its evolution may then be described by the following dynamical model:

$$\tau_0 \dot{\sigma}(\vec{r},t) = [\epsilon - d(q_c^2 + \nabla^2)^2]\sigma(\vec{r},t) - v\sigma(\vec{r},t)^2 - u\sigma(\vec{r},t)^3 - \vec{v}(\vec{r})\vec{\nabla}\sigma(\vec{r},t) \qquad (1)$$

When the active medium is restricted to a thin horizontal layer (such that its thickness l is smaller than the critical wavelength λ_c), the concentrations of the active species may be represented as a Fourier sum, and in the case of divergence free boundary conditions, the order parameterlike variable is written as $\sigma(\vec{r},t) = \sum_n \sigma_n(x,y,t)\cos n\pi z/l$ (a similar expansion may of course be performed in the case of other boundary conditions). Since $\pi/l \ll q_c$ the only unstable component of σ corrresponds, in the absence of velocity field, to $n=0$ while the other ones remain stable unless $\epsilon \geq d(n\pi/l)^4$. As a result the only stable patterns correspond, in this case, to rolls, hexagons or triangles in the horizontal plane. In the presence of the imposed flow however, the slow mode dynamics may be rewritten as:

$$\tau_0 \dot{\sigma}_0 = [\epsilon - d(q_c^2 + \nabla_\perp^2)^2]\sigma_0 - v\sigma_0^2 - u\sigma_0^3 - \sum_n (v_n + u_n\sigma_0)\sigma_n^2$$

$$- \sum_n (\vec{v}\vec{\nabla})_{0n}\sigma_n$$

$$\tau_0 \dot{\sigma}_n = [\epsilon - d(q_c^2 - (\frac{n\pi}{l})^2 + \nabla_\perp^2)^2]\sigma_0 - (\vec{v}\vec{\nabla})_{n0}\sigma_0 + \sum_{m \neq 0}(\vec{v}\vec{\nabla})_{nm}\sigma_m$$

$$+ \text{nonlinear terms} \qquad (2)$$

where $\nabla_\perp^2 = \nabla_x^2 + \nabla_y^2$ and $(\vec{v}\vec{\nabla})_{nm} = \int_0^l dz \cos(n\pi z/l)(v_{0x}(\vec{r})\nabla_x + v_{0y}(\vec{r})\nabla_y + v_{0z}(\vec{r})\nabla_z) \cos(m\pi z/l)/\int_0^l dz \cos^2(n\pi z/l)$.

320

As we are considering thin layers, the $n = 0$ mode only remains critical and the other ones may be adiabatically eliminated. One has :

$$\sigma_n \simeq \frac{l^2}{dn^2\pi^2}(\vec{v}\vec{\nabla})_{n0}\sigma_0 \tag{3}$$

and, consequently :

$$\tau_0\dot{\sigma}_0 = [\epsilon - d(q_c^2 + \nabla_\perp^2)^2]\sigma_0 - v\sigma_0^2 - u\sigma_0^3 - \sum_{n\neq0}(v_n + u_n\sigma_0)\frac{l^4}{d^2n^4\pi^4}((\vec{v}\vec{\nabla})_{n0}\sigma_0)^2$$
$$- (\vec{v}\vec{\nabla})_{00}\sigma_0 + \sum_{n\neq0}\frac{l^2}{n^2\pi^2}(\vec{v}\vec{\nabla})_{0n}(\vec{v}\vec{\nabla})_{n0}\sigma_0 \tag{4}$$

Hence the effect of an imposed space-dependent flow field on the patterning instability is twofold. On one side there appears an advection term with a material derivative associated with the mean value of the mean velocity, the mean being taken over the layer thickness. The other effect appears for space-dependent flows and corresponds to an additional anisotropic diffusion term. The influence of these terms on the pattern selection and stability properties of course depend on the characteristics of the imposed flow.

For example, in the case of plane Poiseuille or shear flows in convective systems, $\vec{v} = v_0(z)\vec{1}_x$ with, respectively, $v_0(z) = v_M 4\frac{z}{l}(1 - \frac{z}{l})$ and $v_0(z) = az$. Equation (4) then becomes:

$$\tau_0\dot{\sigma}_0 = [\epsilon - d(q_c^2 + \nabla_\perp^2)^2]\sigma_0 - \bar{v}\sigma_0^2 - \bar{u}\sigma_0^3 - \bar{v}_0\nabla_x\sigma_0 + D_x\nabla_x^2\sigma_0 \tag{5}$$

where the nonlinear couplings $\bar{v}$ and $\bar{u}$ correspond to the renormalization of v and u resulting from the elimination of the noncritical modes and where D_x is proportional to $v_M l^2$ for Poiseuille flows and to $a^2 l^4$ for shear flows, while $\bar{v}_0$ is the vertical mean of the velocity field.

As a consequence of the anisotropic diffusion effect triggered by the flow, the roll patterns which appear first have their wavevector perpendicular to the flow velocity (in the present case $\vec{q} = \vec{q}_c\vec{1}_y$) and their amplitude equation may be written as ($\sigma_0 = Aexpiq_cy + A^*exp - iq_cy$):

$$\tau_0\dot{A} = (\epsilon + D_y\nabla_y^2 + D_x\nabla_x^2)A - \bar{v}_0\nabla_x A - 3\bar{u}|A|^2A \tag{6}$$

The phase dynamics associated to a pattern having a wavenumber q slightly different from critical is then:

$$\tau_0\dot{\phi} = (D_\parallel\nabla_y^2 + D_\perp\nabla_x^2)\phi - \bar{v}_0\nabla_x\phi \tag{7}$$

where

$$D_\parallel = D_y\frac{\epsilon - 3D_y(q - q_c)^2}{\epsilon - D_y(q - q_c)^2}, \quad \text{and} \quad D_\perp = D_x + D_y\frac{q - q_c}{q_c}$$

We see that at the leading order the Eckhaus instability is not affected by the flow while the zig-zag instability is shifted towards smaller values of the wavenumber, and the stability domain of the pattern is increased compared to the isotropic case. Furthermore, if the zig-zag instability triggers modulations along the axis of the rolls, these modulations will propagate due to the presence of the advection term $\bar{v}_0\nabla_x\phi$. As in other symmetry-breaking instabilities the nucleation of defects is expected. From the phase dynamics we see that as a matter of fact dislocations may be triggered by phase singularities and that the advection term should induce a climbing motion for these dislocations.

The patterns with hexagonal symmetry which may appear via a first orderlike transition are also affected by the anisotropic diffusion terms triggered by the flow field and are either deformed or destabilized. On the other hand, if the system is very narrow in the y direction and cannot accomodate rolls with wavevectors in that direction, rolls perpendicular to the flow may still exist but their amplitude is reduced to $|A| =$ and the preferred wavelength becomes $\bar{\lambda} =$. In that case the evolution of phase variations along the flow is given by :

$$\tau_0 \dot{\phi} = D_{\parallel} \nabla_x^2 \phi - \bar{v}_0 (\bar{q} + \nabla_x \phi) \tag{8}$$

We recover here an expression already proposed by Brand [18] and analyzed by Pocheau [16] to interpret his experiment where a through flow induces wavelength variations of Rayleigh-Benard rolls in an annulus. It is also interesting to note that, in this case, the transverse phase diffusion coefficient is negative for rolls of critical wavelength, leading to a bending of these rolls.

The conclusion of this discussion is that the effect of an imposed flow in a thin horizontal layer where pattern forming occur is twofold. Besides the advection term associated to the mean flow, an anisotropic renormalization of the diffusion coefficients is induced by the vertical velocity profile. As a result the pattern selection is modified since roll structures tend to be oriented with the roll axis orthogonal to the flow. Cellular patterns and two-dimensional waves are deformed or even destabilized. In the case of quasi one-dimensional systems, a better physical insight is obtained for the different types of phase dynamical behavior proposed by Brand [18]. On the other hand the above analysis may provide useful tools for the understanding of chemical patterning in fluid solutions where the separation between chemical and hydrodynamical effects is usually not obvious, but also in the deformation of solids where the plastic flow may play a determinent role in the orientation and propagation of slip bands, or other dislocation microstructures but also induce a diffusive behavior for mobile dislocations [20].

SPATIAL FORCING OF WAVE PATTERNS

As shown recently, pure spatial or temporal modulations imposed on Hopf bifurcations leading to wave patterns modify the selection and stability properties of the resulting spatio-temporal structures [21-25]. For example, it was shown that, in one-dimensional systems, beyond spatio-temporal Hopf bifurcations, left and right traveling waves are linearly coupled by uniform oscillations or steady spatial modulations provided the frequency of the oscillations or the wavenumber of the modulations are close to two times the critical ones. Hence, a spatially uniform forcing may restore the left-right symmetry and transform traveling waves into standing waves in regimes where the latter are otherwise unstable.

In two-dimensional systems, new possibilities occur and I will discuss here a particular. Consider an isotropic system invariant under space and time translations and parity transformations which undergoes a Hopf bifurcation with finite wavenumber k_c and frequency ω_c. The growth rate of the unstable modes is written as :

$$\omega(\vec{k}) = \epsilon - \xi_0^2 (k^2 - k_c^2)^2 + i(\omega_c + \omega_1 (k^2 - k_c^2) + \omega_2 (k^2 - k_c^2)^2) + O((k^2 - k_c^2)^3) \tag{9}$$

and the slowly varying envelopes of left and right travelling waves propagating in, say, the x-direction, satisfy, after appropriate scaling, the following amplitude equations [26-28] :

$$\dot{A} + c\nabla_x A = \epsilon A + (1 + i\alpha)[\nabla_x - \frac{i\nabla_y^2}{2k_0}]^2 A + i\eta[\nabla_x^2 + \frac{\nabla_y^2}{\chi}]A - (1 + i\beta)A|A|^2$$
$$- (\gamma + i\delta)A|B|^2$$
$$\dot{B} - c\nabla_x B = \epsilon B + (1 - i\alpha)[\nabla_x + \frac{i\nabla_y^2}{2k_0}]^2 B - i\eta[\nabla_x^2 + \frac{\nabla_y^2}{\chi}]B - (1 - i\beta)B|B|^2$$
$$- (\gamma - i\delta)B|A|^2 \tag{10}$$

When $\gamma > 1$, only travelling are stable, while for $\gamma < 1$ standing waves are the only stable structure, and I will consider the first case in the following. Of course, patterns built on travelling waves which propagate in different directions may also be expected. The stability of these structures also depends on the coefficient of the nonlinear couplings between the different waves. On assuming that this coefficient is scalar and written as $\kappa + i\lambda$ with $\kappa > 1$, we are in a situation where the only stable structure of the system corresponds to waves travelling in one direction.

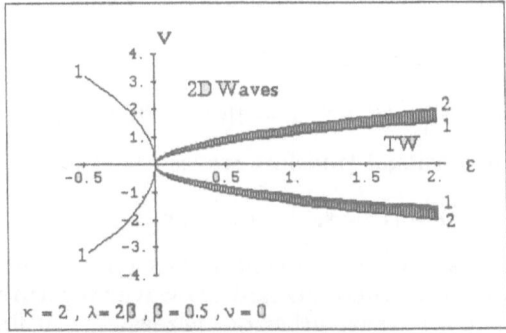

curve 1 : stability limit of 2D wave pattern
curve 2 : stability limit of 1D traveling wave

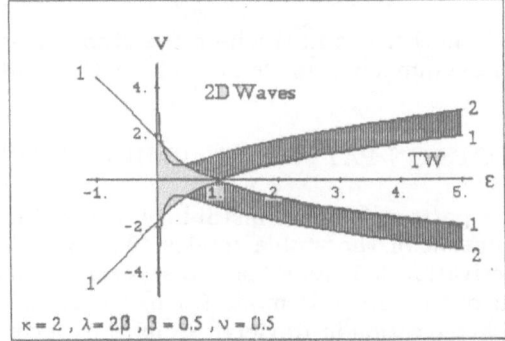

Figure 1 . Phase diagram for a Hopf bifurcation forced by temporal modulations of frequency close to three times the critical frequency, at zero and non zero detuning.

The effect of a purely spatial forcing of hexagonal symmetry has been discussed in [25]. It was shown that, when the wavenumber of the forcing is close to $3k_c$, travelling waves with wavevectors parallel to the basic vectors of the imposed hexagonal modulation are nonlinearly coupled. This leads to a stabilization mechanism for two-dimensional wave patterns. Similar effects may also be obtained with purely temporal forcings. For example, temporal modulations of frequency ω_0 such that $n\omega_0 \neq 3\omega_c$ couple triplets of waves travelling in directions making $2\pi/3$ angles (e.g. $A \exp i(\vec{q}_1\vec{r} - \omega_c t)$, $B \exp i(\vec{q}_2\vec{r} - \omega_c t)$, $C \exp i(\vec{q}_3\vec{r} - \omega_c t)$, with $\vec{q}_1 + \vec{q}_2 + \vec{q}_3 = 0$ and $|\vec{q}_1| = |\vec{q}_1| = |\vec{q}_1| = q_c$) . The corresponding uniform amplitude equations are:

$$\dot{A} = (\epsilon + i\nu)A + \upsilon\bar{B}\bar{C} - (1 + i\beta)A|A|^2 - (\kappa + i\lambda)A(|B|^2 + |C|^2)$$
$$\dot{B} = (\epsilon + i\nu)B + \upsilon\bar{A}\bar{C} - (1 + i\beta)B|B|^2 - (\kappa + i\lambda)B(|A|^2 + |C|^2)$$
$$\dot{C} = (\epsilon + i\nu)C + \upsilon\bar{B}\bar{A} - (1 + i\beta)C|C|^2 - (\kappa + i\lambda)C(|A|^2 + |B|^2) \qquad (11)$$

where $\upsilon \propto h^n$, h being the amplitude of the external modulation, and $\nu = \frac{n\omega_0}{3} - \omega_c$, the frequency detuning between the external field and the waves .

Hence, from the fixed point condition ($A = R \exp i\Phi_A$, $B = R \exp i\Phi_B$, $C = R \exp i\Phi_C$) :

$$[(1 + 2\kappa)^2 + (\beta + 2\lambda)^2]R^4 - [2\epsilon(1 + 2\kappa) + 2\nu(\beta + 2\lambda) + \upsilon^2]R^2 + (\epsilon^2 + \nu^2) = 0 \quad (12)$$

it may be deduced that hexagonal wave patterns exist when

$$\upsilon^4 + 4\upsilon^2[\epsilon(1 + 2\kappa) + \nu(\beta + 2\lambda)] - 4[\nu(1 + 2\kappa) - \epsilon(\beta + 2\lambda)]^2 > 0 \qquad (13)$$

On the other hand, unidimensional traveling waves are stable when

$$\upsilon^2\epsilon - (\nu - \lambda\epsilon)^2 - (\kappa - 1)^2\epsilon^2 > 0 \qquad (14)$$

The corresponding phase diagram is presented in fig.1 for zero and non-zero frequency detuning ν, showing the regions where 2D and 1D wave structures are individually or simultaneously stable. One sees that, either on increasing ϵ at fixed field intensity, or on increaising the field intensity at fixed ϵ, one crosses a region where the only stable structure corresponds to hexagonal wave patterns.

Hence we have here another example where the stabilization of a pattern by an external field of a different symmetry is made possible by the non-variational character of the dynamics.

FLUCTUATION EFFECTS ON PATTERN FORMING INSTABILITIES

In the vicinity of a pattern forming instability, the reduction of the dynamics via the adiabatic elimination of the stable modes [29], via multiple scale analysis [30] or normal forms derivation [31] may lead to several types of dynamical models describing the evolution of the unstable mode (or order parameter-like variable) σ . For example, the onset of convection in Rayleigh-Benard experiments or the formation of spatial patterns in reaction-diffusion systems may be described by the so-called Swift-Hohenberg model which is written as [12]:

$$\tau_0\partial_t\sigma(\vec{r}, t) = [\epsilon - (q_c^2 + \nabla^2)^2]\sigma(\vec{r}, t) - u\sigma^3(\vec{r}, t) \qquad (15)$$

where $\epsilon = (\lambda - \lambda_c)/\lambda_c$, λ being the bifurcation parameter and λ_c its critical value; u may be explicitely computed for any explicit dynamical model. This equation results from a series expansion in powers of σ and limited to the third powers by assuming that we consider regions of the parameter space near the instability where the amplitude of the slow mode remains small (of the order of $\sqrt{\epsilon}$) and that u is positive. If this is not the case, one has of course to consider higher order nonlinearities. Furthermore,

u may also be space dependent, but let us consider it first as a constant, which is the case for most reaction-diffusion systems and several convection problems. In this case, the selected planforms should correspond to steady roll structures defined by :

$$\sigma(\vec{r}) = 2Re\ Aexp(i\vec{q}\vec{r}) \qquad (|\vec{q}| = q_c) \tag{16}$$

and which appear via a supercritical bifurcation :

$$|A| = \begin{cases} 0 & \lambda < \lambda_c \\ \sqrt{\dfrac{\epsilon}{3u}} & \lambda > \lambda_c \end{cases} \tag{17}$$

When u is space-dependent, different types of patterns may be selected [32-33] and when it is negative, one has to consider higher nonlinearities which could favor more complex planforms corresponding for example to quasi-periodic patterns such as Penrose tilings or icosahedral lattices.

The effect of fluctuations on this type of dynamics has been extensively analyzed in the past [10-12], by using the methods developed for the study of equilibrium phase transitions. In particular, it was shown that, near the transition point, critical slowing down and divergence of correlation lengthes should occur, leading to the breakdown of the mean field or deterministic approximation.

The effect of internal fluctuations on the formation of steady state cellular patterns may be studied with the stochastic Swift-Hohenberg equation [34]:

$$\tau_0 \partial_t \sigma(\vec{r}, t) = [\epsilon - (q_0^2 + \nabla^2)^2 + \Delta \nabla_\perp^2] \sigma(\vec{r}, t) - u\sigma^3(\vec{r}, t) + \eta(\vec{r}, t) \tag{18}$$

where η is a Gaussian white noise of intensity Γ and where Δ takes care of an eventual anisotropy in the system. The equation of state of the preferred roll structure may then be written, in the Hartree approximation [10] :

$$0 = \epsilon < \sigma_{q_0} > -u < \sigma_{q_0} >^3 -u < \sigma_{q_0} > \int d\vec{k} < \sigma_{\vec{k}} \sigma_{-\vec{k}} > \tag{19}$$

The correlation function $< \sigma_{\vec{k}} \sigma_{-\vec{k}} >$ may be evaluated at the same approximation through the self-consistent scheme :

$$< \sigma_{\vec{k}} \sigma_{-\vec{k}} > = \frac{\Gamma}{r + d(k^2 - q_0^2)^2 + \Delta k^2 cos^2(\phi)}$$

$$r = -\epsilon + 2u < \sigma_{q_0} >^2 + u \int d\vec{k} < \sigma_{\vec{k}} \sigma_{-\vec{k}} > \tag{20}$$

where ϕ is the angle between the wavevector $\vec{k}$ and the easy axis. As a result, the equation of state may be rewritten as :

$$\epsilon = r_1 + \frac{2u}{\sqrt{d}} \frac{\Gamma}{\sqrt{r_1 + \Delta q_0^2}} K(\frac{\Delta q_0^2}{r_1 + \Delta q_0^2}) \tag{21}$$

where $K(m)$ is the complete elliptic integral of the first kind and with

$$r_1 < \sigma_{q_0} > -u < \sigma_{q_0} >^3 = 0 \tag{22}$$

Therefore the corresponding modulated structure arises with a finite amplitude (i.e. via a *first-order* transition) :

$$< \sigma_{q_0} > = \sqrt{\frac{r_{1c}}{u}}$$

325

at

$$\epsilon_1 = r_{1c} + \frac{2u}{\sqrt{d}} \frac{\Gamma}{\sqrt{r_{1c} + \Delta q_0^2}} K\left(\frac{\Delta q_0^2}{r_{1c} + \Delta q_0^2}\right) \tag{23}$$

where $\epsilon_1 = \epsilon(r_{1c})$ corresponds to the minimum of the curve $\epsilon(r_1)$ defined by equation (23). In the strong anisotropy limit ($\Delta q_0^2 >> r_{1c}$), one has :

$$\epsilon_1 = \frac{3u\Gamma}{\sqrt{d}\Delta q_0^2} ln(\Delta q_0^2) \tag{24a}$$

while in the weak anisotropy limit ($\Delta q_0^2 << r_{1c}$), one has :

$$\epsilon_1 = 3\left(\frac{\pi u\Gamma}{2\sqrt{d}}\right)^{2/3} \tag{24b}$$

Hence fluctuations modify qualitatively the character of the bifurcation, as first shown by Brazovskii in the case of equilibrium phase transitions [35] .

When structures of different symmetries are stable beyond the same supercritical bifurcation, a similar analysis may be performed for each type of structure, and it may be shown that the corresponding effective transition points are different [36]. Hence, the fluctuations also lift the degeneracy of the deterministic bifurcation point.

Due to the symmetry breaking properties of pattern forming instabilities, long range phase fluctuations may develop in all the ordered phase. Since the evolution of amplitude fluctuations is relaxational while the evolution of phase fluctuations is diffusive, the long time behavior of roll patterns with $\vec{q}||\vec{1}_x$ is given, in two-dimensional systems, by its phase dynamics [37]:

$$\tau_0 \partial_t \phi = D_{||} \nabla_x^2 \phi + D_\perp \nabla_y^2 \phi \tag{25}$$

By representing fluctuatiion effects by a gaussian white noise of amplitude Γ, the asymptotic probability of the phase fluctuations of the layered structure is [11]:

$$\mathcal{P} \propto exp - \frac{R^2}{\Gamma} \int d\vec{r} [D_{||}(\nabla_x \phi)^2 + D_\perp (\nabla_y \phi)^2] \tag{26}$$

where R is the amplitude of the pattern. Hence the static correlation function of the order parameter-like variable, $< \sigma(\vec{r})\sigma(\vec{r}') >$, behaves at large separations as [11]

$$< \sigma(\vec{r})\sigma(\vec{r}') > \propto \left(\frac{D_{||}}{D_\perp} x^2 + y^2\right)^{-\eta}, \quad with \ \eta = \frac{\Gamma}{4\pi R^2 \sqrt{D_{||}D_\perp}} \tag{27}$$

This algebraic decay is characteristic of quasi long range order [38] and has been observed in convective instabilities [14]. When $D_\perp = 0$, one recovers the exponential decay of short range order [39]. Furthermore, the phase fluctuations may manifest themselves by the presence of topologic defects (dislocations, disclinations) in the structure, and which are associated with phase singularities. These defects may be thought to play here a similar role as in the melting theory of two-dimensional solids proposed by Kosterlitz and Thouless [40] and Nelson and Halperin [41].

Effectively, the potential associated with an isolated dislocation embedded in a layered structure is given by

$$\mathcal{V} \simeq \frac{m^2 R^2}{\Gamma q^2} \sqrt{D_\parallel D_\perp} \ln(\frac{L}{a}) \tag{28}$$

where L is the size of the system, a the core diameter, q the wavenumber and m the topological charge of the singularity. Hence the probability for the nucleation of isolated dislocations in large systems is vanishingly small. However, since the potential associated with a neutral array of dislocations behaves as

$$\mathcal{V} = \mathcal{V}(\text{core}) + \frac{\pi R^2}{2\Gamma q^2} \sum_{i,j} m_i m_j \sqrt{D_\parallel D_\perp} \ln[\frac{D_\parallel}{D_\perp}(x_i - x_j)^2 + (y_i - y_j)^2], \tag{29}$$

there should be a finite density of dislocation pairs or clusters within the structure. This density increases for decreasing phase diffusion coefficients and should thus be important near the phase stability limits. Furthermore, by varying the bifurcation parameter, the amplitude R and the noise intensity Γ also vary and the destruction of the quasi long range order should occur for [40]

$$\frac{\pi R^2}{2\Gamma q^2} \sqrt{D_\parallel D_\perp} < 2 \tag{30}$$

However, the noise intensity is usually too small in far from equilibrium systems to allow experimental checks of the existence of the fluctuation effects described above. It is why the recent observation of ptretransitional fluctuations in the EHD instability of nematics[15] triggers new hopes for the study of fluctuation effects in the ordered regime aswell. Of course, in this case one has a Hopf bifurcation to wave patterns and the dynamics of the oder parameter-like variable is not potential. Nevertheless, as in Hopf bifurcations to limit cycles, one may expect a separation between potential and gradient parts of the dynamics [42] and, at least for the potential part, fluctuation effects should be qualitatively the same as the ones discussed here.

References

1. P.Coullet and P.Huerre, "New Trends in Nonlinear Dynamics and Pattern Forming Phenomena : The Geometry of Nonequilibrium.," Plenum, New York, 1990.
2. F.Busse and L.Kramer, "Nonlinear Evolution of Spatio-temporal Structures in Dissipative Continuous Systems," Plenum, New York, 1990.
3. D.Walgraef, in "Nonlinear Phenomena in Materials Science," G. Martin and L.P. Kubin, eds., Transtech, Aedermannsdorf (Switzerland), 1988, p. 77.
4. P.C.Hohenberg and M.C.Cross, in "Fluctuations and Stochastic Phenomena in Condensed Matter," Lecture Notes in Physics 268, L.Garrido ed., Springer, New York, 1987.
5. M.Dubois, P.Bergé and J.E.Wesfreid, J.Physique **39** (1978), p. 1253.
6. S.Ciliberto, E.Pampaloni and C.Perez-Garcia, Phys.Rev.Lett. **61** (1988), p. 1198.
7. H.Bercegol, E.Charpentier, J.M.Courty and J.Wesfreid, Phys.Lett. **A121** (1987), p. 311.
8. A.Joets and R.Ribotta, in "Cellular Structures in Instabilities," J.E.Wesfreid and S.Zaleski, eds., Springer, New York, 1984.
9. D.Walgraef and N.M.Ghonlem, Phys.Rev. **B39** (1989), p. 8867.

10. D.Walgraef, G.Dewel and P.Borckmans, Adv.Chem.Phys. **49** (1982), p. 311.

11. G.Dewel, D.Walgraef and P.Borckmans, J.Physique Lett. **42** (1981), p. L-361.

12. J.Swift and P.C.Hohenberg, Phys.Rev. **A15** (1977), p. 319.

13. D.Walgraef, G.Dewel and P.Borckmans, Z.Physik **B48** (1982), p. 167.

14. R.Ocelli, E.Guazzelli and J.Pantaloni, J.Physique Lett. **44** (1983), p. L567.

15. G.Ahlers, these proceedings.

16. A.Pocheau, "Structures Spatiales et Turbulence de Phase en Convection de Ray leigh-Bénard," These de Doctorat, Université de Paris 7, 1987.

17. S.C.Müller, T.Plesser and B.Hess, in "Physichochemical Hydrodynamics: Interfacial Phenomena," NATO ASI Series, M.G. Velarde and B. Nichols, eds., Plenum, 1987.

18. H.Brand, Phys.Rev. **A35** (1987), p. 4462.

19. D.Walgraef, in "Instabilities and Nonequilibrium Structures," E.Tirapegui and D.Villaroel, eds., Reidel, Dordrecht, 1987, pp. 197-216.

20. D.Walgraef and E.C.Aifantis, Res Mechanica **23** (1988), p. 161.

21. P.Coullet, L.Gil and D.Repaux, in "Instabilities and Nonequilibrium Structures II," E.Tirapegui and D.Villaroel, eds., Kluwer Acad. Publ., Dordrecht, 1989, p. 189.

22. H.Riecke, J.D.Crawford and E.Knobloch, Phys.Rev.Lett. **61** (1988), p. 1942.

23. D.Walgraef, Europhys.Lett. **7** (1988), p. 485.

24. I.Rehberg, S.Rasenat, J.Fineberg, M.De La Torre Juarez and V.Steinberg, Phys. Rev. Lett. **61** (1988), p. 2449.

25. P.Coullet and D.Walgraef, Europhys.Lett. **10** (1989), p. 525.

26. M.C.Cross, Phys.Rev.Lett. **57** (1986), p. 2935.

27. P.Coullet, S.Fauve et E.Tirapegui, J.Physique (Paris) **46** (1985), p. 787.

28. H.R.Brand, P.S.Lomdhal et A.C.Newell, Phys.Lett. **118A** (1986), p. 67.

29. H.Haken, "Advanced Synergetics," Springer, Berlin, 1983.

30. A.C.Newell, in "Lectures In Applied Mathematics," vol.15, M.Kac, ed., American Mathematical Society, 157, 1974.

31. V.I.Arnold, "Geometrical Methods in the Theory of Ordinary Differential Equations," Springer, New York, 1983.

32. E.Guazzelli, P.Borckmans, G.Dewel and D.Walgraef, Physica **D35** (1989), p. 220.

33. B.Malomed, A.Nepomnyanschii and M.Tribelskii, Sov.Phys. JETP **69** (1989), p. 388.

34. D.Walgraef, G.Dewel and P.Borckmans, Phys.Rev. **A21** (1980), p. 397.

35. S.A.Brazovskii, Sov.Phys. JETP **41** (1975), p. 85.

36. D.Walgraef, *Transitions between Patterns of Different Symmetries*, preprint.

37. Y.Pomeau and P.Manneville, J.Physique Lett. **40** (1979), p. L609.

38. B.I.Halperin, in "Physics of low dimensional systems," Y.Nagaoka and S.Hikami, eds, Publication Office, Progr.Theor.Phys., Kyoto, 1979.

39. J.Toner and D.Nelson, Phys.Rev. **B23** (1981), p. 316.

40. J.M.Kosterlitz and D.J.Thouless, J.Phys. **C6** (1973), p. 1181.

41. D.Nelson and B.I.Halperin, Phys.Rev. **B19** (1979), p. 2457.

42. H.Hentschel, Z.Physik **B31** (1978), p. 401.

PANEL DISCUSSIONS

PANEL DISCUSSION:

DIRECT NUMERICAL SIMULATION OR EXPERIMENTS?

Moderator: A. Roshko, Caltech

Panel Members: L. Kleiser, DLR, Göttingen
P. Moin, Stanford University

Roshko: I have been tapped as the moderator for this panel, and I hope that everybody else is prepared to say a few things. I think that our job as panellists is to provoke a little discussion from the audience, and I would like to say a few words to get things started. First, to clarify the subject of the discussion a little bit, I think that what is implied in the title by "experiments" are laboratory experiments. I think that the direct numerical simulations are, in fact, also experiments, and that we should be talking about using direct numerical simulation in the same way that we use the laboratory, to produce flows that can be studied and brought into the research picture.

I think that it is not really one *or* the other. I think that it is appropriate to say one *and* the other, at least at this period of time, in 1990. What I think that is really implied by the question in the title is: how long will the situation last? I thought about that a little and I will try my conclusions out on you. It may be appropriate to say that, up to about 1980, it was laboratory experiments only. There were plenty of attempts at DNS, but I think that, until about ten years ago, or maybe even less, we were not learning anything from the direct numerical simulation that was not already known from experiments. This situation had existed back since about 1890 - a hundred years ago - which was the time of Reynold's experiment and which, in a way, marks the beginning of the modern era of research experiments. It has been now roughly a hundred years that we have been doing laboratory experiments, and it has been only in the last ten years, I think, that the DNS has joined in that activity.

Looking down the line in the other direction, I wonder what will happen in the next ten years, by the year 2000. I am pretty sure that it will still be one and the other, operating side by side. From what I have heard at this meeting, a Reynolds number of 10^3 is roughly where DNS is now, even if for some flows it is higher and for others it is lower. For example, I was astonished to find out that, for the flow over a circular cylinder at a Reynolds number of 100, there is still a great deal of disagreement amongst different people producing numerical simulations. On the other hand, boundary layers apparently can be computed up to $Re = 1000$ or 2000 fairly confidently, and for some other flows you can go perhaps a little higher. So maybe by the year 2000 we will be on a similar situation to where we are in now, but the Reynolds number will be higher. Maybe we will push it up by an order magnitude.

How about a hundred years from now? where will we be in the year 2090? Will 10^7 be a good guess? Because if it is so, at that time it will be DNS only. The limit for experiments in the laboratory is about 10^7. If you want to get higher Reynolds numbers you have to go into the ocean or the wind or somewhere else but, for laboratory experiments in turbulence, that is about as high as we can go.

I hope that some of this will provoke responses either from my panel colleagues or from the floor. All the panellists will first make some statement and then we will pass the baton to the floor.

Moin: Roshko basically said what I had in mind. I did not think that there was any controversy on the point that numerical simulations are like experiments, but it turns out that we may have to clarify that. I agree with Roshko that, at this day and age, that should not be an issue any more, but yesterday Mike Gaster told me that he does not believe a thing that comes out of computations, so maybe it is. So, let me qualify the statement that numerical simulation is like experimentation. To be able to make that statement, the first thing one has to do is to raise the issue of validation. How correct are these numerical simulations? To what extent are the numbers that computers produce physically realistic?

I think that in the last ten years or so it has been conclusively demonstrated that, in some of the flows that have been simulated, numerically generated data bases are as good, if not better, than experiments. One example that I am very familiar with are boundary layers, channel flows, etc... The accuracy of that simulation data is not questioned any more, at the Reynolds number that we are talking about. Furthermore, early discrepancies that existed between the simulation results and some of the earlier experiments, have been removed now by additional experiments that showed that those discrepancies were due to experimental errors. I think that, without making blanket statements about which method is better, the adequacy of each method is based on its merits, and on its checks and balances. We have good experiments, and bad experiments. We have good computations, and bad computations. For each case one has to do the internal consistency checks as well as the external checks with other computations, to be sure that the results are correct. And I think that both sides have to do that, the experiments as well as the computations. Once these checks are made, I do not think that we should have a controversy on whether we believe it or not.

I think that what simulations provide, and what any method of simulation of data should provide in the future, is raw, unprocessed data. The history of turbulence research, which has been dominated by experiments, has been characterised by data generated in a given laboratory and processed according to the taste of that laboratory. The data that were generated in one place were very difficult to access in raw form by investigators in another place, so as to be able to test their theories, etc...

What simulations provide, and what I hope modern experiments would provide also, is data in raw, unprocessed form. Instantaneous space-time information on the velocity field. Once these data are available, we can use them to validate a theory, which is what all experiments, including numerical ones, should do. I think that the function of all of us is to verify theories. I think that this is a new development in turbulence research. The availability of raw data for anyone, anywhere. Of course we are not at that stage yet. We don't have magnetic tapes or optical disks to send around for people to process them, but I think that we should move in that direction. Certainly, the calculations that I mentioned the other day, which took six months, are a very expensive investment that should not be done by everybody. They should be done once - spend the taxpayers money once - and then be made available to everybody else.

After the availability of data, the next issue is: what are the right questions to ask? The future of turbulence research depends on that. I think that we will not get paid forever to generate data, just for the sake of generating them. We will get paid to ask the right questions and hopefully to find the answers. And I think that having a data base in front of us, on a workstation or on some other electronic medium, will help us to ask the right questions and then to be able to quickly evaluate the answers. For example, among the questions that I think have been left out in the study of turbulence, or at least in the study of coherent structures in boundary layers, is the issue of dynamics. I think that too much emphasis has been spent on kinematics, on looking at pictures of the flow fields and on identifying different kinds of coherent structures. A good portion of the boundary layer research has been spent doing that, and I think that we want to spend more time looking at dynamics.

To make a more general comment about experiments, I think that something that will come up in the next years - or something that I hope to see - are experiments of discovery. Not just general purpose tunnels which can be used to generate a flow situation and measure it, and then move to a different problem and just make more measurements. I mean experiments designed especially to answer specific questions. Of course, when I say experiments, I mean numerical simulations as well as laboratory ones. Workers in computational fluid mechanics spend too much time making computations to get a good agreement with a given experiment, and a good portion of the Government sponsored experimental research, at least in United States, is devoted to running mundane experiments to validate codes and vice versa. I think that we need a lot more experiments of discovery in both categories.

I move now to the question that Roshko raised on what is the relevance of numerical simulations to actual situations. He pointed out that there is a Reynolds number hierarchy. Experiments can do Reynolds numbers of 10^7. Calculations can do Reynolds numbers of 10^3. If our objective is going to be to simulate as high a Reynolds number flow as possible, I don't think that we will succeed, at least for flows encountered in practical applications. I do not think that we will be able, in our lifetimes, to calculate the flow around a 747 aircraft at cruise speed in the detail that the direct numerical simulation requires. But I wonder whether that is the relevant thing to worry about. I wonder whether there is a critical threshold beyond which the Reynolds number effects are not important. Or whether we can come up with scaling laws to absorb the Reynolds number dependence. There is some evidence that such laws exist in many situations. The indications are that we don't have to go to a Reynolds number of 10^7. Roshko, in his talk this morning, spoke about the mixing transition. In that case, at least up to the highest Reynolds number for which experiments have been done so far, there is a flat region, beyond $Re = 5,000 - 6,000$, where it does not appear to be a Reynolds number dependence.

The same is true of the strength of the wake profile in boundary layers as a function of the momentum thickness Reynolds number. At the very beginning there is a Reynolds number dependence, but beyond $Re_\theta = 3,000$ or maybe $5,000$, at least for this one parameter, there is no dependence. People have worried about that issue in the past, and have gone to higher Reynolds numbers and, in going to $Re_\theta = 50,000$, there are indications that there maybe some Reynolds number effect again. One can raise doubts about the validity of those experimental data, at least of the measurement of skin friction at such high Reynolds numbers. But even if it is true, and the dependence is there, to what extent is it important when it comes, for example, to designing algorithms for turbulence control? I am not sure that the issue of Reynolds numbers will be important once we go above a certain threshold.

Just to wrap up, I address again the question that was put to this panel, *Experiments or Simulations?* I think that the method of choice is whichever one generates the data that we need, and the rest of the time and energy should be spent in finding the right questions to ask. Let me give an example in the case of coherent structure research in boundary layers, or channel flows, at low Reynolds numbers. These are the flows that we have been computing, and also where the experimental research has concentrated. I don't think frankly that there is any point in doing experiments at these Reynolds numbers to study coherent structures any more. For that problem, I think that it has been established over and over again that the direct simulations provide the accurate data that are needed; three dimensional, time dependent data. Experimentalists - Ron Blackwelder, Ron Adrian, and others - are now using these data to study coherent structures. I would not do another experiment of turbulence in a channel flow at a Reynolds number of five to six thousand. I think that is done.

Roshko: I am going to take some exception to the remark that, eventually, the highest Reynolds numbers will not be of interest, because I can think of quite a few examples where that is not true. The outstanding one is the flow over a circular cylinder, where there is a very big change in what is going on between Reynolds numbers of 10^6 and 10^7, and where we would really like to see what happens when we get to 10^8. We even talk about towing a cylinder between two ships out in the ocean, or some other way to get a Reynolds number of 10^8. It is very difficult to do that experiment, and it will certainly be difficult to do it by DNS for some time. So I don't think that there is a finite horizon out there yet.

Moin: In the case of cylinder, if scientists believe that there is a mechanism that occurs at Reynolds number of 10^8, then that is a motivation to perform an experiment to reveal that phenomena. I agree.

Roshko: Or even a trend with Reynolds number. Even the trend is not clear.

Lumley: I just want to point out that the Reynolds number is the least of your problems. You are leaving out all the reacting flows, the flows with moisture, the flows with heat transfer. If you try to compute a flow in a gas turbine combustor, for example, the ratio of the size of the combustor to the flame thickness is of the order of five hundred, which is far beyond the capabilities of even the best imaginable machine at the moment. You are decades and decades away from being able to do that. Also it is a totally inhomogeneous situation. You are not just interested in isothermal, non reacting, homogeneous turbulence.

Moin: I agree with that comment. Whenever we do research we don't touch those situations, but when it comes to discussing the relevance of simulations, those issues always come up. And I agree with you. I think that flows with phase changes, boiling, etc. are best done experimentally. In the cases where you have length scales that are generated in a specified, fixed, location, or at least in locations that do not vary too much during the calculation, then I think that there are techniques of adaptive meshing that you can use to resolve those scales, rather than trying to put the fine mesh everywhere.

Kleiser: I do not really have to add very much to what has been said already, both in the two previous presentations and in Moin's talk yesterday. I want just to emphasise a few more points, some of them from the viewpoint of the simulation of transition to turbulence, which has also been covered in some of the contributions here. One of the advantages of direct simulation, which was mentioned already yesterday by Moin, is what he called conceptual experiments. With numerical simulation you can very easily control various terms, and introduce certain effects or take them out of the system, which would not be possible in the laboratory. Another point which has just been raised by Moin is the access to the technique of direct numerical simulation. First, of course, is the access to the computers to generate the data. There are only a few places in the world right now where you can do a channel flow simulation. Second is the access to the data that have been produced, which is not as widespread as it should be. These are two points which certainly have to evolve in the next ten years or so.

Now, if you look briefly at the problem of transition - and much of this is also relevant for turbulence research - I think that what we want to do can be classified roughly in four groups in terms of research: One is to describe the phenomena; a second one is to explain the mechanisms behind them; a third one is to control these various types of flows or phenomena; and, finally, to have prediction methods, not only for engineering tasks but also for qualitative behaviour. For all of these tasks, of course, we can use experiments to some extent but, for others, direct simulation would certainly be of help. From the point of view of explaining the mechanisms, I think that neither of them provides the answers directly, as Moin has pointed out. You just generate raw data. In order to get to the mechanisms you have to use a further tool, your brain, to analyse the data and to investigate the mechanisms behind them, to derive simplified models and to study their behaviour, and to perhaps get some hints on the complexities of the full system.

For the direct simulations, an important point, raised by Moin already, is that it is important to provide reliable data and, as he said, standards for verification have evolved very much during the past ten years. The flows that we can cover now with direct simulations are, in my view, a tiny fraction of all the problems that are being treated in the field of flow engineering and, as it has been pointed out already in this discussion, there are many problems which have not been tackled by simulations. That does not mean that they cannot be done in the future. I think the future of direct simulation is bright, and that the field will develop as dramatically as it has done in the past ten or twenty years.

The use of data bases has already been mentioned, and is important for the documentation of the data, for the investigation of phenomenology, and for the improvement of quantitative predictions. For example, in turbulence modelling of the Reynolds averaged equations, but also for higher order equations.

Another point that is very important for laminar wing flows, or for hypersonic re-entry vehicles, for example, are the models of the transitional zone, and there is a hope now that numerical simulations can be used, at least to some extent, to provide data in that respect, which in some fields are actually not available at all.

Finally, my last remark is that the answer to the question *"Simulation or experiment?"*, or *"What use for each technique?"* is a strong function of time. If you look back in time, it has been influenced very strongly by how the computer hardware, and the computer architecture, have evolved. If you look ahead, parallel processing may add an order of magnitude to the capabilities of numerical simulation techniques. Of course, the experimental techniques are certain to also evolve further, and both will certainly go hand in hand. I will stop here and ask for more contributions from the audience.

Metais: I want to go back to the issue of Reynolds number in direct numerical simulation. We know that we have to use low Reynolds numbers, but even so we have an advantage over some laboratory experiments right now because, for example, when you look at the available numerical experiments on homogeneous turbulence, they are the only way of checking the structure of the dissipative range. I would think that a proper laboratory experiment of three dimensional isotropic turbulence with a passive scalar, for example, would be of great help in understanding the dissipative structure and to test what we heard before about, for example, the validity of intermittency models, like fractals or multifractals. I think that, without speaking about combustion or even channel flow, it would be very nice to have a proper three dimensional isotropic turbulence laboratory experiment with a very big tunnel.

Zabusky: The title posed by the organisers, *"DNS or experiments?"*, reminded me of a remark that a professor made when I first entered graduate school. Professor Richard Leighton, who taught me modern physics, said the first day that he walked into class: "I am technologically unemployed". And what he meant was that he was working on cloud chambers, and cloud chambers had seen their best day at that time because big high energy machines were more convenient, more controllable, and producing data that was more accessible. So that is what the question is. *"DNS or experiments?"* means that somebody is technologically unemployed.

I think that everything that I would have said has already been said already, but for one point, which is the rapid development of experimental techniques, which I want to link with the use of data bases. I think that the way in which people conduct their experiments now, except for perhaps the laser sheet, is the same one in which Reynolds did. In other words, I think that experimentalists have fallen very far behind in the process of generating the enormous data bases that computers routinely spit out. There is a reason for it. They have not instrumented that way because of the problems implied by the storage of information and by the access to that information. We have to get enormous work-stations or hierarchical memories so that we can store these data. We can generate these runs and soon we are tying up enormous resources, just for storage. In fact some times it becomes easier to do the run over again. What I think is the challenge to experimentalists is to develop their laser diagnostics, to generate enormous numbers of sheets so they can see a time development and then to rapidly be able to discretise them and - the most important thing - to juxtapose them. That is, to move data very quickly from the experimental domain and to put them in parallel to the computation domain. In doing this we will actually begin to answer some of the questions that John Lumley brought up, mainly: when is the model that we are using wrong? Right now, all you have to do is to raise the Mach number, ionise something, and you know that you may as well stop doing your calculation.

I think that the ability to generate enormous data bases is what is changing the attitude towards experiments. I am not saying everybody is going to do that, but there should be a certain number of experimentalists who should produce data just like the NASA-Ames people generate data on their computers and I think that, when that happens, experimentalists will never lose their opportunity to work.

Roshko: I would just like to take a little exception to what you say. I think that if you have too much data coming at you, you can't see the forest for the trees. I think that has been demonstrated by the great success of the hot wire anemometer, which was a marvellous instrument that could spit out tremendous amounts of data. Too much, because you could not see what was behind it. So there is nothing wrong with having the capability to produce huge amounts of data but you have to be careful that that is not all you do.

Papailiou: I am not a computer man. I do experiments and I am always reading that the only factor limiting the capability of computers to give us all the answers is their capacity of storage and calculation. It seems that all we really need is a bigger computer. Could I ask whether this is really the feeling of the people here?

Moin: I don't know whether this is actually what you had in mind but I want to comment on whether the computer is the only thing that we need or not. To begin with, we are using the governing equations, the Navier-Stokes equations, but in solving them we need boundary conditions. In the case of open boundaries, I think that there is a great deal that needs to be done in that respect. We do not really know how to prescribe turbulence on open boundaries. This is an *ad hoc* input that is going to be with us for some time, and considerable research is going on right now on it. Experimentalists face the same issue. At the end of their wind tunnel there is a boundary condition of some kind, and it changes from one laboratory to another, although it might not be such a serious problem in that case.

On the question that I think you had in mind: is the computer going to give us all the answers? No; that is the whole point. We never said that it was going to give us an answer. Nor do hot wires give us an answer. It is the power of analysis that ultimately is going to do that. What computers have been doing for us during the past ten years has been to give us the data easily. To generate data so that we can analyse them and perform conceptual experiments with them. The computer is not going to solve the problem by itself.

On that respect, I would like to make one point that I did not mention before, and which has perhaps a little to do with the work of the physics community that I saw here. I think that, in Fluid Mechanics, we are very fortunate to have the Navier-Stokes equations. It is a gift that we have, and I think that, as much as we can, we should not throw it away. This, or approximations to this, is what the numerical simulations have been working with. I know that, in most of theoretical Physics, a great deal of effort is put into coming up with the governing equations, but Fluid Mechanics is not one such area. The equations are known, and they have been known for a long time. Of course we can make approximations to them, but only rational approximations.

Wygnanski: I very much agree with you in that what is important are the questions that you ask yourself when you do either experiments or numerical computations but, on the issue of this discussion, I think that there are many examples that show that experiments are going to be with us for a long time. Some have already been given here, but let me give you others. Suppose that you ask yourself the question of feasibility regarding control experiments. I can give you the example of the experiment that I talked about today. I can ask myself what I think is the right question, or, at any rate, a question. I can then construct a small, cheap facility, and check the answer. I could not use the analogue of a little wall jet, say a mini computer, in order to answer that question. I think, although I may be wrong, that I would need a huge computer to check it. I think that laboratory experiments provide a certain flexibility on the feasibility level - particularly on questions such as: can I control certain phenomena or not? - which the computer does not. What I see in the computer is the analogue of the largest and most perfect wind tunnel. But you do not always get access to the largest wind tunnel when you need a small thing. You do not shoot at a fly with a nuclear cannon. And that is why I think that this duality will coexist for quite a long time.

Kleiser: I agree with that statement and I think that the important question that you have to ask yourself is: what result do I want? If you are happy with the results that your small wind tunnel gives you, there is no point in making an expensive direct simulation which gives you all the data of the 3-D velocity field in space and time. If you just want to measure, let us say, some averaged velocity distribution, or some skin friction distribution in the streamwise direction, then your experiment may be sufficient, but what we are talking here is about what happens if you are interested in the details, in the structure, in the mechanisms of turbulence physics. Then I think that not even the biggest wind tunnel is able to provide you with the amount and quality of data that the simulation gives you. The main point of the simulation is not just generating the flow, but replacing the data acquisition. I think that that is the biggest advance that has been made in Fluid Dynamics in comparison with the experimental approach.

Moin: I have no objection to what Wygnanski said. I don't know why the conversation always drifts towards whether experiments are going to remain with us or not. That is not the point. I think experiments are going to be with us for as long as the computers are going to

336

be with us. Just take an example: tomorrow I have flat plate calculation and I want to check the effect of inserting a rod in it, or a sharp edge to generate streamwise vorticity; it is going to take me a long time to set up the calculation to do it and, in that case, experiments are much better. I do not think this debate has anything to do with whether experiments are going to be with us or not. Of course they are going to be with us, as far as I am concerned. There are a lot of things that you can do better experimentally and there are some things that you can do better computationally.

Wygnanski: I fully agree with you. All I actually wanted to plead for was an interaction on a continuous basis. I think that feasibility studies are preferably done in a laboratory. Once a few basic questions have been asked, and once certain hope, in terms of feasibility, has been achieved, the next step probably belongs to the computer and to the simulation, where you can provide answers for which a laboratory is limited; and the right thing to do is perhaps some sort of interactive or iterative system between the DNS and the experiments.

Saffman: I just want to comment briefly that there is another aspect to this *"or"* question which affects the universities, where we are concerned not only with the research, but also with education and training, and with preparing the future generations. I think that, for the universities, the *"or"* is an important question which has to be faced now, because we have finite, or even vanishing, resources, and universities have to face the question of where are they going to spend their money. Are they going to spend what little money they have on a new computer or in modernising the laboratories? Are they going to hire a professor of Computational Fluid Dynamics or are they going to hire professors who are experts in experiments and instrumentation? And this is a question and a problem which is with us now, and we have to make decisions and, if we make the wrong decisions, it is going to impact unfavorably on the future research for the next twenty or thirty years.

Reynolds: I have been trying to figure out why are we having this discussion, and I think that there has been a lot of territorial defence going on here. What we probably ought to be doing is asking ourselves what Zabusky started saying. What is going to cause the big change in this? I do not think that it is the biggest, fastest, computer that is going to make the big difference in the computations in the future. I think that it is the smart computer. The computer that has an algorithm in it that can handle small scales as well as large scales. The small scales probably by some kind of analysis that results into some credible type of algorithm - not what Moin would call "glue", as the current algorithms are for handling sharp discontinuities - but something that can handle calculations with a Schmidt number of 10^4. That computer would have a much bigger impact than another decade in speed, and I think that is the kind of thing that we ought to be thinking about. What could we do that would push us orders of magnitude ahead either in the laboratory or in the computers? Zabusky said what he thought it was in the laboratory, and I would agree with that. In the computers I think it is being smarter in their use, not just bigger.

Lesieur: I want to change the subject for a moment. It seems to me that an important aspect in this discussion is that we have been talking about the relationship between laboratory and numerical experiments, but mainly from the point of view of a computational person. I told already yesterday that Fluid Mechanics contains some essential quantities that allow us to understand the dynamics of the flow much more easily than others, and that vorticity is an example. An I hope that you all agree that this is much easier to get from a numerical simulation than from the laboratory. In numerical simulation you can have all the fields, pressure, velocity, temperature and vorticity. It seems to me that most of the major progress that has been done by Computational Fluid Dynamics in understanding the physics of flows has been because we can follow the evolution of vorticity, the straining of vortex lines, the motion of vortex tubes; we can see that big billows correspond to low pressure tubes and things like that. This is very easy to investigate numerically but - I don't know if the experimentalists would agree - the severest limitation of the probes right now is that you cannot get these quantities. You can put fifty hot wires in a turbulent flow, or you can build a probe with several hot wires oriented in all directions to measure vorticity, but it is very difficult, and you don't get many results; and if you spend two years building a probe and then you break it, you have to start again. That is a problem, and something that we might do is to try to derive some experiments, maybe some simple shear flow for which we are starting to understand the vorticity dynamics, and do exactly the same experiment numerically and in the laboratory. We could then correlate global things that can be observed in the laboratory, like passive scalars and dyes, with the detailed vorticity distributions that we obtain from the computations, and use them to calibrate the laboratory measurements using the numerical results. Maybe in that way we may be able to develop non intrusive experimental probes to measure vorticity with reasonable precision.

Roshko: Let me summarise that for you. I had it here in writing, actually. I had prepared this little list of things that I envy of DNS, and one of them is the possibility of measuring everything. That has been implicit in what the speakers have been saying here, but nobody has quite said it. As you say, you can measure vorticity, you can measure fluctuating pressures and so on, and that is certainly one of the main attributes of DNS. Another thing that nobody has mentioned, although it has kept coming up in the discussions, are two dimensional flows. I feel a little strange saying this - being an experimentalist - but I want to encourage my computer friends not to overlook two dimensional flows, because I think they are very interesting objects in themselves, and we cannot do two dimensional flows in a laboratory above some Reynolds number.

I think it is also fair to say some of the things which I do not envy, and one of them is staring at a screen all day, but there is also a couple of other things, not so important, that have been discussed already today: Reynolds number limitations, mentioned by John Lumley, and the geometrical limitations of various kinds.

Fauve: This is an answer to Marcel Lesieur. I think that the main advantage of experiments is probably the fact that we can only use a small number of probes and, sometimes, only one probe. It is a great advantage, because you must think about the right quantity to measure, and if you succeed, it is an important point. Also, about the Navier-Stokes equations, the comment is very similar. I think that another equations which are well known are the Newton equations of motion and, if Boltzmann had had the possibility of using a large computer, he would probably have followed the motion of each particle, and it would have been a pity for Thermodynamics.

Ahlers: I would like to say first of all that, of course, we all recognise the great virtues of numerical simulation. Roshko just made the main point: you find out everything about the flow that you are studying. You have all the parameters, while the experimentalists just cannot record or measure simultaneously that much information. That is one of its great advantages as I see it. But nonetheless I must take exception to calling it a experiment. I really and honestly don't believe that it is one. It may sometimes, even quite often and over a wide parameter range, give results that are identical, within known errors, to an equivalent experiment. But there are parameter ranges where it will fail, and it will fail for various reasons. The obvious one, of course, is that you don't study a partial differential equation, but a finite difference equation, or whatever numerical technique you may be using. You have to be conscious of the fact that there are parameter regions where the finite difference equation no longer corresponds to the partial differential equation. You can say that you may change the number of grid points, and do all kinds of tests, but the tests only look at statistical measures of the flow. I think that Moin showed us a beautiful example in his talk this morning, where there were two flows that looked visually very similar and yet, when one looked at certain aspects of them, they were different, and had different drags on the wall. I wonder if one could not be similarly mislead in one's tests of the numerics.

Then, at the next level, and the previous speaker already put his finger on it, we say that we know the governing equations, but, what does that mean? The Navier-Stokes equations are not Nature. They are a very good approximation to Nature, but there are certain cases under which we have to expect the Navier-Stokes equations not to contain everything that the experiments would contain. They are of course a gradient expansion, although people are not too worried about that aspect of it. But they also involve coarse graining, and that smooths over the stochastic effects. We also know that, when one has situations where there are two attractors which are very close to each other, so that their attractive basins don't have much of a barrier between them, noise can play a very important role, and noise can cause shifting from one type of flow to another. This would be completely absent in the numerical calculation, except for the noise caused by the numerical round-off. So I would be worried about referring to direct numerical simulation as a experiment.

338

I fully recognise its virtues. I think that it is a great thing to do; it tells us things that we can not get out of experiments, but I think that we always have to be aware that there is the possibility that it does not necessarily correspond to the real physical world, and I believe that only measurements on the real physical world should be called experiments. Now, if you want to call them numerical experiments, that may be all right, but it distinguishes a bit between the two approaches.

Moin: First of all, I would like to go back to the question of the equations of motion. Any theory has to go in some limit to the Navier-Stokes equations or, if you prefer, to the Boltzmann equation. I think that the limit has to be preserved because you cannot throw terms in there and study different equations. For example, I believe in amplitude equations. I understand the sort of work that starts with the Navier Stokes equations, simplifies them, and derives an amplitude equation, but I am concerned when I see new terms arbitrarily thrown in, absolutely with no justification, just because they produce certain results that look like something that has been measured before.

I raise the same objection to my colleagues in computational fluid dynamics when, just to get smooth solutions in their calculations, they throw in artificial viscosity. That is as bad as throwing in additional terms. It is just playing some kind of mathematical trick.

Now, going back to the reliability of computations. The same things that you said about fine tuning numerical methods can be said about experiments. If you put one brand of hot wire in there, or a different one; or if you buy one kind of plexiglass versus another, your detailed flow situation is going to change. Also, even when you validate experiments, by default you have to refer to statistics; there is no guarantee that experiments that are done at USC or at Caltech are the same physical phenomena. They have different boundary conditions, different roughness elements, different screens, and so forth. So the great resolution of calculations, the fact that they may lead to different instantaneous flow situations, I frankly think is irrelevant.

Then there is the field of numerical analysis. A great many people are devoting their lives to numerical analysis, and to try to make sure that, in some limiting process, these numerical calculations converge to the equations of motion. And as long as those precautions are taken, I do not think that we should worry about whether the computations correspond precisely to one particular physical experiment or not, because the experiments are not unique either.

Kleiser: I would just like to emphasise Moin's last point. I sharply disagree with the statement that what computer simulations provide is some solution to a finite difference equation which might or might not agree with the solutions of the partial differential equation. I think that the direct simulations that we are talking about are the ones in which it has been verified that there is no difference in the limit between the two solutions, by doing the proper checks. For example, convergence studies or comparisons with the experimental data, including time dependent behaviours. I think that what you may be referring to are some numerical results published about ten or twenty years ago, which were not always reliable, as Moin has also pointed out. But I think that what we now mean by direct simulation produces numerical results which are as reliable as experimental measurements.

Moin: In both cases there has to be a limit process. We can talk here about equations and the numerical analysis that goes with them. You can prove that, for a convergent numerical method, the results converge to the solution of a certain equation under certain limiting criteria. In every scientific discipline this assurance has to exist - otherwise we can't trust anything - and numerical analysis has it.

Roshko: I think that this a good time to stop, unless somebody still has some burning statement that he wants to do. There is no way to sum up and no need to. The session is over.

PANEL DISCUSSION:

WHAT ARE TURBULENT PUFFS?

Moderator:	C. Van Atta, U. California, San Diego
Panel Members:	Y. Pomeau, ENS, Paris
	I. Wygnanski, U. Tel Aviv and U. Arizona, Tucson

Van Atta: A turbulent puff is an organised structure that occurs in pipe flow. To set the scene for those of you who are not familiar with it, let me refer to figure 1, which is a map of the different types of flows that you get when you perturb a pipe, near its entrance, with a finite disturbance. The Reynolds number in the abscissae are based on the pipe diameter and on the average velocity. The other axis in the figure is the strength of the initial disturbance. In the lower left part of the graph you have laminar flow and in the right upper corner, you generate fully turbulent flow, but there is a region in between $Re = 2,000$ and $Re = 2,700$ in which a finite perturbation which is strong enough will generate a puff. As the Reynolds number grows higher we start moving into the slug region, for reasonably weak perturbations. The original distinction that Wygnanski and co-workers did in the past was that puffs are generated by finite amplitude perturbations, while slugs grow from a boundary layer instability that generates the disturbances by itself. Today we are going to concentrate on puffs.

The structure of a puff was first obtained by Wygnanski and co-workers and digested later by Coles, and looks roughly like a vortex ring. There is a fairly well defined trailing interface, concave upstream, and a much less well defined leading interface, and the flow is turbulent in between, with the vortex ring lying near the sharp trailing end (figure 2a). The general picture is that of a vortex ring propelling itself down the pipe, and extracting the energy from the mean flow. The interesting thing is that, once this structure gets generated, it comes to an equilibrium or statistically steady state, and moves along the pipe without changing its length. In the slug region, on the other hand, the front moves faster than the back, and the slug grows.

This is the outline of the physics of the turbulent puff, and the interesting thing is that the flow is self organising. If you hit it hard with a stationary disturbance, and you really have to hit it hard to make it work in the first place, it generates a train of puffs, while a momentary disturbance generates a single one. Various forcing techniques have been tried and they all seem to work. The point is to make a big mess near the entrance and the flow organises itself into a structure looking something like a vortex ring, and then stays that way.

The interesting problem is to decide how a turbulent flow gets organised into such an well defined structure as a vortex ring using the mechanisms of what we usually think of as turbulent diffusion and vortex interactions, etc. A possible point of view is that the turbulent puff is a prototype for a large scale structure in a fully turbulent flow but, in another sense, it is a very distinct transitional feature of pipe flow.

I do not think that direct numerical simulation has been too successful up to now in describing flows like this. Leonard has tried to model a turbulent spot in a boundary layer. It is a tough job and it takes a long computation and it doesn't look realistic when you are finished.

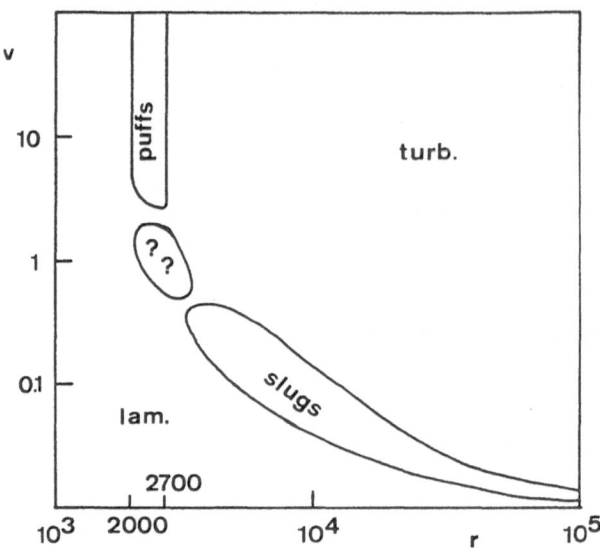

Figure 1. Flow regimes in pipe flow, as a function of Reynolds number and initial perturbation amplitude (percent). Adapted from [1].

Gaster: Is there actually a region of amplitude between the puff and the slug, where neither are produced? I find that extraordinary.

Wygnanski: I don't think that there is a region where there is neither. What happens at a little bit higher Reynolds number, in the gap between the two regions, is that puffs tend to grow slightly at first and then split. After the splitting, they rejoin and start creating a slug. It was shown later that, in fact, you can look at the slug as a conglomeration of puffs and you can recognise them quite easily at least near the leading and the trailing interface. I could show that as part of the data that we have.

Van Atta: I think that those are some of the empirical facts, and now we should try to explain why this happens. Not: what is a turbulent puff? but: why is a turbulent puff? That is one possible approach.

Wygnanski: I will take a couple of minutes to show what we know about puffs and what we don't. What Van Atta has described are really the very first measurements that we took [1,2], but we have more data. Let me talk first about the structure. If you use a rake of hot wires near the exit of the pipe, most of the high frequency turbulence is near the centre of the pipe, and there is hardly anything near the wall. There is a characteristic length of the order of 25 to 30 diameters and, if you look carefully, you can see that it contains from 3 to 4 very large coherent structures (figure 2b). I disagree with the one vortex model. That single vortex appears when you ensemble average over 500 events and you align them at the sharp trailing edge. Then the rest of the structures smear out but the one at the trailing edge sticks out. It is only then that you see the model in figure 2a.

Another interesting fact is that initially we did perturb the inlet flow and saw the puff peeling off and moving into some equilibrium downstream. We wondered whether they would be generated in a fully developed parabolic velocity profile. We took data in which the perturbation was introduced 400 diameters downstream, where the parabolic profile was fully developed, and still the shape and the length of the puff was more or less identical. So, puffs are not an inlet phenomenon that remains a long time, they are a characteristic of the fully developed pipe flow. Another aspect of interest, which was tried by Frank Champagne, but never published, is that you can take a small diameter pipe, with $Re \approx 10^4$, expand it through a one degree diffuser and

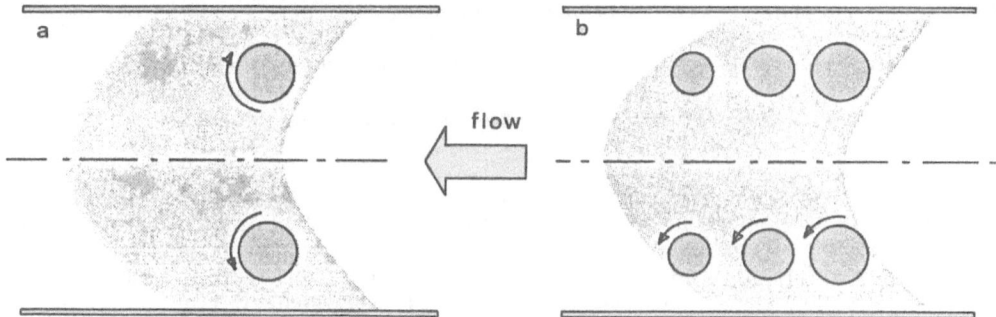

Figure 2. Schematic structure of a puff. See text.

get to the critical Reynolds number of 2000, and you find out that puffs are generated in spite of the fact that what you have is some sort of fully turbulent decaying flow. All that changes is that instead of having a clean velocity signal, you get the same structure on a noisy background.

In all cases, when you ensemble average, the average flow is axisymmetric and you can draw streamlines, and what you see is that the puff looks schematically like the picture in figure 2b, with two, or maybe three, large toroidal vortices. You can also measure azimuthal correlations, which are quite high, suggesting that the structure is actually quite axisymmetric, although we cannot rule out that it might be partly helical. The moment that you ensemble average, you loose all that information.

There was a recent paper by Breuer and Haritonidis [3] in which they actually counted the number of vortices and made a histogram, and they observe that most puffs have two vortices, while about half as many have three and a few have one or even four. So, what may happen is that the second vortex in the row moves forward through the first one, and in the process it gets squeezed and, since the production in the centre is very small, it dissipates very slowly, and that is why the leading interface is hardly visible. In this process, it maybe that a local inflectional instability is generated and a new vortex is produced, and the whole structure moves that way. At a given Reynolds number the dissipation of the old vortices compensates exactly the production of new ones. There are other possibilities, but this is something about which we don't know too much.

There is something else I wanted to talk about. If you look at the slug, it is a structure that grows. However, if you look very carefully at its trailing edge and at its leading edge, paste them together and forget everything in between, you see that it is essentially a puff. So we suspect that a slug is nothing but a conglomeration of puffs. Going back to the previous model, it is possible that what happens is that, at a slightly higher Reynolds number, the vortex ring that moves forward does not dissipate through the process and remains there, and that is how the splitting is generated. This keeps reoccurring and, by rejoining, you get the slug.

With respect to the splitting, what the recent observations of Breuer and Haritonidis show is just that something happens in the middle part of the puff that seems to be tearing it apart.

Saffman: All the experimental work I am familiar with is for circular pipes, but we know that, if we have non-circular pipes, the structure of the turbulence can be a little different because of the secondary motions which are set up. What happens if you have a slug or a puff in a non-circular pipe? Has anybody looked at that?

Wygnanski: Not to my knowledge.

Jiménez: I compute puffs in plane channels and they look quite similar. They are probably a different thing, but they also have the three or four structures and a front and a tail.

Saffman: The plane channel and the circular pipe are two special cases. You want something with an elliptical structure so that the difference in the normal turbulent Reynolds stresses produces secondary motions, and mean flows in the direction perpendicular to the main flow.

Pomeau: What I want to speak about are some remarks that we made together with Hakim and Jakobsen [4], and which are related to the behaviour of solutions in a problem which is intermediate between a system which has a purely dissipative behaviour, in the sense that it goes to an equilibrium state, and a system which is completely conservative. I know that it is an act of faith to believe that it has some connection with experiments in Fluid Mechanics, but nevertheless I think that it is an interesting point.

I will not write the equation, because it is not crucial, but consider figure 3. In the abscissae we have a parameter which allows us to interpolate between $r = 0$, which is a purely dissipative case, and $r = \infty$, which is purely conservative. We can abstract things a little and think of r as a Reynolds numbers, which gives the Stokes limit when it goes to zero, and the Euler limit when it goes to infinity. The other parameter, u, is related to some other term in the equation and measures the difference in depth of a two-well potential, which may be derived, if you wish, from some expansion in powers of some amplitude. You can consider it as measuring, in some vague sense, the difference in energy between two states of the system, one laminar and one turbulent. When this parameter is equal to zero (point **a**), and in a purely dissipative case, you can have a equilibrium between the two phases because they have the same energy.

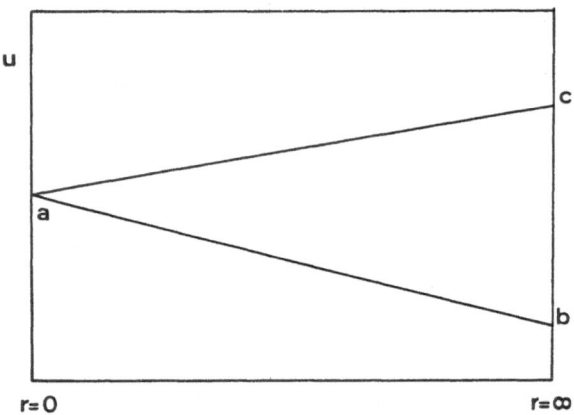

Figure 3. See text.

Even if it is not immediately obvious you can think of this as a general property. In fact, this is a property which is structurally stable, which means that if you start with a system which has a potential energy, and perturb it a little to systems without a potential, you can still retain this property that two states can coexist at equilibrium. This, by the way, maybe an answer to some people who are worried by the use of potential energies in turbulence.

In any case, when you do this extension, there is a line (**ab**) in which you have a steady coexistence between the two states of the system while, if you are on one side of the line, one part of the system will tend to invade the other and, if you are on the other part of the line, you have the reverse situation. But, if we had only this property, it would not be very interesting because this would not explain the formation of the Thual-Fauve solitons [5], which are localised and stable structures. Localised structures are in general unstable entities, and have no chance of being observed in a experiment, but what happens is that in the region bac, at least near the point a, you can do a small parameter expansion and prove that a new stable localised structure appears. The general behaviour is that below the line ab there is only one unstable localised structure, in between ab and ac there are one stable and one unstable localised solution, and above ac there are no localised equilibrium solutions at all. At the line ab, the stable localised wave grows infinitely wide and becomes a steady front.

So I think that this may be what is happening in the case that Wygnanski was talking about a moment ago; as you increase some parameter of the system, you first have no stable localised solution at all, and then you move into a region in which there is a stable solution. You need a finite perturbation to force it but, once you do it, you form a puff and it stays stable forever. Next, and this is again not immediately obvious, as you keep increasing your parameter, the puff becomes larger and larger and, as you cross the lower boundary, you eventually get a structure that expands continuously, as I explained in my presentation two days ago. I think that this may be a general explanation of what is happening here.

If I may now comment on a different point, consider a system that undergoes a subcritical bifurcation in terms of some parameter, say Reynolds number, and think of the point of onset of the linear instability. I think that it is an interesting remark that, at that point, one can predict the speed of expansion of a perturbation by using a method which is well known to people in combustion, which is the Zeldovich theory for the KPP problem. So, at that point, there is a chance of predicting from first principles, by solving the Navier-Stokes equation in a pipe by the usual expansion in powers of the amplitude, some things that could be checked by experiments. I think that the same thing could be used for instance in Couette flow, or in plane Poiseuille flow, which have also been known for a long time to be subcritical. In some of these cases, this may require a slight extension of the Zeldovich theory to slightly nonlinear perturbations, which is not trivial, and has not been done, but I think that the results would be very interesting.

Jiménez: I have a question. Nobody has told anything about spiral turbulence. Is that a puff? Do you think that it is the same phenomenon?

Pomeau: No, I don't think so, I even wrote that in a paper. Spiral turbulence happens at some specific regime of the Taylor Couette flow between counter-rotating cylinders. And at some rotation speed, a spiral band of turbulence appears and rotates as a unit, also without a change of shape. This spiral cover the whole length of the cylinder, and that means that when the flow goes around the cylinder, it always has to cross the turbulent band. That means that, when one does the nonlinear stability analysis of the flow, one has to impose some constrain on it, which in this case is constant pressure drop, and I think that it is because of the interaction between the growth of the perturbation and this constrain that the perturbation stops growing at a definite size.

Liñán: Going back to Saffman's hint that perhaps there are no puffs in channels, the secondary flows do not persist in the laminar flow, which is just a parallel flow. So that the secondary motions, if they exist, should be just imbedded in the three or four vortex rings in the puff, but the stability of the laminar flow should not change.

Saffman: It was my understanding that the turbulence inside the puffs is almost indistinguishable from the fully developed turbulence that you have in a pipe when the intermittency is zero. So, if you have the fully developed turbulence in the puffs, you are going to have normal Reynolds stresses which are unequal, and that will drive a secondary helical flow. Then, if one has a flow in a pipe that is not circular and not planar you should have some kind of helical structure and the question is: do the puffs then exist? Because there is a possibility that this extra secondary motion might destroy the structures or mechanisms which keep the puffs going.

Van Atta: If there are no further comments from the panel or the floor, we can consider this session closed.

REFERENCES

[1] I. Wygnanski and F. Champagne, J. Fluid Mech. **59**, 281-335 (1973).

[2] I. Wygnanski, M. Sokolov and D. Friedman, J. Fluid Mech. **69**, 283-304 (1975).

[3] K.S. Breuer and J.H. Haritonidis, Proc. 3rd. European Turb. Conf. (A. Johansson, ed.), Stockholm, July 1991 (to appear, Springer).

[4] V. Hakim, P. Jakobsen and Y. Pomeau, Europhys. Lett. **11**, 19-24 (1990).

[5] O. Thual and S. Fauve, J. Phys. France **49**, 1829-1833 (1988).

PLENARY DISCUSSION:

IS THIS THE THEORY FOR THESE EXPERIMENTS?

Moderator:	W.C. Reynolds, Stanford University
Panel Members:	J. Lumley, Cornell University
	P.G. Saffman, California Institute of Technology

Reynolds: This discussion comes at the end of the Conference. I have had some problems deciding what the subject was supposed to mean, but I propose that we use it to discuss whether all the various theoretical ideas presented during the Conference are relevant to the experimental observations of turbulence in fluids. Also, there have been two different positions along the Conference. One one side there have been those who are willing to write equations to fit particular phenomena, be it amplitude equations or something else, just because they seem to work. This is what I would call the method of Moses, in which equations are treated as direct revelation. One the other side are those who would not admit anything that can not be explicitly derived from the Navier Stokes equations. I suggest that we try to discuss that also. I will let the members of the panel make individual opening statements, and the floor will then be open to the general discussion.

Lumley: First I want to identify myself for the purpose of this discussion. Some of my colleagues may disagree with this, but for today I will say that I am an engineer, and engineers have certain characteristics. We are monomaniacs about the Navier-Stokes equations. That does not mean that I can not understand the attraction of a lot of other interesting equations in the world and that, even if you are married to the Navier-Stokes equations, you may see another interesting equation go by and have the temptation to follow it. But we have taken holly orders and we can not be seduced. Also, from our point of view, we do not feel that turbulence is a mystery. There are certainly a lot of things we do not know about three dimensional fluid turbulence, but *grosso modo* I think that we understand what is going on in a rough physical sense. What we can not do is to compute it, and we want to do it because we want to be able to design aeroplanes and do things like that. That is what we are trying to do, and very crude rough things are no longer of an enormous interest to us.

Several people have been using the word turbulence to describe all sorts of different chaotic situations and, even if it is just semantics, it is not going to help things much if we keep using the same word for a number of physically different, distinct, situations. We have found in the real world that fluid turbulence is something that is essentially three dimensional, essentially non linear, essentially vortical, and strongly interacting. We found that two dimensional and one dimensional flows, such as for example plasma turbulence, or ocean surface wave turbulence, have entirely different mechanisms. Atmospheric two dimensional turbulence also has different physical mechanisms and it does not help to lump everything under the same heading. I think that it would be helpful if we could try not to introduce verbal turbulence by using the word in a loose way. The subject is hard enough without these other problems.

There appears to me to be a natural progression in science, by which you see a new phenomenon and try to model it, usually very crudely at first and then in a more sophisticated way. The initial method is what Reynolds has called the method of Moses, which is essentially

The Global Geometry of Turbulence
Edited by J. Jiménez, Plenum Press, New York, 1991

all you can do if you are trying to describe the stock market or storm surges in the Adriatic, or something where you have no idea of what the equations are. But if you know what the equations are, you have to move to an exact theory, in which you have to show that the exact equations have in them the characteristics of the model that you have found to work. On that connection, for example, we know that turbulence does not have potentials and that it is not a Hamiltonian system. As a consequence, either the potentials or the Hamiltonian nature are not essential, and you can obtain the same properties without making those assumptions, or the results are not relevant for three dimensional fluid turbulence. It is one or the other. You can not have it both ways. So I think that it is very important to start working to show that the results can actually be obtained without making those assumptions, or they will have to be ignored, even if I admit that they are very seductive.

I agree that amplitude equations often describe with great generality systems close to a bifurcation point, so that they have universality and are not just arbitrary models. I think it is also a good idea to remember that dynamical systems theory describes primarily closed low dimensional systems, and temporal chaos, and that there are relatively few results in dynamical systems theory for open systems, spatio-temporal chaos and large numbers of dimensions.

One thing that has come out of this meeting, for me anyhow, is that Mackay gave a definition of a strange attractor which applies to turbulence. He essentially said that anything is a strange attractor which is ergodic. He did not use that word, but he said that time averages have to be the same as phase averages with an appropriate measure, which is what we would call ergodic. He also said that solutions have to diverge exponentially, and a number of calculations have shown that solutions do diverge exponentially in three dimensional fluid turbulence. That says to me that real fluid turbulence does have a strange attractor, which I think is a step in the right direction.

Finally I would like to say that, from our experience, it is very helpful to put together interdisciplinary groups. Mathematicians and physicists and engineers have a great deal to contribute to each other; when we put our group together at Cornell, it took about two years of a learning process before we found what the other guy thought was important, but it has been a very productive thing. Mathematicians often see things that engineers are willing to accept but that they feel need proof, and by the time they have tracked down the proof and found out what assumptions need to be made, they have uncovered something that is important. The engineers are very good at keeping the mathematicians headed in the right direction, putting blinders on them to keep them from running off after interesting equations, so I think that we all need each other.

Saffman: Like Bill, I was a little puzzled by what was meant by the title *"Is this the theory ...?"*. Also, having reached the age in which you find that, when you can no longer solve the equations, you can do something easier, like doing philosophy, I am going to consider some philosophical aspects of the question: *"What is the theory ...?"*. One of the things that I always found troubling in the study of the problem of turbulence is that I am not quite sure what the theoretical turbulence problem actually is. There is a number of questions about which I will talk in turn. First of all: what actually is the theoretical turbulence problem? One reason I think we have so much difficulty in solving it, is that we are not really sure what it is. If one tries to become more precise, you can ask: what are the theoretical questions of turbulence? What does the engineer want the theoretician to do? what should the theoretical physicist do when he leaves the world of high energy physics, or the other exotic areas of physics, and decides to come and solve the turbulence problem? What are the precise theoretical questions for which answers are required?

Then of course, since no questions are ever fully solved - they are usually just forgotten and left unsolved, and people go on to other things - there is the question of what is an acceptable answer. This is subjective. I think that there is really no objective criterion for what an acceptable answer is, in turbulence or in any branch of science. I think the answers can be separated, although with some overlap, into several kinds, the first of which is what I will call botany. Here, I use botany as an euphemism for cataloguing, or classification. That, I think, can be extremely important, and we learn a great deal from it. We take different flows or different physical situations and we try to find ways of classifying them, which means that we look for the things that they have in common. A recent example are fractals. I would say that the attempts to represent or describe turbulent flows, turbulent structures, or the evolution of chaotic systems, using fractals is essentially a botanical description, unless you can do some prediction, or make some theory to say what is happening, but it is basically, I would say, botany.

Then there is also postdiction, which is a word that you don't find in the dictionary, unfortunately, and should be there. The term was introduced to my knowledge by Feynman, I suppose in connection with some problem in theoretical physics, and means developing a theory, or a model, which is then tested by seeing whether it agrees with what has been measured. It is also, I think, a valid process. I am not trying to be derogatory towards it. I think that it is also a valid, and very useful, scientific process by which we gain understanding of the phenomena, and I would say that the dynamical system models would come into the category of postdiction. You are introducing some general ideas and you are trying to see whether they work, by seeing whether the results of calculations agree with what is known or what is observed.

Then, of course, after these two, if they are really successful, would come prediction, which means that you say what is going to happen. You take knowledge gained from your botanical investigation, or from the postdiction phase and you say that if somebody did such and such, then such and such would happen. And I think that at the present stage the only thing we have for this is the direct numerical simulation, however limited it is. Parviz Moin was very good and very honest in the way he pointed to its disadvantages as well as to its advantages. Simulation is at the moment, unfortunately, our only method for predicting what is going to happen in a turbulent flow. But this also has its problems: there is the question of how much detail you want; you can't predict everything, and you do not want to predict everything, and this comes back to what are the theoretical questions. Predictions can be of two forms, again without a totally clear distinction: there are the qualitative ones, in which the ideal answer is yes or no, and there are also the quantitative.

After putting this down I decided to stop to think, and to go through my memory for how many predictions there actually have been in the study of turbulence. I came up with a depressingly small number. Anybody can make predictions, but I am limiting this to predictions confirmed by experiment.

First are the small scale isotropic relations, which were given by G.I. Taylor around 1937, when he wrote the four classic papers on homogeneous turbulence, and which were later confirmed by his experiments. Then is the Kolmogorov scaling, which is the most noted, I think. Then, although I may be wrong, are the shear flow similarity scalings; the fact that the angle of spread of a jet is independent of the Reynolds number, and the decay laws for turbulent jets, wakes, mixing layers, etc. They may have been predictions, although I am not sure of which came first, whether predictions were made before the experiments. My recollection is that they were about simultaneous.

After that I just can not think of anything where a genuine prediction for the dynamics of turbulent flow has been confirmed by an experiment. So we have a big vast empty field. I hope that when the next meeting takes place, maybe in 5 or 10 years, it will be possible to add something to this list.

Reynolds: I would just like to make a couple of points. I would like to look at three characteristics of turbulence, as I understand it, and then discuss what I see in dynamical systems theory, and whether both things can be related to each other. First, I think turbulence is characterised by local instabilities occurring over a very wide range of scales, and when I see theories based on global modes I get worried. I have never seen an eddy that looked like a sine wave. I am much more interested in the things that we saw with solitons. That is a very interesting area and I would urge things to be pushed in that direction. There is also, I think, some evidence that, in turbulence, small scales get organised into large clusters, into things that get names like puffs and things like that, specially high Reynolds numbers. The mean deformation on the flow probably plays a role on that organisation. It seems to me that the main focus in the dynamical systems theory has been in the creation of disorganisation and not on the creation of organisation, and it may be useful to have more work, more attention on the process of organisation.

Finally let me echo John Lumley in that we are sure that turbulence is the solution of the

Navier-Stokes equations. I get criticised pretty badly when I invent turbulence models using k-ε ideas and things like that which are sort of hooky. I was surprised by how much of that I saw at this meeting. I think that there ought to be more reliance on the Navier-Stokes equations as the bases for the simpler equations that everybody is using.

I would like to conclude by posing two questions. When I came to the meeting I hoped that I might get some information that could help me answer the question of what is the nature of the anisotropy of the small scales of turbulence, in a flow that is subjected to, say, persistent mean deformation. There is some evidence from numerical simulations and other things that the small scales may be more organised than we really thought, at finite Reynolds numbers. I did not see any theory here that would help me address that question, which leads me to question number two: did I miss something? With that let me open the floor to a general discussion about what has gone on at this meeting: comments, responses, ... whatever you like.

Saffman: I just want to say that the panel is not unanimous in its desire to stay wedded to the Navier-Stokes equations. Again I think that, as Feynman pointed out, what we may be needing in the next generation is a qualitative theory of differential equations, and it may be that turbulence is not dependent on the Navier-Stokes equations. There may be other equations, and not even necessarily differential equations, whose properties have the same kind of structure as the turbulent structure of the Navier-Stokes equations. And these other equations may be easier to solve, or the required properties may be more transparent. So I think there is a great deal to be said for exploring other equations in trying to find out what are the generic properties of their solutions. And it therefore makes sense to study a class of equations rather than to stay wedded to any particular one.

There is also a very powerful tool, which I think that we have all experienced at one time or another, and which can be described as Serendipity, which is finding a dollar when you are looking for the cent that you have dropped under the chair. It can very well be that, in studying these equations, the mental processes which are stimulated by those other equations will suddenly lead you to have insight into the properties of the Navier-Stokes equations. So I think that, from the point of view of the theoretician of the applied mathematician, of the theoretical physicist, and I think that it should also be that of the engineer, one should be prepared to consider systems of equations other than just the Navier-Stokes equations.

Huerre: I am not the best qualified person to answer some of the questions that were raised on amplitude equations but I will try nevertheless. I think that almost everybody agrees that amplitude equations close to bifurcation points are fairly well defined objects that we can derive asymptotically, so there may be a minimum of controversy on that point. Where things become a little more muddled is when we try to apply them as generic models, by applying Moses law as stated by Bill at the beginning of this conference. I think that there is a great virtue in playing with such models, even away from bifurcation points, because they are in the class of postdiction models that Phil Saffman talked about. We include certain effects, based on either extensions of amplitude equations close to bifurcation points or on phenomenological considerations, see what kind of answers we get from the evolution equations, either numerically or analytically, and try to compare with the experimental findings.

I think that this strategy has been more successful in closed systems than in open ones until now. I think that in the Taylor Couette flow and in Rayleigh Benard convection there have been significant achievements in this regard. I think that in open flows the difficulty may come from the fact that they are inhomogeneous in space, and highly non parallel. But I think there are possible extensions to this strategy by taking into account the inhomogeneity, and this is where I come to what was mentioned by Bill Reynolds on global modes. There we try to deal with objects which take into account the inhomogeneity of the flow and we can derive equations from this global modes in an asymptotic sense. We do not have to rely on phenomenological considerations. For instance in the case of the Kármán vortex street at very low Reynolds numbers, we can learn a great deal about how the Kármán vortex street frequency is being selected, from instability ideas and amplitude evolution ideas. Another example is what has been done by Gaster and Wygnansky over the years in the case of mixing layers, where you can apply the instability wave approach, even linear evolution equations, to try to get a handle on the large scale structure evolution in shear flows. So I think that these avenues are hopeful.

Mackay: I would like to follow Phil Saffman's point because, when I see something with structure on a whole range of scales, I think of renormalization. Is there some way of taking the system and doing a scale transformation, and representing the new system as a different system of the same form? It maybe that, as you shrink the scale, you see a universal behaviour described by a fixed point on a space of possible systems. If you go to really small scales, you probably go to the Stokes flow, but in between there may be some saddle point on the space of systems that may describes the mixing layer transition. This is totally off the wall, but I would like to ask what people here think about the potential for renormalization theories. I know that there have been some attempts, by Orzsag and others, but I am not familiar with how far they have got.

Lesieur: The RNG theory was developed for turbulence by these people that you mentioned, and by Frisch, Foster and others before. The problem is that, although it gives quite good results for forced turbulence in the infrared limit, if you want to reproduce the Kolmogorov cascade, you have to assume that the small expansion parameter is equal to 4, which is not so small. So I do not think that it is very useful for small scale turbulence.

Reynolds: We have also been looking at the Orzsag and Yakhot work. The perturbation analysis, which is really the key of it, is a perturbation about a balance between a modified viscous force and the time dependent term, so that the inertial terms, which are so crucial in turbulence, are not in the lowest order approximation to the perturbation analysis. I think that is the root of the problem. Turbulence is inherently inertial.

Simó: Let me take the point of view of the dynamical systems people. I think that there is a big difference between the problems that we have been discussed during these days and the dynamical systems approach. The main difference is the following: In dynamical system theory, one tries to explain not the particular behaviour of a given solution, but the global behaviour of all the solutions, or at least of all the solutions in a significant portion of the phase space. And I have not seen that this is the point of view in fluid mechanics.

Personally I consider that it is impossible to understand the dynamics, or the main characteristics of the dynamics, without understanding a significant portion of the phase space. In particular, neither in the laboratory experiments, nor in the numerical experiments, have I seen any result concerning unstable solutions, for instance unstable periodic orbits. If somebody finds some periodic solution in the laboratory, or in the computer, they are looking at stable periodic solutions, stable tori or something like that, but not at unstable ones. In dynamical systems theory, it is those unstable solutions the ones which give the key to understanding the global behaviour in some significant parts of the phase space.

Reynolds: I think that is a good point. I would just mention that Laurence Keefe, who was working with Parviz Moin, has done a cluster of some 3000 solutions, and then looked at the Liapunov exponents generated by their exponential divergence. I think that is along the line of what you are looking for

Mackay: I would like to defend Hamiltonian systems. I was somewhat disappointed that Carlos Simó did not respond to the question about what use they might be, but they have come under fire again. People who work in turbulence are interested in high Reynolds numbers, in the limit where viscosity is very small, so that the flow is very close to a Hamiltonian system. There is also other stuff going on, some dissipation, but it is small. I think that it is interesting to understand the behaviour of the Euler equations, and then to look at what happens when you add a small dissipation. I think that is the main problem.

Lesieur: I am sorry to contradict you, but three dimensional turbulence is characterised by a finite rate of dissipation of kinetic energy when viscosity goes to zero. It is a strongly dissipative system. It is not at all a Hamiltonian system.

Reynolds: Well said.

Pomeau: I think it is precisely this question that we don't understand. Why is it that in flows at large Reynolds numbers there is finite dissipation.

Lesieur: That is the Kolmogorov theory

Pomeau: But why is it true? There is no theory in Kolmogorov. It is just a scaling law.

Lesieur: Kolmogorov theory says that, in the limit of zero viscosity, you have a finite flow of energy, ϵ, towards infinitesimal scales, and this corresponds to a finite rate of energy dissipation in the limit of viscosity going to zero. I think that anybody here who has been interested in at least three dimensional turbulence will confirm what I say. Take Bachelor's book on turbulence. Of course you can not demonstrate it without solving the Navier-Stokes equations, but anybody who has manipulated turbulence knows it. John Lumley said that you can not use the word turbulence for everything. Well, I mean three dimensional turbulence, and this is an essentially dissipative system at, if not zero viscosity, at least infinitesimal viscosity. This is a very important point.

Saffman: Has anybody here looked carefully at the experimental evidence for the assumption of Kolmogorov that the energy dissipation is independent of Reynolds number? I did. The only evidence that I could find is in Batchelor (1953) book, figure 6.1, in which, since this fact - he does not call it an assumption, but a fact - is so important, he has a little appendix to chapter 6 to give the experimental evidence. Now, the experimental evidence in Batchelor's book does *not* support the hypothesis. In fact, if you plot ϵ against Reynolds number, the fit is closer to the $Re^{-1/4}$. Bachelor's book was written in 1953, and these are old data taken prior to 1950. These data are all at very low Reynolds number, and extrapolating them to high Reynolds numbers is completely unjustified. But I know of no other experimental evidence, and this is why I ask people here to show that ϵ is independent of viscosity in what is supposed to be homogeneous turbulence.

Reynolds: This is going to start a good discussion, and I am going to give priority to people who have books. Is there anybody who has a book and who would like to respond?

Lesieur: Well, I do not claim to show it in my book, but I am sure that everybody who has worked on the decay of isotropic turbulence in a wind tunnel behind a grid, for instance Van Atta or John Lumley, would agree that, when you increase the Reynolds number, you see a finite dissipation. I do not claim that we can be sure that it is independent of the Reynolds number, because the experimental Reynolds numbers are not even high, but you have a strong dissipation of kinetic energy even when you increase the Reynolds number as much as you can. And you can also ask to the people who are developing three dimensional numerical simulations of isotropic turbulence in a box, that, when they increase the Reynolds number, they see a finite rate of decay of kinetic energy. You can not tell whether the decay exponent is t^{-1}, or $t^{-1.3}$ or $t^{-1.6}$, because, up to now, the Reynolds numbers are too moderate, but nobody who has looked at these experiments, either in the laboratory or in numerical experiments, would say that you don't have dissipation of kinetic energy at infinite Reynolds number.

Reynolds: That question simply can not be answered by numerical simulations in this century.

Roshko: I am not surprised that you can not get much out of the example that you mentioned. It would take a tremendous amount of leverage in Reynolds number to settle that question from detailed measurements of the fluctuations and of their relation to dissipation. But one example which I think may indicate that the dissipation reaches a certain limit with Reynolds number is the simple problem of the round jet. It has a rather well defined spreading angle under a wide range of laboratory conditions and, if you look at something like pictures of a rocket jet on a test stand, it seems to have pretty much the same spreading rate. this is not a tight scientific experiment but there is a tremendous spread of Reynolds number and you do not have to measure the fluctuations. The spreading rate itself tells you that the energy is being dissipated at a certain rate, which does not seems to vary much in a range of Reynolds numbers from 10^4 to 10^8.

Mackay: I remember looking in some engineer's handbook at the friction factor of flow through a pipe, and it levels off at high Reynolds number, but it depends on the roughness of the pipe. If you use a smoother pipe it levels off at a lower level and, if you imagine an infinitely smooth pipe, the friction factor actually keeps going down. It is not clear to me whether there is a non-zero limit to dissipation.

But I also want to point out that there is a lot more to Hamiltonian systems than energy conservation. There is conservation of the Poincaré invariants, and I think that the simplectic structure could play a very important role in understanding the behaviour of high Reynolds number flows. If you imagine the Poicaré invariants as projecting $P\,dQ$ on the canonical planes, the ones corresponding to the contribution from the macroscopic scales will probably be quite well preserved, even if down at the microscopic scales they are not. So that the simplectic structure will still play an important role.

Reynolds: Who is next?

Lumley: I would just like to say that I do not want to get involved in the question of whether the dissipation goes to zero as the Reynolds number goes to infinity. I would rather look at it from an engineering point of view and say that, over the range of Reynolds numbers that are encountered in geophysics and in technology, the picture that the dissipation remains fixed as the viscosity goes to zero is adequate. I think it would be a waste of time to pay much attention to systems in which the dissipation goes to zero. The best evidence, from my point of view, is the fact that, if you take the dissipation times L/U^3, it is essentially constant over a wide range of Reynolds numbers. That means that it can been parameterised by the energy containing range and not by the viscosity.

Reynolds: Any more discussion on this point?

Simó: Another comment I want to make is the following. I am convinced that turbulence is a very difficult subject, but there is some hope the following: some 30 or 40 years ago the state of knowledge on Hamiltonian systems with two or three degrees of freedom was almost zero. Now, with the help of some ideas of dynamical systems, with computers, with experiments and so on, the state of knowledge is fairly complete. Not only can we use dynamical systems for postdiction, but also for prediction. We can do a few numerical computations of some relevant quantities, and using normal forms, unfolding of singularities, computing some periodic orbits, some tori and separatrices, we can predict the global picture of the dynamical system. And we can even do quantitative predictions by just looking at some specific points. In Hamiltonian systems with 4, 5 or 6 degrees of freedom this is much harder; if we add dissipation, or go to an infinite number of degrees of freedom, even more, but let us hope that in the next, say, 60 or 100 years, this approach of studying, not just local orbits, but the neighbourhood of orbits, or a significant set of orbits, can give us some idea about global behaviour.

Lesieur: I don't want to keep up this discussion on dissipation, but I wanted to make clear that I was talking about three dimensional turbulence. If you accept the concept of 2-D turbulence - even if I know that John Lumley does not like it - then you have conservation of kinetic energy when the viscosity goes to zero. It is the enstrophy that is dissipated at a finite rate.

Papailiou: I would like to state two things from the experimental point of view. To begin with, I would like to remind you that after all the research that we have devoted to the subject, we still have to predict transport phenomena in turbulence. The average engineer still uses the old Prandtl model, which seems to me to indicate that we are not very far from where we started. The second point is that, as experimentalists, we would like to know more about vorticity, rather than about the velocity field. We don't know how to measure it, and velocity is a bad substitute for vorticity. If we were able to describe statistically the small vortices, in the same way that we describe the velocity field, and if we were able to predict the dynamic behaviour of the larger vortices as they interact with the mean velocity field, I believe that we would have a much better insight of turbulence than we have today.

Reynolds: I think you would get a lot of agreement on that, but you may not know that Juan Agüí here has developed a neat optical technique for measuring vorticity components. Other questions or comments?

Mackay: Can Philip tell us now what the questions are?

Saffman: I wish I knew what the questions are. I had made a list of what I personally thought would be some interesting questions. But I did that before the meeting and I am not sure now of whether I agree with that list any longer. I will show them to you, but I am not longer convinced that they are particularly relevant. First I will give some qualitative questions, which are really too vague to be considered questions at all.

- Which is the structure of the cascade?

- Which is the lifetime of coherent structures? This is presumably or possibly related to the fission processes which take place in the vortices.

- Is there an inverse cascade? Can we have generation of large scales from small scales in turbulent motion? Might this be identified with the fusion processes in vortices?

Then of course there is the question, that has been discussed by many people,

- Transition.

And lastly there is this question which is perhaps related to our arguments about dissipation and Hamiltonian systems:

- Is the three dimensional structure of turbulence really important? Is the stretching of vortex lines crucial? a *sine qua non* feature of the evolution of turbulence?

  It is clear that, if we drop the three dimensions and the stretching of vortex lines, something has gone but, is what we have lost important? What we are loosing is the generation of vorticity fluctuations on a small scale and, for some applications of turbulence, that would clearly be important. But there are many more applications, such as diffusion, dispersion, or the large scale convection of pollutants or passive scalars, where there really is no evidence that the three dimensional aspect of turbulence are important. And there is no argument that two dimensional turbulence is a Hamiltonian system, where the energy dissipation vanishes when the Reynolds number becomes large. Perhaps we don't loose too much by forgetting about the dissipation of 3-D turbulence. There may be some similarities there.

  With regard to quantitative, or semi-quantitative questions, there is the one that we have been arguing about.

- Is there a non-zero limit to dissipation at high Reynolds numbers?

  I was deliberately a little bit provocative there, because I actually do believe that ϵ is independent of viscosity. I was just commenting that, when you look at the hard evidence, it is not there. I admit Anatol's point, I fully appreciate that there is nothing to show otherwise, but that is not evidence, and nobody here seems to have been able to come up with hard evidence.

- Why is the law of the wall universal?

  The interaction of turbulence with solid boundaries is obviously a concept of fundamental dynamical importance. By some miracle, I think it was Kármán who discovered that the mean velocity profile near the wall has a logarithmic behaviour. This has been postdicted by simple similarity arguments, by Millikan, and I think by somebody else, around the 1930's. It is almost universal. This log law the wall is one of the most fundamental facts of turbulence, and none of the theories other than the general, similarity type of, analysis seems capable of saying anything about it.

- Why is the spreading rate of a jet independent of viscosity? Anatol was talking about that. We know that this is true, or that perhaps it depends very weakly. One could say that there is some power law. If so, what is it?

  Another question, and this is one where the Kolmogorov theory went wrong.

- Why is the flatness factor not a constant?

  The Kolmogorov theory, in its original form, is right in maybe its most important feature, the prediction of the energy spectrum, but it is also very wrong in other predictions and, in particular, in the statistics of the high order moments of velocity and vorticity fluctuations. The flatness factor is the mean value of $(du/dx)^4$, normalised on the mean of $(du/dx)^2$ squared. According to Kolmogorov it should be constant, independent of the viscosity, by the same type of argument which he uses to say that ϵ is independent of viscosity. Experimentally it is found that this is a very high power of the Reynolds number. The question mark here is not small. There have been models trying to postdict this but, to my knowledge, they all seem to fail. They are all wrong in the numbers that they predict.

  And finally there is another question:

- How does the intermittency factor depends on Reynolds number? People have mentioned it throughout this meeting over and over again, and quite rightly so, because the spatial and temporal intermittency of turbulence is one of its most important features, and the question of its dependence of Reynolds number is a very puzzling one to the theoreticians.

These are some of the questions which I regard as important, but I want to stress the *"I"*. This is very much a list of my personal preferences, of my personal taste, of the things in which I happen to be interested. I am sure that other people could come out with other lists, which may be more interesting. There are many problems which I have just not mentioned.

Reynolds: That seems a pretty good note on which to end this meeting.

REFERENCES

Batchelor, G.K. 1953 "The theory of homogeneous turbulence", Cambridge University Press, Cambridge, pp. 105-106.

ORGANISING COMMITTEE

J. Jiménez (Director)
Departamento de Mecánica de Fluidos
Esc. Técn. Sup. Ing. Aeronáuticos
Pl. Cardenal Cisneros 3
28040 -Madrid
SPAIN

Y. Pomeau
Groupe de Physique Statistique de l'ENS
24, Rue Lhomond
75231 Paris Cedex 05
FRANCE

P. Huerre
Dept. Mecanique
Ecole Polytecnique, Departamentale 36
91128 Palaiseau cedex
FRANCE

P. Saffman
Applied Mathematics (217-50)
California Institute of Technology
Pasadena, CA 91125
USA

A. Liñán
Departamento de Mecanica de Fluidos
Esc. Técn. Sup. Ing. Aeronáuticos
Pl. Cardenal Cisneros 3
28040 -Madrid
SPAIN

LECTURERS

G. Ahlers
Lehrstuhl fur Theoretische Physik IV
Universitat Bayreuth
Postfach 101251
8580 Bayreuth
GERMANY

J.C. Antoranz
Dpto. Física Fundamental
Univ. Educación a Distancia
Apto. 50487
28080 Madrid
SPAIN

Roberto Benzi
Universita di Roma
Dip. di Fisica
Via Orazio Raimondo
I-00173 Roma
ITALY

T. Bohr
Niels Bohr Institute
Blegdamsvej, 17
2100 Copenhagen
DENMARK

H.R. Brand
FB 7 Department of Physics
University of Essen
D 4300 Essen 1
GERMANY

H. Chaté
Institut de Recherche Fondamentale
DPh-G/PSRM, CEN Saclay
F 91191 Gif-sur-Yvette cedex
FRANCE

J-M. Chomaz
CNMR
42 Av. G. Coriolis
31057 Toulouse
FRANCE

P. Clavin
Laboratoire de Recherche en Combustion
Université de Provence
Centre Saint Jérome
13397 Marseille Cedex 13
FRANCE

R.J. Deissler
Center for Nonlinear Studies
Los Alamos National Laboratory
Los Alamos, N.M. 87545
USA

Stephan Fauve
Ecole Normale Superieur Lyon
46 allée d'Italie
69364 Lyon Cedex 07
FRANCE

H. Fiedler
Technische Universitat Berlin
Hermann-Fottinger Institut
Strasse des 17 Juni
D-1000 Berlin 12
GERMANY

M. Gharib
Dept. of AMES, R-011
Univ. California San Diego
La Jolla, CA. 92093
USA

J.D. Gibbon
Mathematics Department
Imperial College
London, SW7 2AZ
UNITED KINGDOM

J.A. Hernández Ramos
Departamento de Mecánica de Fluídos
Esc. Técn. Sup. Ing. Aeronáuticos
Pl. Cardenal Cisneros 3
28040 -Madrid
SPAIN

E. Hopfinger
Institut de Méchanique de Grenoble
B.P. 68
38402 S.-Martin d'Héres Cedex
FRANCE

M. Jensen
Nordita
Blegdamsvej, 17
2100 Copenhagen
DENMARK

J.C. Lasheras
Department of Mechanical Engineering
University of Southern California
Los Angeles, CA 90089-1191
USA

M. Lesieur
Institut de Méchanique de Grenoble
B.P. 53X
38041 Grenoble Cedex
FRANCE

J.L. Lumley
Sibley School of Mechanical
and Aerospace Engineering
Upson and Grumman Hall
Cornell University
Ithaca, NY 14853-7501
USA

L. Kleiser
DFLVR
Institute for Theoretical Fluid Mechanics
Bunsenstrasse 10
D-3400 Gottingen
GERMANY

R. MacKay
Mathematics Institute
University of Warwick
Coventry CV4 7AL
UNITED KINGDOM

J.M. Massaguer
E.T.S. Ingenieros Telecomunicación
Universidad Politécnica de Cataluña
Jordi Girona Salgado 31
08034 Barcelona
SPAIN

O. Metais
Institut de Méchanique de Grenoble
B.P. 53X
38041 Grenoble Cedex
FRANCE

P. Moin
Mechanical Engineering Department
Stanford University
Stanford, CA 94305
USA

D. Papailiou
Dept. Mechanical Engineering
University of Patras
Rio 26001
Patras
GREECE

C. Pérez García
Dpto. Física
Facultad de Ciencias
Universidad de Navarra
E-31080 Pamplona
SPAIN

A. Pumir
Laboratoire de Physique Statistique
Ecole Normale Supérieure
24, Rue Lhomond
F-75231 Paris Cedex 05
FRANCE

W. Reynolds
Mechanical Engineering Department
Stanford University
Stanford, CA 94305
USA

Renzo Ricca
Dept. AMTP
Silver Street
Cambridge, CB3 9EW
UNITED KINGDOM

Jean-Pierre Rivet
Observatoire de Nice, BP139
06003 Nice Cedex
FRANCE

Erika Roesch
Max-Planck Inst. für Stromungsforschung
Bunsenstr. 10
D-3400 Gottingen
GERMANY

A. Roshko
Aeronautics Department
California Institute of Technology
Pasadena, CA 91125
USA

Miguel A. Rubio
Dpto. Física Fundamental
Univ. Educación a Distancia
Apto. 50487
28080 Madrid
SPAIN

C. Simo
Dept. Matemáticas Aplicadas y Análisis
Facultad de Matemáticas
Universidad de Barcelona
Plaza de la Universidad
Barcelona
SPAIN

C. Van Atta
Department of Applied Mechanics
and Engineering Sciences (R-013)
University of California, San Diego
La Jolla, Ca. 92093-0413
USA

M.G. Velarde
Facultad de Ciencias
Univ. Educación a Distancia
Apto. 60141
28071 Madrid
SPAIN

D. Walgraef
Service de Chimie-Physique
Universite Libre de Bruxelles
Campus Plaine, C.P. 231
B-1050 Bruxelles
BELGIUM

I. Wygnanski
Faculty of Engineering
Tel-Aviv University
Ramat-Aviv
Tel-Aviv, 69978 Israel
ISRAEL

N. Zabusky
Dept. Mechanical and Aerospace Eng.
College of Engineering
Rutgers University, P.O. Box 909
Piscataway, NJ 08855-0909
USA

W. Zimmerman
Lehrstuhl fur Theorische Physik
Universitat Bayreuth
Postfach 101251
8580 Bayreuth
GERMANY

Stéphane Zaleski
Lab. Physique Statistique
Ecole Normale Supérieure
24 Rue Lhomond
75231 Paris cedex 05
FRANCE

PARTICIPANTS

J.C. Agüí
IBM Scientific Centre
Paseo Castellana 4
28046 -Madrid
SPAIN

Roque Corral
Departamento de Mecánica de Fluídos
Esc. Técn. Sup. Ing. Aeronáuticos
Pl. Cardenal Cisneros 3
28040 -Madrid
SPAIN

Owen E. Cote
Air Force Office of Scientific Research
European Off. Aerospace Res. & Dev.
233 Old Marylebone Rd.
London NW1 5TH
UNITED KINGDOM

M. de la Torre
Dpto. Física Fundamental
Univ. Educación a Distancia
Apto. 50487
28080 Madrid
SPAIN

C. Dopazo
Dept. Mecánica de Fluidos
E.T.S.I. Industriales
María Zambrano 50
Polígono Actur
50015, Zaragoza
SPAIN

Miguel A. Fernández Sanjuán
Dpto. de Física
ETS Arquitectura
Universidad Politècnica de Madrid
28040 Madrid
SPAIN

M. Gaster
Engineering Department
Cambridge University
Cambridge, CB3 9EW
UNITED KINGDOM

F. Higuera
Departamento de Mecánica de Fluídos
Esc. Técn. Sup. Ing. Aeronáuticos
Pl. Cardenal Cisneros 3
28040 -Madrid
SPAIN

P. Juvet
Mechanical Engineering Department
Stanford University
Stanford, CA 94305
USA

Carlos Martel
Departamento de Mecánica de Fluídos
Esc. Técn. Sup. Ing. Aeronáuticos
Pl. Cardenal Cisneros 3
28040 -Madrid
SPAIN

Roy E. Reichenbach
Aeronautics and Mechanics Branch
US Army European Res. Off.
233 Old Marylebone Rd.
London NW1 5TH
UNITED KINGDOM

Ezequiel del Río-Fernández
Dpto. Física Fundamental
Univ. Educación a Distancia
Apto. 50487
28080 Madrid
SPAIN

Maurice Rossi
Mecanique Theorique
Université de Paris VI
Tour 66- 4, Place Jussieu
75230 Cedex 05, Paris
FRANCE

Jaume Timoneda
Matemáticas Aplicadas i Análisis
Univ. de Barcelona
Gran Vía 585
08091 Barcelona
SPAIN

Douglas A. Varela
15 Flag St.
Massachusetts Institute of Technology
Cambridge, MA 02139
USA

P.D. Weidman
Department of Mech. Engineering
University of Colorado
Boulder, Co.
USA

J. Zufiria
IBM Scientific Centre
Paseo Castellana 4
28046 -Madrid
SPAIN

INDEX

Toroidal chaos, 242
Toroidal fields, 252-253, 256, 343
n-Torus, 235, 237
2-Torus, 235, 237, 239, 240, 241
3-Torus, 235, 236
    uniform flows on, 233, 238-243
Translative instabilities, 149
Truly incompressible method, 177
Turbulence memory, 33, 34, 38
Turbulent coupled map lattices, *see* Coupled
    map lattices
Turbulent patches, *see* Vortex patches
Turbulent puffs, *see* Puffs
Turbulent scalar, 155-164, *see also*
    Large-eddy simulation
Turbulent wall jets, *see* Wall jets
Two-dimensional boundary layers, 128-130
Two-dimensional parallel flows, 248
Two-dimensional shear layers, 167-177, *see*
    also Rayleigh-Benard convection
Two-dimensional turbulence, 309, 347, 353
Two-dimensional wakes, 33-49, *see also* Large
    scale vortices

Unforced flows, 67, 68-75, 76, 78, 82, 84, 85,
    86, 143
SUB = behind a backwards-facing step, 143,
    149-150, 152
Uniform flows, 233, 238-243
Uniform state instabilities, 135-137
Uniform wavetrains, 138

van der Pol oscillators, 111, 287
Velocity
    bubble formation and, 133, 135, 136
    DNS of, 124, 125, 126, 127, 129, 147, 336
    external excitation and, 81, 82
    intermittency and, 222
    in Karman vortex streets, 113-114
    in laminar, co-flowing forced jets, 97
    L.E.S. of, 155, 156-157, 158, 159, 160-163
    in round jets, 89, 90, 91, 92, 93
    as vorticity substitute, 353
    in wall jets, 68-75, 76-79, 84, 86
Vertical channels, 139
Viscosity, 354
    in dynamical systems theory, 219
    intermittency and, 222
    in Kolmogorov theory, 352, 355
    large scale vortices and, 43, 49
    L.E.S. of, 155, 156, 157, 158, 162, 163-164
    mixing transitions and, 9
    RBC and, 167, 169
    stationary boundary conditions and, 59
    vortex line stretching and, 266, 269
    in wall jets, 67, 68, 70, 73, 84, 86, 169
Vortex decay, 33, 38
Vortex dipoles, 43, 44
Vortex dynamics, 201-207
    defined, 201
Vortex filaments, 203, 205, *see also*
    Filamentation

Vortex filaments (*cont'd*)
    Betchov-Da Rios equations and, 257-260
    DNS of, 145-146, 149
Vortex generation, 33, 38, 43, 45, 46, 48, 49
Vortex lines, 337
    stretching of, 265-269, 354
Vortex loops, 101, 104, 105, 106, 107
Vortex pairing
    DNS of, 143, 144
    Gortler instability and, 28
    in large scale vortices, 33, 38, 43
    mixing transitions and, 3, 8
Vortex patches, 207, 273, 274-275, 277, 285
    filamentation of, 202, 206-207
    fusion and fission of, 202-203
Vortex rings, 99-100, 203, 343
    Euler equations and, 262
    in laminar, co-flowing forced jets, 95, 96,
        101, 102, 103, 104
    puffs and, 341, 345
    reconnection and, 204
    in round jets, 89
Vortex shedding, 13-21
    in bluff body wakes, 51-53
    in homogeneous fluids, 13, 14-17
    instability mode frequencies in, 17-18
    in large scale vortices, 34, 38
    in stratified fluids, 14, 19-21
Vortex sheets, 204
Vortex splitting, 51-53
Vortex streets, 38, *see also* Karman vortex
    streets
Vortex tubes, 99-100
    core bursting in, 202, 207
    DNS of, 337
    Euler equations and, 261, 262-263
    in laminar, co-flowing forced jets, 96, 101
    reconnection in, 202, 204-206
    singularity formation and, 204
Vorticity
    DNS of, 150
    in dynamical systems theory, 219
    in laminar, co-flowing forced jets, 95-107,
        see also Laminar, co-flowing forced
        jets
    in round jets, 91
    velocity as substitute for, 353
Vorticity links, 260-263

Wakes, 68, 117, 203, *see also*
    Two-dimensional wakes
    in homogeneous fluids, 13, 14-17
    Karman vortex streets and, 115
    in stratified fluids, 19-21
Wall jets, 67-86, 336, *see also* External
    excitation
Wall stresses, 80, 81, 86
Wave breaking, 206, 207
Wavelengths, 138, 139
Wave patterns, 320, 322-324
Waves, 202, 203, *see also* Billows
Wavetrains, 138